Hemoglobin

T0177649

Hemoglobin

Insights into Protein Structure, Function, and Evolution

Jay F. Storz

Professor, School of Biological Sciences,
University of Nebraska

OXFORD
UNIVERSITY PRESS

OXFORD
UNIVERSITY PRESS

Great Clarendon Street, Oxford, OX2 6DP,
United Kingdom

Oxford University Press is a department of the University of Oxford.
It furthers the University's objective of excellence in research, scholarship,
and education by publishing worldwide. Oxford is a registered trade mark of
Oxford University Press in the UK and in certain other countries

© Jay F. Storz 2019

The moral rights of the author have been asserted

First Edition published in 2019
Impression: 1

All rights reserved. No part of this publication may be reproduced, stored in
a retrieval system, or transmitted, in any form or by any means, without the
prior permission in writing of Oxford University Press, or as expressly permitted
by law, by licence or under terms agreed with the appropriate reprographics
rights organization. Enquiries concerning reproduction outside the scope of the
above should be sent to the Rights Department, Oxford University Press, at the
address above

You must not circulate this work in any other form
and you must impose this same condition on any acquirer

Published in the United States of America by Oxford University Press
198 Madison Avenue, New York, NY 10016, United States of America

British Library Cataloguing in Publication Data
Data available

Library of Congress Control Number: 2018950707

ISBN 978–0–19–881068–1 (hbk.)
ISBN 978–0–19–881069–8 (pbk.)

DOI: 10.1093/oso/9780198810681.001.0001

Printed in Great Britain by
Bell & Bain Ltd., Glasgow

For my kids, Jessie and Cody

Preface

"More may have been written about hemoglobin than about any other molecule. Physicists, crystallographers, chemists of all kinds, zoologists, physiologists and geneticists, pathologists, and hematologists have all contributed to a vast literature. In the erratic ways that scientific research shares with other human endeavors, the multifarious work of that great throng has provided us with an enormous store of knowledge from which one can extract data on subjects as diverse as the quantum chemistry of iron and the buoyancy of fish."

—Perutz (2001)

In the chapters to follow, we will explore questions about protein structure and function, biochemical adaptation, and molecular evolution by focusing on lessons learned from research on a single, paradigmatic protein. There are several reasons to focus specifically on hemoglobin (Hb) and other members of the globin superfamily; one is historical importance. During the last half century, Hb has played a starring role in research efforts to understand relationships between protein structure and function, and in efforts to identify the molecular underpinnings of physiological adaptation and pathophysiology. Hb and its cousin, myoglobin (Mb), were the first proteins to have their crystal structures solved, Hb serves as a paradigm for understanding principles of allosteric regulatory control, and clinical research on mutant Hbs ushered in the modern era of molecular medicine.

As stated by Dickerson and Geis (1983): "with this one family of macromolecules [Hb and related globin proteins] one can illustrate nearly every important feature of protein structure, function, and evolution: principles of amino acid sequence and protein folding, a mechanism of activity that resembles that found in enzymes although not itself catalytic, specificity in the recognition and binding of large and small molecules, subunit motion and allosteric control in regulating activity, gene structure

and genetic control, and the effects of point mutations on molecular behavior." Dickerson and Geis also emphasized the merits of familiarity: "People for whom alpha-ketoglutarate has no charms immediately recognize hemoglobin as the essential constituent of blood, without which human life would be impossible."

The other rationale for focusing specifically on Hb is based on the time-tested scientific practice of using a model system to extrapolate general principles. As foundational knowledge accumulates, the value of the model continues to increase. For example, solving the crystal structure of a protein can provide insights into the stereochemical basis of observed functional properties. These insights can then suggest new hypotheses about structure-function relationships which, in turn, motivate further experimental work. Research on Hb structure, function, and evolution illustrates how a well-chosen model system can enhance our investigative acuity and bring key questions into focus.

This book is aimed at an interdisciplinary audience, including evolutionary biologists with an interest in how mechanistic studies of protein function can be used to address fundamental questions about evolution, and biochemists and physiologists with an interest in how evolutionary approaches can broaden and enrich their field of study. Most previous book-length treatments of Hb structure and function have devoted much space to discussions of Hb disorders such as sickle-cell anemia and various forms of thalassemia (Dickerson and Geis, 1983, Bunn and Forget, 1986). The volume by Steinberg et al. (2009) provides an especially authoritative and complete compendium of information about Hb-related diseases. The present volume has a different aim, so I have not devoted much space to the discussion of pathophysiology except in cases where understanding the etiologies of particular Hb disorders helps illustrate a broader point. Relative to previous books about Hb structure and function, I have sharpened the focus on conceptual issues of relevance to questions about biochemical adaptation and mechanisms of protein evolution.

To lay the foundation, Chapter 1 reviews basic principles of protein structure—the nature of proteins as polymers of amino acids, the variety of amino acids, and the way in which the physicochemical properties of amino acid side chains influence the folding of a polymer into a three-dimensional protein with specific functional properties. Chapter 2 then provides an overview of Hb function and its physiological role in respiratory gas transport. Much of the chapter is devoted to explaining the physiological significance of cooperative O_2 binding by Hb and therefore provides a point of departure for Chapter 3, which provides a brief overview of allosteric theory. Chapter 4 provides an overview of Hb structure and explains the mechanistic basis of allosteric effects. Chapter 5 provides an overview of the evolutionary history of the globin gene superfamily and places the evolution of vertebrate-specific globins in phylogenetic context. The remaining chapters explore the physiological significance of gene duplication and Hb isoform differentiation (Chapter 6), the evolution of novel Hb functions and physiological innovation (Chapter 7), and mechanisms of biochemical adaptation to environmental hypoxia (Chapter 8). Finally, Chapter 9 discusses conceptual issues in protein evolution and provides a synthesis of lessons learned from studies of Hb.

During the preparation of this book, I have been fortunate to receive helpful suggestions from friends and colleagues over the world: Andrea Bellelli (Sapienza University of Rome, Italy), Michael Berenbrink (University of Liverpool, UK), Colin J. Brauner (University of British Columbia, Canada), Thorsten Burmester (University of Hamburg, Germany), Kevin L. Campbell (University of Manitoba, Canada), Angela Fago (Aarhus University, Denmark), Chien Ho (Carnegie Mellon University), Federico G. Hoffmann (Mississippi State University), David Hoogewijs (University of Duisburg-Essen, Germany), Frank B. Jensen (University of Southern Denmark, Denmark), Amit Kumar (State University of New York-Buffalo), Hideaki Moriyama (University of Nebraska), Chandrasekhar Natarajan (University of Nebraska), John S. Olson (Rice University), Juan C. Opazo (Universidad Austral de Chile, Chile), William E. Royer (University of Massachusetts Medical School), Graham R. Scott (McMaster University, Canada), Anthony V. Signore (University of Nebraska), Jeremy R. H. Tame (Yokohama City University, Japan), Tobias Wang (Aarhus University, Denmark), Roy E. Weber (Aarhus University, Denmark), and Mark A. Wilson (University of Nebraska). I am also very grateful to F. G. Hoffmann, A. Kumar, and C. Natarajan for their help with figures.

I thank Angela Fago and Roy Weber for their part in maintaining a fun and productive transatlantic collaboration built on strong friendship. Our work together inspired me to write this book.

Finally, I thank my wife, Eileen, our kids, Jessie and Cody, and our dog, Gwydion, for helping to ensure that time spent writing was well-balanced with other pursuits!

Jay F. Storz
Lincoln, Nebraska

References

Bunn, H. F. and Forget, B. G. (1986). *Hemoglobin: Molecular, Genetic and Clinical Aspects*, Philadelphia, PA, W. B. Saunders Company.

Dickerson, R. E. and Geis, I. (1983). *Hemoglobin: Structure, Function, Evolution, and Pathology*, Menlo Park, CA, Benjamin/Cummings.

Perutz, M. F. (2001). Molecular anatomy and physiology of hemoglobin. *In:* Steinberg, M. H., Forget, B. G., Higgs, D. R., and Nagel, R. L. (eds.) *Disorders of Hemoglobin: Genetics, Pathophysiology, and Clinical Management*, pp. 174–96. Cambridge, UK, Cambridge University Press.

Steinberg, M. H., Forget, B. G., Higgs, D. R., and Weatherhall, D. J. (2009). *Disorders of Hemoglobin: Genetics, Pathophysiology, and Clinical Management*, 2nd edition, Cambridge, UK, Cambridge Medicine.

Contents

CHAPTER 1

Principles of protein structure

1.1 Introduction

In the chapters to follow, we will explore the oxygenation properties of Hb and its physiological role in respiratory gas transport. Hb is a complex and exquisitely constructed molecular machine. The circulatory conveyance of chemically bound O_2 is its *raison d'être* and it also has physiologically important interactions with carbon dioxide (CO_2) and nitric oxide (NO). In each of the four subunits of the tetrameric Hb protein, a single O_2 molecule binds reversibly to an iron atom that is coordinated by four coplanar nitrogens at the center of a flat porphyrin ring (the heme group). Each heme is enclosed in a folded protein chain (the globin), to which it is bound via a coordinate covalent bond. How does the protein chain modulate O_2 binding by the iron atom? And what is the benefit of enclosing the iron atom in a protein cage in the first place? To provide a foundation for addressing these questions and others, we will first briefly review relevant principles of protein structure.

1.2 The hierarchy of protein structure

Proteins are polymers of amino acids that are linked in a specific linear sequence by peptide bonds. The 20 standard amino acids have different side chains with different physicochemical properties (Fig. 1.1), and these properties dictate how the polymer folds into a three-dimensional structure—the native conformation. The way in which the linear chain of amino acids spontaneously folds into a functionally intact, three-dimensional molecule, and the rules that govern this origami-like process, are foundational problems in the field of structural biology.

The primary structure of a protein refers to the linear sequence of amino acids comprising a single polypeptide chain. This is the only level in the hierarchy of protein structure that can be directly predicted from gene sequence. The genetic code indicates which particular amino acid is specified by each triplet of mRNA nucleotide bases (Fig. 1.2), and therefore serves as a Rosetta Stone for translating DNA sequence into amino acid sequence. The secondary structure of a protein refers to the spatial arrangement of residues that are close together in the linear sequence, and which give rise to regular, repeating patterns of hydrogen-bonded main chain conformations such as α-helices and β-sheets. The tertiary structure of a protein refers to the spatial assembly of helices and sheets and the interactions between them. It describes the three-dimensional folding pattern of the polypeptide chain. Finally, quaternary structure refers to the spatial arrangement of individual subunit polypeptides, and is therefore only applicable to multimeric proteins like Hb.

1.3 The peptide bond

"A protein molecule has the advantage of constructional simplicity that comes from being built from backbone parts of standardized dimensions."
—**Dickerson and Geis (1983)**

Each amino acid consists of a central carbon atom (called the α-carbon or "C_α") with an attached amino group (NH_3^+), a carboxylic acid group (COO^-), a

Hemoglobin: Insights into Protein Structure, Function, and Evolution. Jay F. Storz, Oxford University Press (2019).
© Jay F. Storz 2019. DOI: 10.1093/oso/9780198810681.001.0001

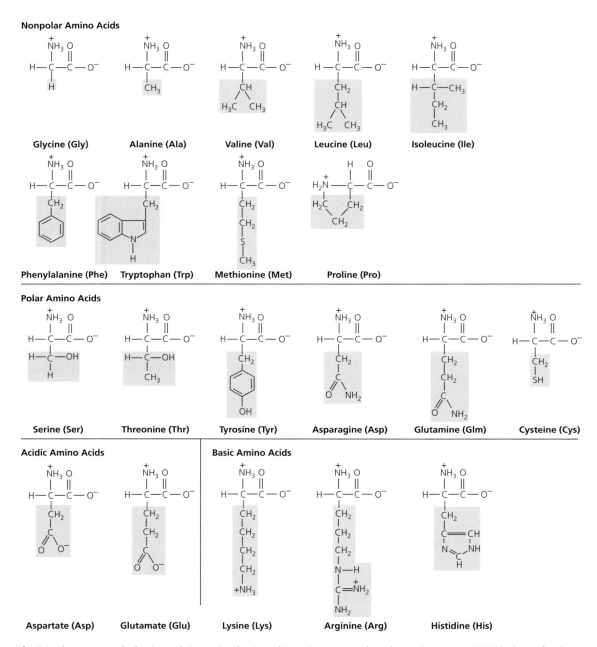

Fig. 1.1. The twenty standard amino acids in proteins. The pictured ionization states are those that predominate at pH 7. Side chains of each amino acid are indicated by shading.

Reproduced from McKee and McKee (2012) with permission from Oxford University Press.

Second base

	U	C	A	G	
U	UUU Phe F UUC Phe F UUA Leu L UUG Leu L	UCU Ser S UCC Ser S UCA Ser S UCG Ser S	UAU Tyr Y UAC Tyr Y UAA Stop UAG Stop	UGU Cys C UGC Cys C UGA Stop UGG Trp W	U C A G
C	CUU Leu L CUC Leu L CUA Leu L CUG Leu L	CCU Pro P CCC Pro P CCA Pro P CCG Pro P	CAU His H CAC His H CAA Gln Q CAG Gln Q	CGU Arg R CGC Arg R CGA Arg R CGG Arg R	U C A G
A	AUU Ile I AUC Ile I AUA Ile I AUG Met M	ACU Thr T ACC Thr T ACA Thr T ACG Thr T	AAU Asn N AAC Asn N AAA Lys K AAG Lys K	AGU Ser S AGC Ser S AGA Arg R AGG Arg R	U C A G
G	GUU Val V GUG Val V GUA Val V GUG Val V	GCU Ala A GCC Ala A GCA Ala A GCG Ala A	GAU Asp D GAC Asp D GAA Glu E GAG Glu E	GGU Gly G GGC Gly G GGA Gly G GGG Gly G	U C A G

First base (left), Third base (right)

RNA Codon   Amino acid

Fig. 1.2. The standard genetic code. The coding sequence of a gene specifies the order in which amino acids are linked together in the encoded protein, with each unique triplet of nucleotide bases (codon) specifying a particular amino acid or a punctuation mark (e.g., a stop codon that signals the end of the protein chain).

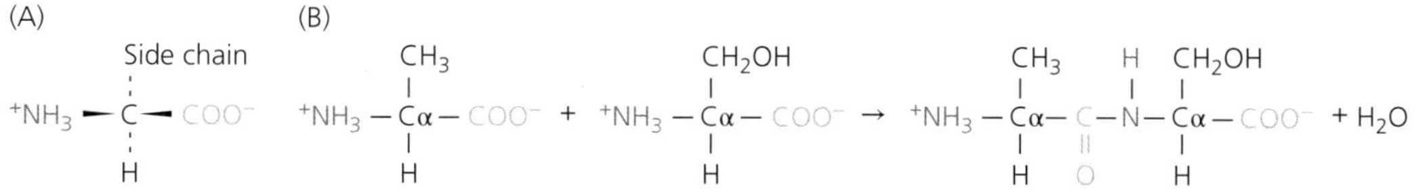

Fig. 1.3. Amino acids and the formation of peptide bonds. (A) General structure of an amino acid. (B) A peptide bond is formed when the α-carboxyl group of one amino acid reacts with the amino group of another, resulting in the elimination of a water molecule. This example shows the formation of a dipeptide involving alanine (side chain = CH_2) and serine (side chain = CH_2OH). In the cell, the synthesis of peptide bonds is an enzymatically controlled process that occurs on the ribosome and is directed by the mRNA template.

Modified from Lesk (2010) with permission from Oxford University Press.

hydrogen atom, and a characteristic side chain (Fig. 1.3A). Amino acids are covalently linked together in a protein chain via peptide bonds that are formed by the reaction between the carbon atom of the carboxylic acid group of amino acid "n" and the amino group of amino acid "$n + 1$," resulting in the loss of a water molecule (Fig. 1.3B). Individual amino acids are linked together end-to-end to form the main chain or backbone of a protein polypeptide with 20 different possible types of side chain protruding from the α-carbon of each residue (Fig. 1.4). The amino group of the first residue in the chain and the carb-

oxylic acid group of the last residue remain intact, so the polypeptide chain is described as extending from the amino (N) terminus to the carboxy (C) terminus. Individual residues in the chain are numbered accordingly, from the N- to the C-terminus. This is also the order in which polypeptides are synthesized at the ribosome, as each new amino acid is added to the free carboxy terminus of the growing chain.

The stability and polarity of peptide bonds is attributable to resonance, the delocalization of electrons over adjoining chemical bonds (Martin, 2001, Voet and Voet, 2011). Because of resonance,

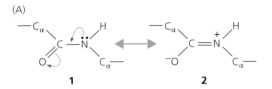

Fig. 1.4. Proteins are polymers of amino acids containing a constant main chain of repeating units with variable side chains. It is the sequence of variable amino acid side chains that gives each protein its distinctive character.

(A)

(B)

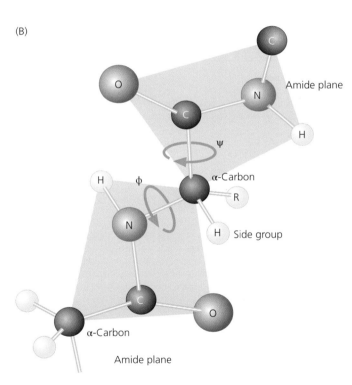

Fig. 1.5. The peptide bond. (A) Resonance forms of the peptide bond. (B) The partial double-bond character of the C-N bond means that all of the atoms connected by the shaded quadrilateral lie in the same two-dimensional plane, called the amide plane. Due to the rigidity of the peptide bond, the polypeptide chain can only rotate around the Cα—C bond (with a rotation angle of ψ) and the Cα—N bond (with a rotation angle of ϕ). This limits conformational degrees of freedom of the polypeptide chain, and therefore dictates allowable folding patterns in three dimensions.

Reproduced from McKee and McKee (2012) with permission from Oxford University Press.

the double-bond character of the C=O bond is shared with the adjoining C-N bond (Fig 1.5A), thereby preventing free rotation about that bond. Consequently, the three non-hydrogen atoms that make up each individual peptide bond (the carbonyl oxygen O, the carbonyl carbon C, and the amide nitrogen N) lie in the same two-dimensional "amide plane." Consecutive amide planes can only rotate about the single N-Cα and Cα-C bonds, so the α-carbons represent swivel points in the polypeptide chain. The angle of rotation around the N-Cα bond is called the Φ (phi) torsion angle and that around the Cα-C bond is called the Ψ (psi) torsion angle (Fig. 1.5B). The allowable angles of rotation around the N-Cα and Cα-C bonds are constrained by steric hindrance between main chain atoms and side chain atoms. Thus, rotatable (but sterically constrained) covalent N-Cα and Cα-C bonds alternate with comparatively rigid peptide bonds along the main chain, and this imposes limits on the number of possible folded conformations a polypeptide chain can adopt (Ramachandran et al., 1963, Richardson and Richardson, 1990).

1.4 Folded proteins are mainly stabilized by weak, non-covalent interactions

The main chain polypeptide is linked together by covalent bonds, but the three-dimensional structure of native state proteins is mainly stabilized by a multitude of non-covalent, weakly polar interactions (Burley and Petsko, 1988, Jaenicke, 2000). These weakly polar interactions depend on the electrostatic attraction between opposite charges. The strength of association between positively and negatively charged atoms or groups of atoms depends on the distance between them and whether the interaction involves full or partial charges.

Van der Waals interactions involve a weak attractive force between atoms caused by fluctuations in electron density around their nuclei (Fig. 1.6A). The interaction is strongest between groups that are the most polarizable, such as the methyl groups and methylene groups of hydrophobic amino acid side chains. Van der Waals interactions are exclusively short range, generally involving atoms less than 5 Å apart.

Hydrogen bonds are formed between atoms of nitrogen (N) and oxygen (O) via an intermediate hydrogen atom (H): For example, N-H...N, NH...O, OH...O, or OH...N, where in each case the hydrogen is covalently bonded to the atom on the left (the donor) and more weakly bonded to the negatively polarized, non-bonded atom on the right (the acceptor) (Fig. 1.6B). The most common hydrogen bonds in proteins involve the N-H and C=O groups of the polypeptide main chain. In this hydrogen bond, N-H...O=C, the typical H...O distance is 1.9–2.0 Å, whereas the covalent N-H distance is ~1 Å. The hydrogen atom that is covalently bound to the more electronegative donor atom has a partial

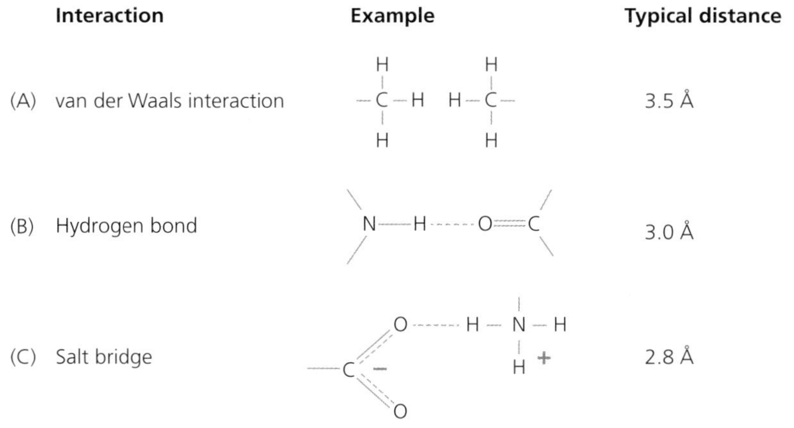

Interaction	Example	Typical distance
(A) van der Waals interaction		3.5 Å
(B) Hydrogen bond		3.0 Å
(C) Salt bridge		2.8 Å

Fig. 1.6. Summary of non-covalent interactions that stabilize polypeptides. Interatomic distances for the different interactions are highly context-dependent, so the values shown here should be regarded as approximate averages.

positive charge, and is therefore attracted to the partial negative charge of the acceptor. Although individually quite weak, hydrogen bonds are great in number. They have been aptly described as "the lacing that stitches a protein molecule together" (Dickerson and Geis, 1983). If the donor and acceptor atoms are both fully charged, then the hydrogen-bonded ion pair is called a salt bridge (Fig. 1.6C). The bonding energy of a salt bridge is much higher than that of a hydrogen bond in which one or both members of the bonded pair are only partially charged.

The peptide bonds that link together the main chain are the only covalent bonds that contribute to the stabilization of most proteins but, in some cases, additional covalent interactions can form cross-links between loops of the polypeptide chain. For example, a disulfide bridge is formed by the connection between the sulfur atoms of two cysteine residues (which have $-CH_2$-SH side chains) that are brought together in the tertiary structure of a protein. Disulfide bridges are generally rare in intracellular proteins because the reduced -S-H groups of cysteine residues are favored over the oxidized -S-S- linkage in the reducing environment of the cytosol.

1.5 The physicochemical properties of amino acids: consequences for protein folding and function

Amino acids can be divided into three main classes defined by the chemical nature of their side chains: non-polar (hydrophobic), polar (hydrophilic), and charged (acidic or basic) (Fig. 1.1). These properties influence how amino acids interact with one another and with water, interactions that have important consequences for protein folding and protein stability in the native state.

The amino acid glycine has a single hydrogen atom as a side chain and therefore confers conformational flexibility; it is often found in tight corners and in the interior of compact, tightly packed proteins. Amino acids with non-polar, hydrophobic side chains are typically buried in the interior of the folded protein and are not exposed to solvent, whereas those with polar or charged side chains typically occur in exterior, solvent-exposed positions where they can participate in hydrogen-bonding interactions. Some of these latter amino acids have

different charge states depending on the solvent pH and the structural microenvironment in which they occur. For example, the carboxylic acid side chains of aspartic acid and glutamic acid have acid dissociation constants (pK_as) of 4–5 in aqueous solution, so they are normally unprotonated and negatively charged at physiological pH (Fig. 1.7A). However, in the hydrophobic interior of a protein or in the presence of a neighboring negative charge, the pK_a may shift higher than 7, allowing such residues to act as proton donors at physiological pH. Three other hydrophobic amino acids (lysine, arginine, and histidine) have side chains with unused electron pairs that attract protons, which confers a positive charge. The amine group of lysine has a pK_a >10 in aqueous solution, so it is normally protonated and positively charged at physiological pH (Fig. 1.7B). However, in a non-polar environment or in the presence of a neighboring positive charge, the pK_a for lysine may shift below 6, so it can act as a proton acceptor. The guanidino group of arginine has a pK_a range of 11.5 to 12.5, so it is permanently protonated and positively charged at physiological pH (Fig. 1.7C). The imidazole side chain of histidine has two titratable –NH groups, each with a pK_a of ~6. When one of the –NH groups is deprotonated, the pK_a of the other group increases above 10. When both –NH groups are protonated, the residue as a whole is positively charged (Fig. 1.7D, right). When only one of the two groups is protonated, the side chain is neutral (Fig. 1.7D, left) and therefore has the capacity to accept or donate protons in response to fine-tuned changes in pH in the physiological range. The remaining amino acids are not excessively hydrophobic (so they are not forced into the interior of the folded protein) nor excessively polar (so they are not forced to the exterior). Serine, threonine, glutamine, and asparagine do not ionize but are able to donate or accept hydrogen bonds simultaneously. Such residues are well-suited to the formation of intermolecular interfaces between different protein domains or subunits.

Whereas hydrophobic (non-polar) amino acids mainly participate in van der Waals interactions, amino acids with charged or polar side chains are able to form hydrogen bonds with one another, with the polypeptide main chain, with polar organic molecules, and with water molecules (Fig. 1.8). The most common form of hydrogen bond in proteins is

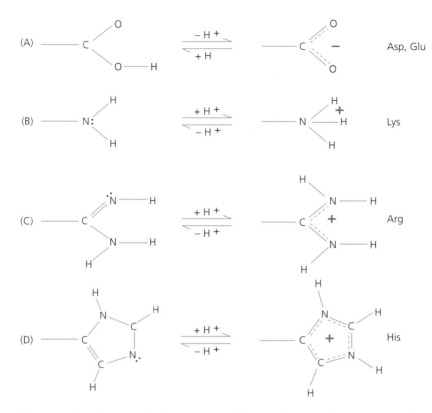

Fig. 1.7. Acidic and basic groups found in protein side chains. Aspartic and glutamic acids contain the carboxyl group (A), which can lose a proton and acquire a negative charge. Lysine side chains have an amine group (B) whose nitrogen electron pair can attract a proton, thereby acquiring a positive charge. Arginine has a more complicated guanadino group (C), but the principle of acquiring an H^+ ion is the same. Histidine (D) can attract a proton to one of its imidazole ring nitrogens. Dashed lines in these drawings represent electrons that are shared or delocalized among multiple atoms, which lowers the overall energy of the group and makes it more stable.

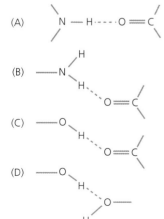

Fig. 1.8. Hydrogen bonds are non-covalent interactions involving one hydrogen atom that has a partial positive charge—because it is covalently bound to a more electronegative nitrogen (N) or oxygen (O) atom—and a nearby atom that has a partial negative charge. Hydrogen bonds are frequently encountered in proteins with NH (A, B) or OH (C, D) as hydrogen atom donors, and CO (A, B, C) or OH (D) as acceptors. When such groups occur in the interior of a globular protein, they are almost always involved in hydrogen bonding.

that between the NH and CO of consecutive amino acids in the main polypeptide chain, but side chains of several amino acids can also form such bonds. The NH and NH_2 of asparagine, glutamine, lysine, arginine, histidine, and tryptophan can serve as H atom donors (Fig. 1.8A,B), as can the OH groups of serine, threonine, and tyrosine (Fig. 1.8C,D). The oxygen atoms of these OH groups can also act as hydrogen-bond acceptors (Fig. 1.8D), as can the CO of asparagine and glutamine (Fig. 1.8A,B,C). All of the above-mentioned amino acids are polar, so they will only be found in the interior of a globular protein if they form hydrogen bonds with other donor or acceptor residues.

In addition to serving as hydrogen-bond donors, amino acid side chains with unused electron pairs can form coordinate covalent bonds with metal ions such as iron, zinc, magnesium, and calcium, which can play functionally important roles in binding reactions or catalysis. The imidazole side chain of histidine is an especially common ligand to iron atoms, and this is how the heme group is held in place in globin proteins such as Mb and Hb.

1.6 Secondary structure

Segments of folded polypeptide chains generally adopt conformations in which the Φ and Ψ torsion angles of the main chain repeat in a regular pattern, forming elements of secondary structure such as α-helices and/or β-sheets (Pauling et al., 1951). These characteristic structural elements keep the main chain in an unstrained conformation and they satisfy the hydrogen-bonding potential of the main chain N-H and C=O groups. Mb and the Hb subunits have an exclusively α-helical folding pattern with consecutive helices linked by short interhelical loops. An α-helix consists of a single-stranded, right-handed coil (like a clockwise spiral staircase) with 3.6 amino acid residues per turn, and a pitch (the distance between corresponding points per turn) of 0.54 nm (Fig. 1.9). The side chains of α-helical residues are oriented outward in a radial fashion. The protruding side chains determine the interactions of the α-helix with other parts of the folded protein or with other proteins. Successive turns of the helix are stabilized by a regular pattern of hydrogen bonding between amino acid residues located one turn away

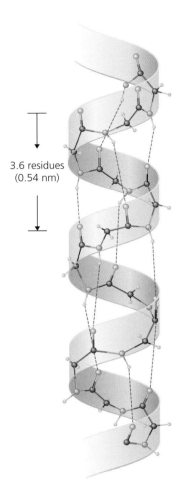

3.6 residues (0.54 nm)

Fig. 1.9. The α-helix is a common element of protein secondary structure, and it represents the modular building block of the globin fold. Hydrogen bonds form between N-H and C=O groups along the long axis of the helix. There are 3.6 residues per turn of the helix, which has a pitch of 0.54 nm.

Reproduced from McKee and McKee (2012) with permission from Oxford University Press.

from one another. If the hydrogen atom acceptor is the CO of amino acid residue n, then the bond donor is the NH of residue $n + 4$, one turn away in the helix. In both Mb and Hb it is common for individual α-helices to be capped by a tightening of the final turn, where the CO of residue n is hydrogen-bonded to the NH of residue $n + 3$ instead of $n + 4$ (a 3_{10} helical turn), or a loosening of the final turn, where residue n is bonded to residue $n + 5$ (a π-helical turn). In compact, globular proteins, the α-helical structure is interrupted at specific intervals by non-helical segments, thereby allowing the polypeptide chain to fold at sharp angles. In many cases, the interruption of α-helix is caused by the presence of a proline residue. The unique feature of proline is that its side chain (attached to the α-carbon) makes an additional connection to its amine nitrogen. This prevents rotation about the α-carbon and can force a contorted bend in the main polypeptide chain. Thus, prolines typically occur at the protein surface and can only be accommodated within the first three N-terminal residue positions of an α-helix.

In polar solvent, the tendency of non-polar groups to self-associate (the "hydrophobic effect") plays an important role in protein folding. The clustering of hydrophobic side chains from disparate parts of the polypeptide chain promotes compact folding by minimizing the total hydrophobic surface area in contact with water and by bringing polarizable hydrophobic groups into close proximity with one another, which promotes the formation of van der Waals contacts. Interactions with water are therefore critically important: proteins fold into their characteristic tertiary structures by simultaneously maximizing the exposure of hydrophilic groups to the polar solvent while minimizing the exposure of hydrophobic groups. In the interior of a globular protein, the polar main chain N-H and C=O groups cannot form hydrogen bonds with water molecules, so they tend to hydrogen bond with one another, which gives rise to the secondary structure that stabilizes the folded state (Lesk et al., 1980, Rose and Roy, 1980, Richards and Richmond, 1997). The various elements of secondary structure fold into a compact and highly specific topological arrangement, giving rise to a tertiary structure that is stabilized by numerous polar and non-polar interactions (Brandon and Tooze, 1999, Petsko and Ringe, 2004).

1.7 Tertiary structure

When a protein is denatured, the flexible main chain can trace out a limitless number of calligraphic patterns in three-dimensional space. In the native state, by contrast, the same conformational freedom is not thermodynamically permissible. Under physiological conditions of solvent and tem-perature, proteins with identical primary structures typically adopt the same three-dimensional folding pattern. This is because there is a single optimal solution to the "thermodynamic jigsaw puzzle" (Lesk, 2010) that requires satisfaction of hydrogen-bonding potential of polar groups, burying of non-polar groups, and dense packing of residues in the interior of the folded protein. Lesk (2010) described the emergent properties of tertiary structure as follows: "The polypeptide chains of proteins in their native states describe graceful curves in space. These are best appreciated by temporarily ignoring the detailed interatomic interactions and focusing on the calligraphy of the patterns. The major differences among protein structures come not at the level of local interactions, such as formation of helices and sheets; in most proteins these are quite similar. The differences appear rather at the calligraphic level, in which similar substructures are differently deployed in space to give different protein folding patterns."

Mb and the individual subunits of Hb have highly distinctive tertiary structures, in which seven to eight α-helices are connected end-to-end by sharp turns and short interhelical loops. The helices are packed together in a criss-crossing arrangement to form a hydrophobic core and a pocket where the heme group is bound. The protruding side chains of each α-helix fit into grooves along the cylindrical surface of adjacent helices—a "ridges and grooves" arrangement that allows dense packing. Hydrophobic interactions in the core help stabilize the globular structure, while polar, hydrophilic residues on the surface make the protein soluble in water.

1.8 Quaternary structure

Proteins or subunits of proteins that are the products of a single gene are called monomers. Multisubunit assemblies comprising two, three, four, or more distinct polypeptide chains are called dimers, trimers, tetramers, and so on. Multimeric proteins containing structurally identical subunits have the prefix "homo" and those containing structurally distinct subunits have the prefix "hetero." For example, the Hb of jawed vertebrates consists of four monomeric subunits—two α-type and two β-type—and is therefore a heterotetramer ($\alpha_2\beta_2$).

In multimeric proteins, different polypeptide subunits are assembled together by non-covalent interactions and—in some cases—coordinated metal ions or covalent cross-links such as disulfide bridges. Protein subunits are generally held together by the same weak bonds that stabilize the tertiary structures of monomeric proteins: van der Waals interactions, hydrogen bonds, and salt bridges. The intersubunit interfaces are typically closely packed to ensure formation of all possible van der Waals contacts between non-polar groups on opposing subunits, and to ensure that hydrogen-bond donors and acceptors form intersubunit pairings instead of forming hydrogen bonds with water molecules. These non-covalent interactions between subunits require a high degree of complementarity and dense atomic packing, with non-polar groups immediately opposite other non-polar groups, hydrogen-bond donors opposite acceptors, and positive charges opposite negative charges. The density of atomic packing at intersubunit interfaces is similar to that observed in the hydrophobic interior of monomeric proteins.

Having now secured the base camp with this brief overview of protein structure, in Chapter 2 we will explore the functional properties of Hb and its physiological role in respiratory gas transport.

References

Brandon, C. and Tooze, J. (1999). *Introduction to Protein Structure, 2nd edition*, New York, NY, Garland Publishing.

Burley, S. K. and Petsko, G. A. (1988). Weakly polar interactions in proteins. *Advances in Protein Chemistry*, **39**, 125–89.

Dickerson, R. E. and Geis, I. (1983). *Hemoglobin: Structure, Function, Evolution, and Pathology*, Menlo Park, CA, Benjamin/Cummings.

Jaenicke, R. (2000). Stability and stabilization of globular proteins in solution. *Journal of Biotechnology*, **79**, 193–203.

Lesk, A. M. (2010). *Introduction to Protein Science: Architecture, Function, and Genomics*, Oxford, Oxford University Press.

Lesk, A. M., Chothia, C., Ramsay, W., Foster, R., and Ingold, C. (1980). Solvent accessibility, protein surfaces, and protein folding. *Biophysical Journal*, **32**, 35–47.

Martin, R. B. (2001). Peptide bond characteristics. *Metal Ions in Biological Systems*, **38**, 1–23.

McKee, T. and McKee, J. R. (2012). *Biochemistry: The Molecular Basis of Life*, Oxford, Oxford University Press.

Pauling, L., Corey, R. B., and Branson, H. R. (1951). The structure of proteins: two hydrogen-bonded helical configurations of the polypeptide chain. *Proceedings of the National Academy of Sciences of the United States of America*, **37**, 205–11.

Petsko, G. A. and Ringe, D. (2004). *Protein Structure and Function*, Sunderland, MA, Sinauer Associates.

Ramachandran, G. N., Ramakrishnan, C., and Sasisekharan, V. (1963). Stereochemistry of polypeptide chain configurations. *Journal of Molecular Biology*, **7**, 95–9.

Richards, F. M. and Richmond, T. (1997). Solvents, interfaces and protein structure. *Ciba Foundation Symposium*, **60**, 23–45.

Richardson, J. S. and Richardson, D. C. (1990). Principles and patterns of protein conformation. *In:* Fasman, G. D. (ed.) *Prediction of Protein Structure and the Principles of Protein Conformation, 2nd ed.*, pp. 1–98. New York, NY, Plenum Press.

Rose, G. D. and Roy, S. (1980). Hydrophobic basis of packing in globular proteins. *Proceedings of the National Academy of Sciences of the United States of America-Biological Sciences*, **77**, 4643–7.

Voet, D. and Voet, J. G. (2011). *Biochemistry*, Hoboken, NJ, John Wiley and Sons.

CHAPTER 2

A study in scarlet

The role of hemoglobin in blood gas transport

"The respirable portion of air has the property to combine with blood and its combination results in its red color..."
—Antoine Laurent Lavoisier (1862)

"Blood is a very special juice"
—Mephistopheles, in *Faust* (Johann Wolfgang von Goethe, 1808)

2.1 Hemoglobin, the O_2-transport protein

Hb is a red blood cell protein that plays an essential role in sustaining aerobic metabolism by transporting O_2 from the respiratory exchange surfaces (e.g., lungs, gills, or skin), where the partial pressure of O_2 (PO_2) is high, to the cells of respiring tissues, where the PO_2 is low and the partial pressure of CO_2 (PCO_2) is elevated by metabolic activity. After unloading O_2 in the tissue capillaries, Hb facilitates the transport of metabolically produced CO_2 back to the respiratory surfaces to get rid of it via expiration.

Hb is one of the most abundant proteins in the human body. Under normal conditions, each milliliter of human blood contains approximately 5 billion red blood cells, each of which is packed with approximately 280 million Hb molecules in virtually saturated solution (Dickerson and Geis, 1983). Complementing the role of Hb in convective O_2 transport, the related Mb protein serves as an O_2 storage protein and facilitates intracellular O_2 diffusion from the sarcolemma (the plasma membrane of muscle fibers) to the mitochondria of cardiac and skeletal muscle cells (Wittenberg and Wittenberg, 2003, Gros et al., 2010, Helbo et al., 2013).

In humans and other jawed vertebrates, the Hb protein is a heterotetramer composed of two α-chain subunits and two β-chain subunits (Fig. 2.1, Plate 1). Each of these subunit polypeptides contains a heme group—an iron atom at the center of a poryphyrin ring—which reversibly binds a single O_2 molecule in the ferrous state (Fe^{2+}) (Fig. 2.2). Whereas free heme binds O_2 irreversibly and is converted to the ferric state (Fe^{3+}) in the process, Hb can reversibly bind O_2 because the valence state of the iron atom is protected by encapsulating the heme in the globin protein fold. Each tetrameric ($\alpha_2\beta_2$) Hb can therefore reversibly bind four O_2 molecules. Oxygenation changes the electronic state of the Fe^{2+} heme iron, which is why the color of blood changes from the dark, purplish hue characteristic of venous blood to the brilliant scarlet of arterial blood (Fig. 2.3). As stated by the English physician Richard Lower, pioneer of seventeenth-century medicine (and the first scientist to perform a blood transfusion): "[T]he blood takes in air in its course through the lungs, and owes its bright color entirely to the admixture of air...[I]t is equally consistent with reason that the venous blood, which has lost its air, should forthwith appear darker..." (Lower, 1671).

Hemoglobin: Insights into Protein Structure, Function, and Evolution. Jay F. Storz, Oxford University Press (2019).
© Jay F. Storz 2019. DOI: 10.1093/oso/9780198810681.001.0001

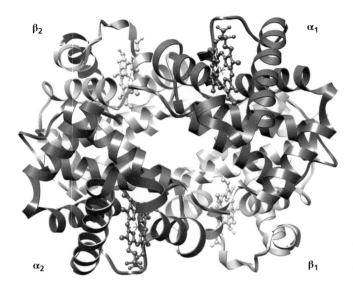

β_2
α_1
α_2
β_1

Fig. 2.1. The Hb tetramer is composed of two α-chains and two β-chains ($\alpha_2\beta_2$), and functions as a pair of semi-rigid $\alpha\beta$ dimers.

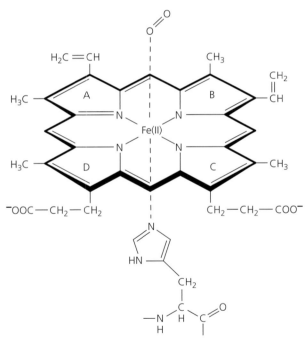

Fig. 2.2. The heme group gives blood its scarlet red color. The heme is a flat porphyrin ring "with an atom of iron, like a jewel, at its center" (Perutz, 1978). The organic component of the heme group—the protoporphyrin—is made up of four pyrrole rings linked by methine bridges to form a tetrapyrrole ring. Four methyl groups, two vinyl groups, and two proprionate side chains are attached. The iron atom at the center of the protoporphyrin is bonded to the four pyrrole atoms. Under normal conditions the iron is in the ferrous (Fe^{2+}) oxidation state. The iron atom can form two additional bonds, one on each side of the heme plane, called the fifth and sixth coordination sites. The fifth coordination site is covalently bound by the imidazole side chain of the globin chain (the "proximal histidine," $\alpha 87$ and $\beta 92$). The sixth coordination site of the iron ion can bind O_2 or other gaseous ligands (CO, NO, CN^-, and H_2S).

In contrast to the tetrameric Hb protein, Mb is a monomer (Fig. 2.4, Plate 2), and is structurally similar to a single heme-bearing subunit of Hb. The monomeric Mb protein and the individual Hb subunits have similar heme-coordination chemistries, but Mb has a much higher O_2 affinity than Hb. This fulfills an important requirement of an efficient O_2-transporting system, as the storage molecule (Mb) should have a higher O_2-affinity than the carrier molecule (Hb) at the low PO_2s that prevail at the sites of O_2 unloading.

2.2 The O_2-transport cascade

Vertebrates primarily rely on aerobic metabolism to meet their energy requirements. Accordingly, a

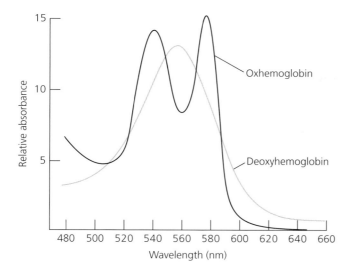

Fig. 2.3. The visible absorption spectra of oxygenated and deoxygenated Hb.

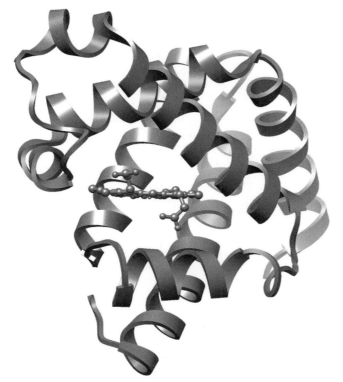

Fig. 2.4. The monomeric Mb is structurally similar to a single heme-bearing subunit Hb.

primary function of the cardiorespiratory system is to extract O_2 from the atmosphere and deliver it to the mitochondria of the cells. Mitochondria then use O_2 as the terminal electron acceptor to produce adenosine triphosphate (ATP) by means of oxidative phosphorylation in the electron transport chain. Since ATP cannot be effectively stored, its continued synthesis requires the O_2-transport system to maintain a balance between O_2 supply (delivery of O_2 to the tissue mitochondria) and demand (chemical utilization of O_2 by oxidative phosphorylation). This balance is maintained by ventilatory and cardiovascular adjustments that jointly ensure an uninterrupted supply of O_2 to the cells of respiring tissues. Increases in O_2 demand (e.g., when a cheetah launches into a sprint or when a hummingbird flits from its perch) trigger a rapid response in convective O_2 transport (increased ventilation and increased blood flow). Conversely, in endothermic and ectothermic vertebrates alike, reductions in O_2 supply can induce metabolic suppression, thereby reducing O_2 demand at the cellular level (Hochachka et al., 1996, Hochachka and Lutz, 2001, Hicks and Wang, 2004).

The organization of the O_2-transport system exhibits tremendous variability among vertebrates. Gas exchange organs include the lungs of obligate air-breathers (most amphibians and all amniotes), the

gills of water-breathers (most fish and larval amphibians), and even skin in the case of some fish and amphibians that rely on cutaneous respiration. There is also considerable variation in cardiovascular anatomy among different vertebrate groups (Burggren et al., 1997). Nonetheless, these varied cardiorespiratory systems have the same primary function of adjusting tissue O_2 delivery in response to changes in O_2 availability and metabolic demand. Whether O_2 diffuses from air that fills the lungs or from water that flows over the gills, it is then transported convectively by the blood to the tissues where it diffuses from the capillaries and is used for respiration in the mitochondria. The transport of O_2 from the respiratory medium (air or water) to the mitochondria in

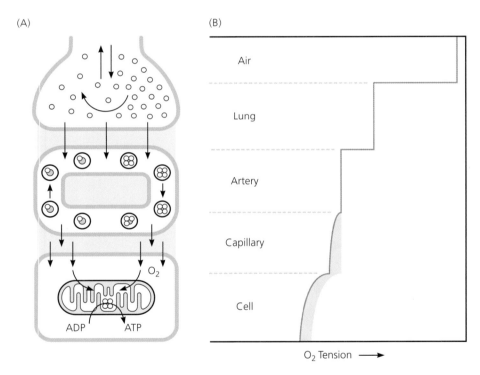

Fig. 2.5. The O_2-transport cascade. (A) Diagrammatic representation of compartments of respiratory gas exchange, with the mammalian cardiopulmonary system shown as an example. O_2 is transported from atmospheric air to the tissue mitochondria along a pathway with several diffusive and convective steps. The rate of O_2 transport in the air compartment of the lungs is determined by the product of ventilation rate and the inspired-alveolar difference in PO_2. The rate of O_2 transport across the blood-gas interface is determined by the product of pulmonary O_2 diffusion capacity and the alveolar-blood PO_2 difference. The remaining steps include convective transport of O_2 by Hb and O_2 diffusion across the blood-tissue interface to the mitochondria, culminating in the utilization of O_2 to generate ATP by oxidative phosphorylation. (B) Diffusive O_2 flux is driven by differences in PO_2 and accounts for the overall reduction in PO_2 across the O_2 cascade. Blood PO_2 declines from the afferent inlet of the tissue capillary bed to the efferent (venous) outlet as Hb unloads O_2 to the perfused tissue. PO_2 also declines with distance from the capillaries, so there should be a gradient of cellular PO_2 reflecting variation in capillary PO_2 and diffusion distance.

Modified from Taylor and Weibel (1981) and Ivy and Scott (2015).

the cells involves four transfer steps that are arranged in series (Fig. 2.5A): (*i*) ventilation of the gas exchange organs (lungs in most air-breathing vertebrates, gills in water-breathers); (*ii*) diffusion of O_2 across the gas exchange surface, from the respiratory medium into the blood; (*iii*) circulation of O_2 throughout the body by blood; and (*iv*) diffusion of O_2 from the blood to the mitochondria of tissue cells, where it is then used to generate ATP via oxidative phosphorylation. The rate of O_2 transfer at each step can be quantified using the Fick principle (for convective steps) and Fick's law of diffusion (for diffusive steps). Differences in PO_2 provide the driving force for O_2 transfer across each of the diffusive steps and account for much of the overall PO_2 gradient between the respiratory medium and the mitochondria (Fig. 2.5B). There is a reverse multistep pathway for the transport of CO_2 from the cells to the respiratory exchange surfaces.

2.3 Convective O_2 transport by the cardiovascular system

In the basic design of the vertebrate cardiovascular system, the heart pumps blood through the arteries to the capillaries, where gas exchange occurs. The deoxygenated blood is then returned to the lungs or gills by the venous circulation. Cartilaginous fishes, ray-finned fishes, and lobe-finned fishes all have a single-cycle circulatory system in which a two-chambered heart pumps blood through the gills and thence to the tissues. In animals with lungs, there is a separate pulmonary circulation that operates in parallel with the systemic circulation. The completeness of separation between the pulmonary and systemic circulations is determined by the anatomical compartmentalization of the heart (Hicks, 1998). In amphibians, turtles, and squamate reptiles (lizards and snakes), the chambers of the heart are not fully divided, so O_2-rich blood returning to the heart via the pulmonary circulation can mix with O_2-poor blood returning from the systemic veins. In crocodilians there is also potential for the recirculation of desaturated blood in the systemic circulation, but for different reasons of cardiac anatomy. Among air-breathing vertebrates, the pulmonary and systemic circulations are completely compartmentalized only in birds and mammals.

Cardiac output (Q) is defined as the product of heart rate (f_H) and the volume of blood that is ejected during each heartbeat (stroke volume, V_S):

$$Q = f_H \times V_S. \qquad (2.1)$$

In accordance with the Fick principle, the rate of O_2 consumption (VO_2) is the product of cardiac output and the difference in the O_2 concentrations of arterial and venous blood ($[O_2]_a$ and $[O_2]_v$, respectively):

$$VO_2 = Q \times ([O_2]_a - [O_2]_v). \qquad (2.2)$$

This Fick equation can also be written in terms of PO_2 and the blood-O_2 capacitance coefficient (βbO_2), which quantifies the amount of O_2 that is unloaded for a given arterial-venous difference in PO_2:

$$VO_2 = Q \times \beta bO_2 \times (P_aO_2 - P_vO_2), \qquad (2.3)$$

where $\beta bO_2 = ([O_2]_a - [O_2]_v)/(P_aO_2 - P_vO_2)$, the arterial-venous difference in O_2 concentration divided by the arterial-venous difference in PO_2.

The blood-O_2 capacitance coefficient, βbO_2, is determined by both the quantity of Hb (its concentration in the blood) and the quality of Hb (its O_2-binding properties). Whereas the O_2 capacitance of air or water is constant for a given temperature, the O_2 capacitance of the blood changes as a function of PO_2. This is due to special O_2-binding properties of Hb, as explained in section 2.7. Changes in the O_2-binding properties of Hb can therefore alter both O_2 loading at the respiratory surfaces and O_2 unloading in the systemic capillaries. Such changes are important for maintaining tissue O_2 supply when arterial PO_2 decreases (hypoxemia) or when O_2 demand increases (e.g., during exercise or digestion).

2.4 Hb-O_2 transport as a key innovation in vertebrate evolution

Vertebrates evolved two key innovations for sustaining O_2 flux to metabolizing cells in support of aerobic ATP synthesis. The first is a closed circulatory system whereby blood is pumped through a system of blood vessels that actively deliver O_2 to cells throughout the body. The second is the use of Hb as an O_2 carrier in red blood cells. Nonvertebrate chordates like amphioxus possess a rudimentary circulatory system that is laid out on the same general pattern as that in vertebrates except that blood is pumped by the coordinated contraction of blood vessels lined with special myoepithelial cells. However, the blood is a colorless plasma that lacks any type of O_2-carrying pigment to augment O_2 content.

The use of Hb as a specialized O_2-transport protein played a key role in the evolution of aerobic energy metabolism in early vertebrates and it alleviated physiological constraints on maximum attainable body size. The low solubility of O_2 results in low O_2 concentrations in blood plasma, $\sim 10^{-4}$ M under normal physiological conditions, whereas whole blood containing ~ 150 g of Hb/L (as is typical in humans) can carry O_2 at concentrations as high as 0.01 M, which is about the same as in air. Thus, Hb dramatically increases the O_2-carrying capacity of the blood. Without Hb to augment arterial O_2 concentration, the fluid convection of physically dissolved O_2 in the blood plasma would not be generally sufficient to meet the metabolic demands of active, free-living vertebrates. A remarkable exception to this rule is provided by icefish in the family Channichthyidae that inhabit the freezing, ice-laden waters surrounding the continental shelf of Antarctica (Plate 3). These physiologically enigmatic fish do not express Hb and a number of species do not express Mb either (Sidell and O'Brien, 2006).

To appreciate the role of Hb in convective O_2 transport, let us consider two alternative solutions to the problem of distributing O_2 to aerobically metabolizing tissues: O_2 diffusion in a tracheal system (as in insects and other arthropods) and convective transport of physically dissolved O_2 in a closed circulatory system (as in the above-mentioned icefish). These have both proven to be viable solutions in particular circumstances, but they involve different sets of physiological constraints that make them untenable as general solutions for vertebrate life.

2.5 The limits of tracheal respiration: lessons from giant insects

A reliance on passive O_2 diffusion to sustain aerobic metabolism is only possible for very small animals with specialized modes of respiratory gas exchange. Many insects and other arthropods do not express any form of respiratory pigment to increase the O_2

content of the hemolymph. Instead, they have elaborate tracheal systems to facilitate the diffusive conductance of O_2 throughout the body, capitalizing on the fact that the O_2 diffusion coefficient is $\sim 8,000$-fold higher in air than in water (Dejours, 1981). In tracheal respiratory systems, the conduits for gas transport are fine tubes that open directly to outside air through spiracles in the exoskeleton. The tracheal tubes ramify into increasingly fine tubules like the bronchioles of our lungs, penetrating every cell in the insect's body. For animals the size of modern-day insects, this is a highly efficient respiratory system because O_2 diffuses through the tracheal tubules in the gas phase and only passes into solution at the last possible moment when it enters the muscle cells.

The maximum distance for passive O_2 diffusion down a blind-ended trachiole is determined by the ambient PO_2. Thus, beyond a certain threshold, an increase in insect size—and the concomitant increase in O_2 diffusion distance—would compromise the oxygenation of muscle cells. This constraint on diffusion distance is especially severe for flying insects because of their high rates of aerobic metabolism.

Crustaceans such as crabs and lobsters are not subject to the same constraints on diffusive gas exchange because—like vertebrates—they make use of circulatory O_2-carrier proteins (functionally analogous to vertebrate Hbs) for convective O_2 transport. As pointed out by the mathematical biologist and population geneticist, J. B. S. Haldane (Haldane, 1927): "If the insects had hit on a plan for driving air through their tissues instead of letting it soak in, they might well have become as large as lobsters, though other considerations would have prevented them from becoming as large as man." Physical constraints associated with diffusion-limited respiratory systems and open circulatory systems ensure that the giant, marauding insects of 1950s horror movies will never become a reality. However, these constraints on O_2 diffusion distances are not immutable; they are expected to change in accordance with changes in atmospheric PO_2. It is therefore noteworthy that atmospheric concentrations of O_2 have changed dramatically throughout the Phanerozoic Eon, the time span extending from the Cambrian Period (541 million years ago) to the present, which brackets the entire evolutionary history of most modern animal phyla. During a period of time roughly 250–350 million years ago during the late Paleozoic Era (spanning the Carboniferous and early Permian periods), the atmospheric O_2 concentration is estimated to have risen to values as high as 35 percent, well above the 21 percent of our modern-day atmosphere (Berner and Canfield, 1989, Berner, 2009) (Fig. 2.6). An increase in atmospheric O_2 to Carboniferous levels of 35 percent is predicted to increase the rate of O_2 diffusion by nearly 70 percent, allowing O_2 to diffuse over a longer distance in the insect tracheal system (Dudley, 1998). Thus, an atmosphere with an increased O_2 concentration would allow insects to grow larger without compromising O_2 delivery to aerobically metabolizing cells. As an aside, it is

interesting to consider that an atmosphere with an O_2 concentration of 30–35 percent would have posed a perennial fire hazard; single lightning strikes could ignite conflagrations that reduced even the wettest rainforests to cinders.

Consistent with the idea that hyperoxia relaxes constraints on maximum attainable body size in insects with diffusion-limited respiration, the O_2-rich atmosphere of the late Paleozoic Era coincided with a period of insect gigantism that is well-documented in the fossil record (Briggs, 1985, Graham et al., 1995, Clapham and Karr, 2012). In select insect groups, late Paleozoic species were up to 10-fold larger than their modern-day relatives (Shear and Kukalovareck, 1990). Representatives of an extinct order of dragonflies (Protodonata)

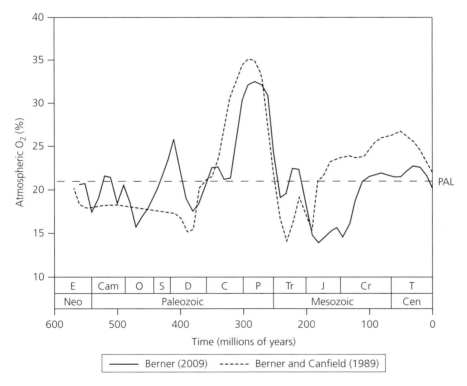

Fig. 2.6. Changes in atmospheric O_2 concentration during the Phanerozoic Eon, estimated by Berner and Canfield (Berner and Canfield, 1989) and Berner (Berner, 2009). The "2009 Berner curve" estimates that there was a roughly 12% increase in O_2 concentration during the 45 million years spanning the interval between the Late Carboniferous and Middle Permian, followed by a subsequent 14% reduction over the 35 million years spanning the interval between the Late Permian and the Early Triassic. Abbreviations of geologic periods are (from most ancient to most recent): *E*, Ediacaran; *Cam*, Cambrian; *O*, Ordovician; *S*, Silurian; *D*, Devonian; *C*, Carboniferous; *P*, Permian; *Tr*, Triassic; *J*, Jurassic; *Cr*, Cretaceous; *T*, Tertiary. Abbreviations of geological eras are: *Neo*, Neoproterozoic; *Cen*, Cenozoic.

Modified from Graham et al. (2016).

attained wingspans of over 70 cm and bodies that were fivefold longer than those of the largest extant dragonflies. Extinct relatives of contemporary mayflies had wingspans of over 50 cm. The trend of Paleozoic gigantism is also evident in fossils of other terrestrial arthropods such as millipedes and scorpions, the extant relatives of which have diffusion-limited tracheal respiration (Graham et al., 1995). In each of these different arthropod groups, ecological factors related to competition and/or predation presumably determined the selective benefits of evolving an increased body size; the higher atmospheric concentration of O_2 simply relaxed constraints on the upper size limit by permitting higher rates of diffusion. The Paleozoic giants all but disappeared by the end of the Permian period 250 million years ago, coinciding with a drop in atmospheric O_2 concentration down to levels below the present atmospheric level, and possibly as low as 15 percent (Berner and Canfield, 1989, Berner, 2009).

The falcon-sized dragonflies that patrolled Carboniferous coal swamps were undoubtedly fearsome aerial predators in their time, but there are good reasons to think that they would not have been able to sustain powered flight or even generate enough lift to become airborne in the comparatively thin air of today's atmosphere. This is due to interrelated biomechanical constraints (the lower air density of today's atmosphere would require higher mechanical power output to stay aloft) and physiological constraints on flight metabolism (the lower PO_2 of today's atmosphere would limit tracheal O_2 diffusion, thereby impairing O_2 delivery to the flight muscle) (Dudley, 1998). Recent experimental work has shed light on the nature of physiological trade-offs between insect body size and diffusion-limited tracheal respiration (Harrison et al., 2006, Harrison et al., 2010). These experimental findings help explain why constraints on the upper limits of insect body size are relaxed under hyperoxic conditions (and why contemporary insects remain small), lending credence to the view that prehistoric changes in atmospheric O_2 concentration played a key role in promoting the initial rise and subsequent decline of the late Paleozoic giants.

In summary, giant flying insects with a tracheal respiratory system can get by without a circulatory O_2 carrier, provided that they live in a hyperdense, O_2-rich atmosphere. However, this same solution for tissue O_2 delivery would not work for pipistrelle bats, peregrine falcons, or most other vertebrate species.

2.6 To have and have not: the curious case of the bloodless icefish

The Norwegian marine biologist, Johan Ruud, first heard reports of pale, nearly translucent "blodløs-fisk" (bloodless fish) in the Antarctic Sea from the crewmen of whaling ships in the 1920s. He initially dismissed reports of fish with colorless blood as fanciful shipboard lore. It sounded physiologically heretical. How would it be possible for a free-swimming, obligately aerobic vertebrate to survive and function without circulatory transport of O_2 by red blood cells? In spite of his initial skepticism, confirmatory reports of clear-blooded icefish from reputable fishery biologists eventually compelled Ruud to voyage to the Antarctic to investigate. After setting up a make-shift laboratory on South Georgia Island, Ruud managed to procure blodløs-fisk specimens of his own, and he set about characterizing the properties of their uniquely colorless blood, which—as he confirmed—was utterly devoid of Hb and red blood cells. Ruud's findings (Ruud, 1954) were met with wide-eyed astonishment by the zoological community.

There are sixteen extant species of Antarctic icefish which together comprise the family Channichthyidae (one of eight families in the perciform suborder Notothenioidei). The Antarctic icefish diversified relatively recently, within the last 6 million years or so (Near et al., 2012). None of today's sixteen icefish species possess Hb, as they descend from a common ancestor in which the underlying α- and β-type globin genes were deleted or inactivated. Six of the sixteen icefish species have also dispensed with Mb (Fig. 2.7), so they have pale white hearts and skeletal muscle in addition to colorless blood. Before exploring the physiological consequences of the Hb-less/Mb-less condition in contemporary icefish, it is useful to consider their history and the unique biogeographical circumstances in which they evolved.

Roughly 35 million years ago in the late Eocene/early Miocene, movement of the Earth's tectonic plates severed the narrow isthmus that connected

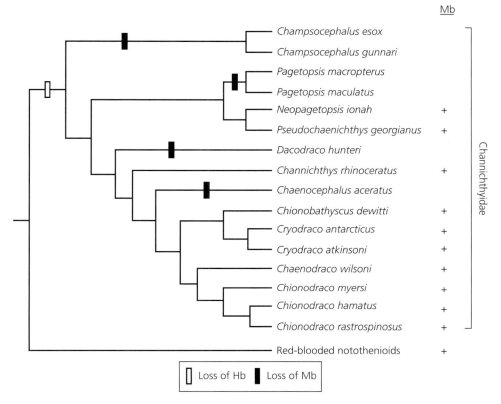

Fig. 2.7. Losses of Hb and Mb in the Antarctic icefish family Channichthyidae. Hb was lost in the stem lineage of the family, and is therefore absent in each of the 16 extant species. By contrast, Mb appears to have been lost four times independently (Sidell and O'Brien, 2006). The tree topology is based on data reported in Near et al. (2003).

the Antarctic Peninsula and the southern tip of South America, opening up the Drake Passage. A few million years later, further seafloor spreading severed the connection between East Antarctica and Australia, opening up the Tasmanian Seaway. Unmoored from its Gondwanan neighbors, the nascent island continent of Antarctica was encircled by a clockwise, circumpolar current that isolated the land mass and its near-shore waters like a castle moat. The establishment of the circumpolar current prevented exchange with warmer tropical waters and resulted in the rapid cooling of the Southern Ocean and the glaciation of Antarctica. Today, near-shore temperatures of Antarctic waters hover just above the freezing point of seawater for most of the year, making it one of the most thermally stable environments on the planet. Against this backdrop of climatic cooling, the Antarctic icefish evolved in

relative isolation and adapted to life in a frigid temperature regime that largely debarred competition from non-notothenioid fish.

To fuel aerobic metabolism without using a circulatory O_2 carrier, icefish rely on the fluid convection of physically dissolved O_2 in their colorless blood plasma. The dissolved O_2 concentration of icefish blood is augmented by their exceedingly low body temperature (due to the inverse relationship between O_2 solubility and temperature), but the O_2 carrying capacity of their blood is still less than 10 percent the capacity of their red-blooded relatives (Holeton, 1970). Importantly, the extremely low body temperature of these ectothermic fish also reduces the metabolic demand for O_2. Since the O_2 capacitance of Hb-containing red blood is far greater than that of Hb-less plasma, a consideration of equation 2.3 suggests that the O_2 demands of icefish could only be satisfied by a compensatory increase in cardiac output. Indeed, research on ice-fish physiology has demonstrated that they have par-tially compensated for the lack of a circulatory O_2 carrier by radically overhauling the entire cardiovas-cular system (Sidell and O'Brien, 2006). Icefish have extraordinarily large hearts for their body size, resulting in a mass-specific cardiac output that is four- to fivefold greater than that of red-blooded teleosts (Hemmingsen et al., 1972). They also have radically increased blood volumes and capillary diameters (Fitch et al., 1984). If the loss of Hb required such dramatic compensatory changes in cardiovascular anatomy, it seems reasonable to expect that the cost was justified by some offsetting energetic advantage. For example, the reduced vis-cosity of Hb-less blood plasma might help sustain adequate rates of blood flow at extremely low tem-peratures. Even red-blooded notothenioid fishes from Antarctic waters have unusually low hemato-crits relative to other teleost fish (the hematocrit is the fraction of blood composed of red blood cells), so perhaps the icefish just went a step further by evolving a hematocrit of zero. Surprisingly, how-ever, comparisons between the clear-blooded ice-fish and their red-blooded notothenioid relatives revealed no evidence that the loss of Hb and red cells confers any offsetting physiological advan-tage. Although the circulatory system of icefish can transport a given volume of blood at lower ener-getic cost, they have to pump a far greater volume of blood to satisfy the same tissue O_2 demand (Sidell and O'Brien, 2006). As stated by Sidell and O'Brien (2006): "In light of these energetic considerations and the rather draconian compensatory alterations in cardiovascular anatomy and physiology seen in ice-fishes, it seems reasonable to conclude … that loss of Hb and red cells did not confer an adaptive advantage to the channichthyids."

In the ancestors of today's icefish, did natural selection favor the loss of Hb (even though the physiological benefits are not discernible in their modern-day descendants)? Or did a relaxation of selection pressure simply fail to prevent the loss of Hb by allowing inactivating mutations to become fixed? Answers to such questions may be lost in the sands of time. For our present purpose, the import-ant lesson is that it *is* possible for relatively large, free-swimming fish to survive and function without a circulatory O_2 carrier, provided that they maintain an exceedingly low rate of metabolism and live in near-freezing, O_2-rich waters. However, this same solution would not work for barracuda, mako sharks, or blue whales. The radiation of vertebrates over the past 500 million years involved the colon-ization of diverse terrestrial and aquatic habitats, and the evolution of exceedingly diverse ways of life, modes of locomotion, and physiological capaci-ties. Vertebrate evolution would have played out very differently without the use of Hb as a circula-tory gas transport molecule. In the following sections we will explore some of the special biochemical fea-tures that make Hb so important as a life-sustaining O_2 carrier.

2.7 The physiological significance of cooperative O_2 binding

Tetrameric Hb is essentially an assembly of four Mb-like subunits, and this multimeric structure is key to its effectiveness as an O_2 carrier. Unlike the monomeric Mb protein, the four subunits of Hb bind O_2 cooperatively, meaning that the binding of O_2 to a given heme iron increases the O_2 affinity of the remaining unliganded hemes and, conversely, O_2 released by each heme reduces the O_2 affinity of the remaining liganded hemes. Max Perutz used a scrip-tural metaphor to describe this all-or-nothing ligand-binding behavior: "For to him who has will more be given, and he will have in abundance; but from him who has not, even that which he has shall be taken away" (Matthew 13: 12) (Perutz, 1978). As a result of cooperative O_2 binding, Hb has a high O_2 affin-ity at the sites of respiratory gas exchange (the alveoli of the lungs in humans and other mammals) where the PO_2 is high, and a reduced affinity at the sites of O_2 delivery in the tissue capillaries where the PO_2 is substantially lower. The physiological sig-nificance of cooperativity is that it permits efficient O_2-unloading over a relatively narrow range of blood PO_2.

The physiological division of labor between Hb and Mb, and their fundamental differences in oxy-genation properties, can be readily grasped by plot-ting fractional saturation values as a function of PO_2. The fractional saturation, Y, is defined as the fraction of potential O_2-binding sites that are liganded with O_2; the value ranges from 0 (no sites are liganded) to 1 (all sites are liganded). The plot of Y vs. PO_2 is known as an O_2 equilibrium curve. For a given protein, the shape and position of the O_2-equilibrium curve provides a graphical summary of O_2-binding properties, and can provide insights into the physiological significance of those properties with regard to tissue O_2 delivery.

In the case of Mb, O_2 binding is described by a simple equilibrium reaction

$$Mb + O_2 \leftrightarrow Mb(O_2) \qquad (2.4)$$

with dissociation constant

$$K = \frac{[Mb]PO_2}{[Mb(O_2)]}. \qquad (2.5)$$

The square brackets denote concentration; since O_2 is a gas, its concentration can be expressed as a partial pressure. The O_2 dissociation of Mb is characterized by its fractional saturation:

$$Y = \frac{[Mb(O_2)]}{[Mb] + [Mb(O_2)]} = \frac{PO_2}{K + PO_2}. \qquad (2.6)$$

If we define P_{50} as the PO_2 at which half of Mb's O_2-binding sites are liganded with O_2 ($Y = 0.5$), then substituting this value into equation 2.6 and solving for K yields the following expression for the fractional saturation of Mb:

$$Y = \frac{PO_2}{P_{50} + PO_2}. \qquad (2.7)$$

This equation describes a hyperbolic curve. Under standard conditions (pH 7.4, 37°C), the P_{50} for normal human Mb is ~2 torr (1 torr = 1 mmHg at 0°C = 0.133 kPa; 760 torr = 1 atm). Plugging this value into equation 2.7 shows that Mb would not release much O_2 over the normal physiological range of blood PO_2, which—in humans—is typically ~100 torr in arterial blood and ~20–40 torr in venous blood depending on the rate of O_2 consumption by the perfused tissue. As a hypothetical example, let us consider a venous PO_2 of 20 torr, which is within the range expected for actively contracting muscle. As shown in Fig. 2.8A, the fraction of Mb with bound O_2 rises steeply as a function of increasing PO_2, and then reaches an asymptote. If Mb were encapsulated in circulating red blood cells, it would be nearly saturated at the high PO_2 prevailing in the

pulmonary capillaries ($Y = 0.98$, meaning that almost all possible binding sites are liganded with O_2) but the problem is that it would remain highly saturated in the tissue capillaries where O_2 needs to be released ($Y = 0.91$). Thus, only $0.98 - 0.91 = 0.07$ of binding sites would contribute to tissue O_2 delivery. Mb is clearly ill-suited to the task of circulatory O_2 transport. In contrast to the hyperbolic curve for the monomeric Mb protein, the O_2 equilibrium curve for tetrameric Hb has a distinctively sigmoid shape (Fig. 2.8B), which is reflective of cooperative O_2 binding (i.e., the binding reactions at individual sites are not independent of one another). Thus, in contrast to the case with Mb, the fractional saturation of Hb has a greater-than-first-power dependence on PO_2.

What are the physiological implications of the different O_2-binding properties of Mb and Hb? Under the same conditions considered here (pH 7.4, 37°C), normal human Hb in red blood cells has a P_{50} of ~26 torr. Thus, in the pulmonary capillaries, Hb would be nearly saturated with O_2 ($Y = 0.98$). When Hb moves to the tissue capillaries and releases O_2 to the cells of working muscle, the saturation level drops substantially ($Y = 0.32$ for a venous blood PO_2 of 20 torr). Thus, a total of $0.98 - 0.32 = 0.66$ possible binding sites contribute to tissue O_2 delivery, resulting in a far more complete unloading of O_2 to the tissues (Fig. 2.9). Is this difference in O_2-transport efficiency simply due to differences in O_2-binding affinity, given that the P_{50} of human Hb is 13-fold higher than that of Mb under physiological conditions? Let us consider a hypothetical non-cooperative O_2-carrying protein with a P_{50} that maximizes the arterial-venous difference in O_2 content for the same situation considered here ($PO_2 = 100$ torr at the site of O_2 loading and 20 torr at the site of O_2 unloading). Under these conditions, the best O_2 transport efficiency that can be achieved would be $0.63 - 0.25 = 0.38$ (Fig. 2.9). Thus, it is the shape of the O_2-equilibrium curve (reflecting subunit cooperativity)—not just its position along the X-axis (reflecting binding affinity)—that determines O_2 transport efficiency. Due to the sigmoid shape of the O_2-equilibrium curve, the amount of O_2 bound by Hb changes significantly over the normal physiological range of PO_2 in the blood. For the arterial-venous difference in PO_2 considered in the aforementioned

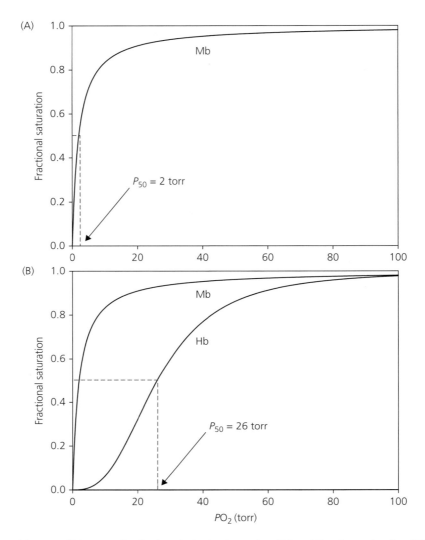

Fig. 2.8. The shape of the O_2-equilibrium curve describes how the fractional saturation of Mb and Hb varies as a function of PO_2. The P_{50} is the PO_2 at which the hemes are 50 percent saturated, so it provides an inverse measure of O_2-affinity. (A) O_2 binding by the monomeric (and, hence, non-cooperative) Mb protein. (B) O_2 binding by tetrameric Hb. The sigmoid shape of the curve reflects the cooperativity of O_2 binding.

example, Hb can transport 9.4-fold more O_2 than the non-cooperative Mb protein and 1.7-fold more than a hypothetical non-cooperative O_2-transport protein with optimal P_{50}. As noted by the Scottish physiologist J. S. Haldane (father of the population geneticist, J. B. S. Haldane, quoted in section 2.5): "A man would die on the spot of asphyxia..." if the sigmoid O_2 equilibrium curve of his blood were suddenly altered to assume the hyperbolic shape of the Mb-O_2 equilibrium curve (Haldane, 1922).

During exercise, the PO_2 of capillary blood perfusing working muscle will decrease in proportion to the rate of cellular O_2 consumption. This scenario further highlights the effectiveness of Hb as an O_2-transport protein because the reduction in venous blood PO_2 (~40 torr for muscles at rest, to ≤20 torr during exercise) occurs over the steepest part of the O_2-equilibrium curve (Fig. 2.10) , which maximizes the amount of O_2 that is unloaded to the cells for a given arterial-venous difference in PO_2.

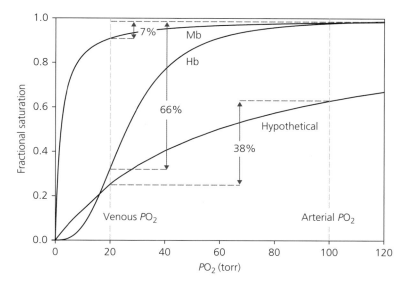

Fig. 2.9. Cooperativity enhances O_2 delivery by Hb. As a result of subunit–subunit interactions, tetrameric Hb unloads a larger quantity of O_2 to the tissues than would be the case for Mb or a hypothetical non-cooperative protein with otherwise optimal P_{50}.

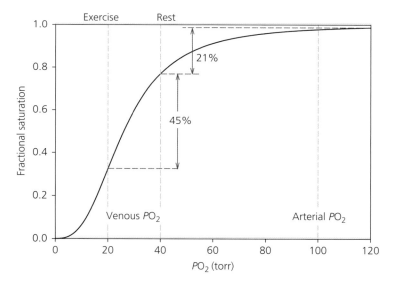

Fig. 2.10. Tissue O_2 delivery by Hb is enhanced during exercise. The drop in PO_2 from ~40 torr in muscle tissue at rest to ~20 torr in working muscle spans the steepest portion of the O_2-equilibrium curve, resulting in a dramatic increase in O_2 unloading.

2.8 The Hill equation as a phenomenological description of Hb-O_2 equilibria

Let us now consider the O_2-transporting properties of a multisubunit globin protein—like Hb—as a function of PO_2. We will follow the general form of Archibald Hill's seminal analysis (Hill, 1913). For a multimeric protein with n subunits (representing n possible O_2-binding sites), O_2 binding can be described by a hypothetical equilibrium:

$$Hb + nO_2 \leftrightarrow Hb(O_2)_n. \qquad (2.8)$$

If n is interpreted as the number of possible O_2-binding sites, then this expression implies that the protein either has zero binding sites liganded or all binding sites liganded. This would be the case if O_2 binding by one subunit of Hb resulted in an infinite increase in the O_2-affinity of the remaining three hemes, in which case partially oxygenated intermediates (i.e., Hbs with one, two, or three sites liganded) would not exist. This infinite cooperativity ($n = 4$ in the case of tetrameric Hb) is not physically possible. However, rather than interpreting n as the number of subunits, it can instead be treated as a non-dimensional parameter that simply quantifies the degree of cooperativity among O_2-binding sites. The Hill equation therefore provides a phenomenological description of the O_2-equilibrium curve, and n is referred to as the Hill coefficient.

The dissociation constant for the equilibrium in equation 2.8 is

$$K = \frac{[Hb]PO_2^n}{[Hb(O_2)_n]} \qquad (2.9)$$

and, as in the case for the Mb monomer, the fractional saturation is expressed as

$$Y = \frac{n\left([Hb(O_2)]_n\right)}{n\left([Hb] + [Hb(O_2)]_n\right)}. \qquad (2.10)$$

Combining equations (2.9) and (2.10) yields

$$Y = \frac{\dfrac{[Hb]PO_2^n}{K}}{[Hb]\left(1 + \dfrac{PO_2^n}{K}\right)}. \qquad (2.11)$$

Algebraic rearrangement and cancellation of terms yields:

$$Y = \frac{PO_2^n}{K + PO_2^n}, \qquad (2.12)$$

which describes the O_2 saturation of a multisubunit protein as a function of PO_2, analogous to equation 2.7.

As with equation 2.7, we can define P_{50} as the PO_2 when $Y = 0.50$, and then substitute this value into the modified equation 2.12,

$$0.5 = \frac{P_{50}^n}{K + P_{50}^n}, \qquad (2.13)$$

such that $K = P_{50}^n$. $\qquad (2.14)$

Substitution of this result back into equation 2.10 yields the following expression of the Hill equation:

$$Y = \frac{PO_2^n}{P_{50}^n + PO_2^n}. \qquad (2.15)$$

If $n = 1$, as is the case for a monomeric protein like Mb or a multimeric protein in which subunits bind O_2 independently, the O_2-binding reaction is non-cooperative, and equation 2.15 describes a simple hyperbola (equation 2.7 is therefore a special case of

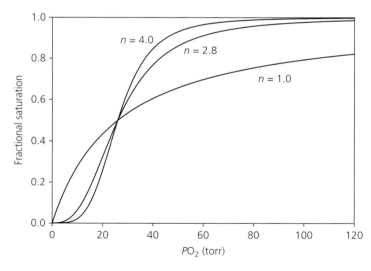

Fig. 2.11. O_2-equilibrium curves for a range of Hill coefficients. A tetrameric protein with $n = 4$ has perfect cooperativity and a protein with $n = 1$ is non-cooperative. The Hill coefficient for human Hb is generally ~ 2.8 under physiological conditions.

equation 2.15 with $n = 1$). A reaction with n greater than 1 exhibits positive cooperativity (as is the case with Hb), and equation 2.15 describes a sigmoidal curve (Fig. 2.11) .

2.9 Hill plots: graphical evaluation of model parameters

The foregoing account explained how the O_2-equilibrium curve can be characterized by two key parameters: the Hill coefficient, n, and the dissociation constant, K. These parameters can be graphically evaluated by rearranging equation 2.15,

$$\frac{Y}{1-Y} = \frac{\dfrac{PO_2^n}{P_{50}^n + PO_2^n}}{1 - \dfrac{PO_2^n}{P_{50}^n + PO_2^n}} = \frac{PO_2^n}{P_{50}^n}, \quad (2.16)$$

and then taking the log of both sides, which yields a linear equation:

$$\log\left(\frac{Y}{1-Y}\right) = n\log[PO_2] - n\log P_{50}. \quad (2.17)$$

A linear plot of $\log [Y(1-Y)]$ versus $\log [PO_2]$ is known as a "Hill plot." The plot has a slope of n and an intercept on the $\log PO_2$ axis of $\log P_{50}$.

As expected, the Hill plot for Mb is linear with a slope of 1.0 (Fig. 2.12A). The intercept of the X-axis at $\log Y/(1-Y) = 0$ gives the log of the equilibrium constant K, expressed in units of torr^{-1}. In contrast to the straight 45° line for Mb, the Hill plot for Hb is qualitatively different (Fig. 2.12B). The plot for Hb is essentially linear for values of Y between 0.1 and 0.9 (10–90 percent saturation) and the Hill coefficient (with notation "n_{50}") is defined as the slope of the line at $Y = 0.5$ ($Y/[1-Y] = 1.0$) where $PO_2 = P_{50}$.

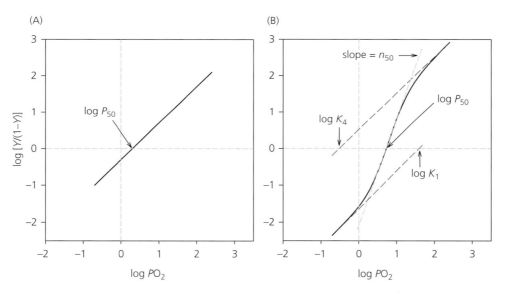

Fig. 2.12. Hill plots for Mb and purified ("stripped") Hb. Y = fractional saturation with O_2, K = equilibrium constant for O_2 binding. (A) The plot for Mb is a straight line with a slope of 1.0, which indicates an absence of cooperativity. The intercept with the horizontal axis at 50 percent saturation ($\log [Y/(1-Y)] = 0$) yields the logarithm of the equilibrium constant K (= $\log P_{50}$, the PO_2 at which half the Mb molecules have bound O_2). If the Mb-O_2 affinity is increased, K is reduced accordingly, as a lower PO_2 is required to attain the same half-saturation value. A decrease in O_2 affinity shifts the line to the right (increase in P_{50}), and an increase shifts it to the left (decrease in P_{50}), but the slope of the line remains unchanged. (B) The sigmoid shape of the curve for Hb reflects cooperative O_2 binding. The upper and lower asymptotes of the curve both approximate a slope of 1.0, similar to the Hill plot for Mb. The lower and upper asymptotes represent O_2 binding to deoxyHb (at low saturation) and oxyHb (at high saturation), respectively. The lower asymptote approximates a slope of 1.0 because—at low PO_2—most Hbs bind no more than a single O_2 molecule, and the hemes therefore react independently, as is the case with Mb. Since O_2 binding at the first site in deoxyHb increases the binding affinity of the remaining unliganded sites, the curve rapidly steepens as O_2 becomes more plentiful. The tangent to the maximum slope of the line is the Hill coefficient (n), which reflects the degree of cooperativity (which, in this example, is 3.0). At higher saturations the curve flattens out again. The upper asymptote of the curve also approximates a slope of 1.0 because at very high PO_2 most Hb molecules will have only a single unliganded site available to bind O_2, and available hemes once again react independently. The intercepts of the lower and upper asymptotes with the horizontal axis at 50 percent saturation ($\log [Y/(1-Y)] = 0$) yields the logarithms of K_1 and K_4, the equilibrium constants for binding the first and fourth O_2 molecule, respectively.

Under physiological conditions, normal human Hb has a Hill coefficient in the range ~2.8–3.0. The numerical value of the Hill coefficient has no physical meaning; a value of 2.8 simply indicates that cooperativity is positive (because it is greater than 1) but not infinite (since it is less than the theoretical maximum of 4, the number of subunits). Some mutant human Hbs have been characterized that have Hill coefficients less than 2.8, which indicates that cooperative O_2 binding is compromised. Mb has a Hill coefficient of 1.0, and the same is true for isolated α- or β-globin monomers and $\alpha\beta$ dimers, demonstrating that the intact $\alpha_2\beta_2$ tetrameric assembly is required for subunit cooperativity.

In Fig. 2.12B, the lower asymptote of the Hill plot for Hb has a slope close to 1.0, indicating an absence of subunit cooperativity at the lowest saturation levels (Y near 0). This reflects the fact that at the initial phase of O_2 binding when Y is less than 0.1, most hemes are unliganded and compete independently for O_2. The upper asymptote of the same Hill plot also has a slope close to 1.0 at the highest saturation levels (Y near 1.0). This reflects the fact that at the final phase of O_2 binding, most Hbs will already have at least three of the four available binding sites liganded; the few remaining unliganded sites will be on different Hb molecules and will therefore bind O_2 independently of one another.

Extrapolation of the lower and upper asymptotes to the X-axis at log $[Y/(1–Y)] = 0$ gives the log of the equilibrium constants for binding the first O_2 molecule to deoxyHb and the final O_2 to oxyHb, respectively. If we extrapolate the lower asymptote in Fig. 2.12B to the X-axis and calculate the exponential function of the intercept, then— according to equation 2.15—the first site in Hb to bind O_2 has an association constant of $K_1 = 0.021$ torr^{-1} (corresponding to $P_{50} = 47.62$ torr). Likewise, extrapolating the upper asymptote shows that the sole remaining unliganded site in Hb binds the final O_2 with an association constant of $K_4 = 3.4$ torr^{-1} (corresponding to $P_{50} = 0.29$ torr). Thus, in this particular example, the last remaining heme binds O_2 with a greater than 160-fold higher affinity than the first.

2.10 The O_2-equilibrium curve as a graphical expression of the Fick principle

We saw earlier that the Fick principle can be used to describe convective O_2 transport by the cardiovascular system (equation 2.3). In this context, the Fick principle simply states that the rate of O_2 consumption by perfused tissue is equal to the product of blood flow to the tissue and the difference in O_2 concentration between the arterial blood entering the tissue capillary bed and the venous blood leaving the capillary bed. The O_2-equilibrium curve provides a way of graphically depicting the Fick principle when the Y-axis is expressed in terms of O_2 concentration, as shown in Fig. 2.13. For given values of P_aO_2 (which is determined by ventilation and O_2 diffusion across the blood gas interface) and P_vO_2 (which is determined by the rate of O_2 consumption by the perfused tissue), the quantity of O_2 delivered

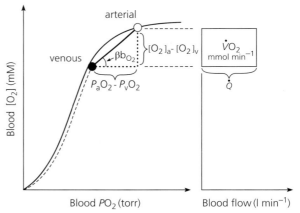

Fig. 2.13. Schematic illustration of blood O_2 transport. The left graph shows an O_2-equilibrium curve under physicochemical conditions prevailing in arterial blood (a, solid curve, open symbol) and venous blood (v, dashed curve, closed symbol). The curve is a plot of blood O_2 concentration (y-axis) vs. PO_2 (x-axis), with paired values for arterial and venous blood connected by a solid line. $[O_2]a – [O_2]v$ denotes the arterial-venous difference in blood O_2 concentration, $P_aO_2 – P_vO_2$ denotes the corresponding difference in PO_2, βbO_2 denotes the blood O_2 capacitance coefficient (see text for details), Q denotes cardiac output, and VO_2 denotes the rate of O_2 consumption. On the right side of the graph, the area of the rectangle is proportional to total O_2 delivery, which can be enhanced by increasing Q and/or by increasing βbO_2. Increases in βbO_2 produce a corresponding increase in the arterial-venous difference in $[O_2]$ through shifts in the shape or position of the O_2-equilibrium curve.

Modified from Storz (2016).

to the tissue is a product of cardiac output (Q) and the blood O_2 capacitance coefficient ($\beta b O_2$). As illustrated in Fig. 2.13, the blood O_2 capacitance coefficient is defined as the slope of the line joining the arterial and venous points on the curve: $\beta b O_2 = ([O_2]_a - [O_2]_v)/(P_a O_2 - P_v O_2)$. An increase in the slope of the line increases the arterial-venous difference in O_2 concentration ($[O_2]_a - [O_2]_v$) for a given arterial-venous difference in PO_2 ($P_a O_2 - P_v O_2$).

In physiological terms, increasing $\beta b O_2$ (via changes in Hb concentration and/or changes in the oxygenation properties of Hb) increases the quantity of O_2 delivered to the tissue for a given difference in PO_2 between the sites of O_2 loading and unloading.

2.11 Cooperative effects of Hb are explained by oxygenation-linked shifts in quaternary structure

The sigmoid shape of the Hb-O_2 equilibrium curve indicates that the ligation state of one heme influences the O_2-affinity of the other hemes, which is why cooperativity is often referred to as "heme–heme interaction." However, given that the hemes do not physically interact, "subunit–subunit interaction" might be a more apt description. What accounts for the indirect coupling between O_2-binding sites in the same Hb tetramer? This question has occupied several highly capable theoreticians and experimentalists since the middle of the twentieth century, and the mechanistic details continue to be refined.

Hb exists in a conformational equilibrium between two semi-discrete conformational states: the deoxygenated form of Hb exists in the so-called T-state (T for "tense"), which has a low-affinity for O_2, and the oxygenated form of Hb exists in the R-state (R for "relaxed"), which has a high affinity for O_2. The terms "tense" and "relaxed" refer to strain on the heme group in the unliganded and liganded states, respectively. T-state Hb is constrained by additional salt bridges within and between subunits (a salt bridge is a bond between a positively charged nitrogen atom and negatively charged

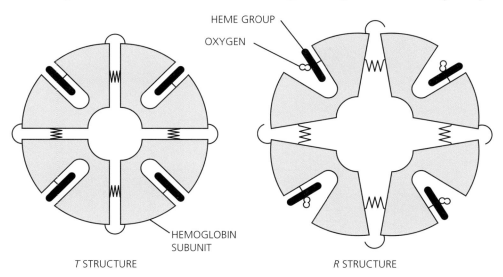

Fig. 2.14. Schematic illustration of the alternative conformational states of Hb. In the T-state molecule on the left, the subunits are constrained by clamps, the springs between them are compressed, and O_2 does not easily gain entry into the tightly shut heme pockets. In the R-state molecule on the right, the clamps have sprung open, the springs are relaxed, and the heme groups are readily accessible to O_2. Binding of O_2 by the T structure confers a degree of strain on the clamps, and once a critical number of sites per tetramer have become liganded, the accumulated strain causes each of the four clamps to burst open in concert, at which point the molecule relaxes into the R structure. Conversely, once O_2 is released from a critical number of sites in the liganded tetramer, each of the four subunits are clamped down in concert, causing the molecule to snap back into the T structure. According to some allosteric models (discussed in Chapter 3), the O_2 affinity of each conformational state is independent of the number of O_2 molecules bound. Modified from Perutz (1978).

oxygen atom), whereas the O₂-binding sites of R-state Hb are subject to less strain and therefore bind O₂ more readily than in the T-state (Fig. 2.14). Thus, to answer the question posed in the preceding paragraph, the binding of O₂ to one heme increases the O₂ affinity of the other hemes by triggering a shift in quaternary structure from the low-affinity T-state to the high-affinity R-state.

In addition to governing cooperative O₂ binding, the oxygenation-linked shift in the T↔R conformational equilibrium is also fundamental to the regulation of heme reactivity by ligands that preferentially bind to deoxy (T-state) Hb at sites remote from the heme active site (Perutz, 1970, Baldwin and Chothia,

1979) (Fig. 2.15). Both types of cooperative interaction occur when the binding of one ligand at a specific site is influenced by the binding of another ligand (an "effector") at a different site on the protein. This is allosteric regulation (allos = "other," steros = "solid" or "space"). In the case of cooperative O₂ binding, the effectors are identical: the binding of O₂ at one heme promotes O₂ binding at the remaining hemes in the same Hb tetramer. This is known as homotropic allostery. By contrast, the binding of non-heme ligands such as H⁺, CO₂, Cl⁻ ions, and organic phosphates can also indirectly influence O₂ binding at the hemes, even though the effector binding sites are structurally remote from

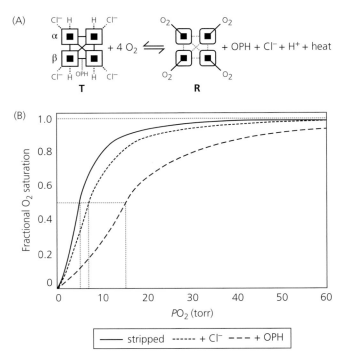

Fig. 2.15. Diagram illustrating the allosteric regulation of Hb–O₂ affinity. (A) The oxygenation reaction of tetrameric Hb ($α_2β_2$) involves an allosteric transition in quaternary structure from the low-affinity T-state to the high-affinity R-state. The oxygenation-induced T→R transition entails a breakage of salt bridges and hydrogen bonds within and between subunits (open squares), dissociation of allosterically bound organic phosphates (OPHs), Cl⁻ ions, and protons, and the release of heat (heme oxygenation is an exothermic reaction). Oxygenation-linked proton binding occurs at multiple residues in the α- and β-chains, Cl⁻ binding mainly occurs at the terminal amino groups of the α- and β-chains in addition to other residues in both chains, and phosphate binding occurs between the β-chains in the central cavity of the Hb tetramer. (B) O₂-equilibrium curves for purified Hb in the absence of allosteric effectors (stripped) and in the presence of chloride ions (+Cl⁻) and organic phosphates (+OPH). The preferential binding of allosteric effectors to deoxyHb stabilizes the T-state, thereby shifting the allosteric equilibrium in favor of the low-affinity quaternary structure. The O₂-equilibrium curves are therefore right-shifted (Hb-O₂ affinity is reduced) in the presence of such effectors. Hb-O₂ affinity is indexed by the P_{50} value—the PO₂ at which Hb is half-saturated. The sigmoid shape of the O₂-equilibrium curves reflects cooperative O₂ binding, involving a PO₂-dependent shift from low- to high-affinity conformations.

Modified from Storz (2016).

the heme pocket. This is known as heterotropic allostery. In principle, heterotropic interactions can be positive or negative (increasing or decreasing reactivity, respectively), but they are typically negative in the case of Hb, as the binding of effector molecules almost always decreases O_2 affinity.

As described in detail in Chapter 4, allosteric effectors such as H^+, CO_2, Cl^- ions, and organic phosphates preferentially bind to deoxyHb at sites located at the N- and C-termini of the globin chains and in the positively charged central cavity of the Hb tetramer (Arnone, 1972, Arnone and Perutz, 1974, O'Donnell et al., 1979, Nigen et al., 1980). The binding of these allosteric effectors stabilizes the T-state through the formation of additional salt bridges within and between subunits (Perutz, 1970, Bettati et al., 1983, Perutz, 1989). This differential stabilization shifts the conformational equilibrium in favor of the low-affinity T-state (i.e., it increases the ratio of the concentration of T-state molecules to that of R-state molecules), thereby reducing Hb-O_2 affinity.

Let us now briefly explore some examples of heterotropic allostery that play an especially important role in the regulation of respiratory gas transport in vertebrates.

2.12 Physiological significance of the Bohr effect for respiratory gas transport

For efficient O_2 transport, Hb-O_2 affinity needs to be jointly optimized for O_2 loading in the pulmonary capillaries and O_2 unloading in the tissue capillaries. A leftward shift of the O_2-equilibrium curve (increased Hb-O_2 affinity) helps maintain a high arterial O_2 saturation even when alveolar PO_2 is below normal, whereas a rightward shift (reduced Hb-O_2 affinity) promotes O_2 unloading to the cells of respiring tissues (West, 1969, Mairbäurl, 1994). The joint optimization of O_2 binding and release is accomplished in part by maintaining a difference in the acid-base status of arterial blood and the capillary (mixed venous) blood in peripheral tissues.

During the circulation of blood through tissue capillary beds, arterial blood is acidified (pH is reduced) due to the diffusion of CO_2 and lactic acid from metabolizing cells. In most vertebrate taxa, H^+

ions are passively distributed across the red cell membrane because the membrane-bound anion exchanger 1 (AE1) protein equilibrates acid-base equivalents (Hladky and Rink, 1977), so changes in plasma pH lead to concomitant changes in red cell pH. In humans, red cell pH is 7.2 when plasma pH is 7.4. Because Hb-O_2 affinity varies as a positive function of pH over the physiological range (6.6–7.6) (Fig. 2.16A), acidosis of capillary blood induces Hb to release O_2 to the tissues that need it most. The pH sensitivity of Hb-O_2 affinity over this range is known as the alkaline Bohr effect, named after the Danish physiologist, Christian Bohr, who—in close collaboration with his fellow Danes, Karl

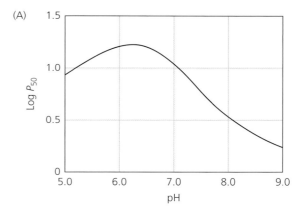

(A)

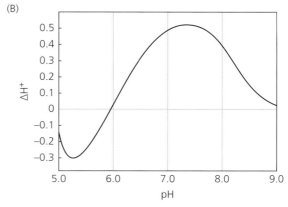

(B)

Fig. 2.16. Changes in pH affect O_2 binding by Hb and, reciprocally, the oxygenation of Hb affects proton binding. (A) The effect of pH on the O_2-affinity of human Hb. The slope of the curve between pH 6.5 and 7.5 provides a measure of the alkaline Bohr effect. (B) The number of protons per heme that are bound (at pH <6.0) or released (at pH > ~6.0) upon oxygenation of Hb.

Modified from Bunn and Forget (1986).

Hasselbalch and August Krogh—first described the phenomenon in 1904 (Bohr, 1904, Bohr et al., 1904). The experimental results presented in these papers significantly advanced the understanding of respiratory gas transport by Hb because they simultaneously revealed the importance of homotropic interaction (in the form of cooperative O_2 binding) and heterotropic interaction (the direct and indirect effect of CO_2 on Hb-O_2 affinity) (Edsall, 1986). Incidentally, Christian Bohr also had a lasting scientific legacy of another sort, as he was the father of Niels Bohr and grandfather of Aage Bohr, both of whom were Nobel laureates in physics.

Whereas the Bohr effect describes the influence of pH on Hb-O_2 affinity, the reciprocal Haldane effect describes the influence of Hb-O_2 saturation on H^+ binding (Christiansen et al. 1914). The Bohr effect and Haldane effect are two sides of the same coin; their interrelationship is expressed by the following linkage equation (Wyman, 1964):

$$\left(\frac{\partial(\log PO_2)}{\partial pH}\right)_Y = \left(\frac{\partial H^+}{\partial Y}\right)_{pH} \quad (2.18)$$

where "H^+" is the number of bound H^+ per heme and Y is the fractional O_2 saturation. The expression for the Bohr coefficient is shown on the left, and that for the Haldane coefficient is on the right. For a symmetrical O_2 equilibrium curve, this relation reduces to

$$\Delta H^+ = -\left(\frac{\partial \log P_{50}}{\partial pH}\right) \quad (2.19)$$

where ΔH^+ (= Bohr coefficient, Φ) is the number of protons per heme that bind upon the full transition from oxy to deoxyHb. The expression on the left side of the linkage equation can be experimentally determined by measuring O_2-equilibrium curves over a range of pHs (Fig. 2.16A), whereas the expression on the right can be determined by means of H^+ titration (Kilmartin and Rossi-Bernardi, 1973, Jensen et al., 1998a, Jensen, 2004) (Fig. 2.16B). The uptake of protons upon deoxygenation of human Hb is known as the alkaline Bohr effect because it only occurs above pH 6 (ΔH^+ is positive). Below pH 6, protons are released upon deoxygenation of human Hb (ΔH^+ is negative). This is known as the acid Bohr effect. In the physiological range of red cell pH (6.6–7.6), the Bohr coefficient for normal

human Hb is approximately –0.6. According to equation 2.19, the binding of 4 moles of O_2 to deoxyHb would therefore promote the release of 0.6 protons per heme, or 2.4 protons per Hb tetramer:

$$Hb(H) + 4O_2 \leftrightarrow Hb(O_2)_4 + 2.4H^+. \quad (2.20)$$

In the hypothetical example considered in section 2.7, we saw how O_2-transport by Hb from the lungs to the cells of working muscle ($PO_2 = 100$ torr and 20 torr, respectively) involves the cooperative release of O_2 amounting to 66 percent of total carrying capacity. Over this same range of blood PO_2, a change in blood pH from 7.4 in the lungs to 7.2 in working muscle results in a release of O_2 amounting to 77 percent of total carrying capacity (Fig. 2.17), further augmenting the O_2 unloading that is attributable to homotropic cooperativity alone.

Diffusion of CO_2 into red blood cells reduces Hb-O_2 affinity via two distinct mechanisms, only one of which is directly related to the drop in pH. Upon entry in the red blood cell, CO_2 reacts with water to form carbonic acid, H_2CO_3:

$$CO_2 + H_2O \leftrightarrow H_2CO_3. \quad (2.21)$$

This hydration reaction is catalyzed by the enzyme carbonic anhydrase, which is abundant inside the red blood cell but not in the plasma. Once formed in the red blood cell, H_2CO_3 rapidly dissociates to form bicarbonate ion, HCO_3^-, and H^+:

$$H_2CO_3 \leftrightarrow HCO_3^- + H^+, \quad (2.22)$$

which causes the drop in intracellular pH (excess of protons). This drop in pH has the effect of reducing Hb-O_2 affinity because protons preferentially bind and stabilize Hb in its low-affinity T-state, thereby promoting O_2 unloading. This proton binding illustrates the buffering power of deoxyHb and indicates that sites for oxygenation-linked proton binding are present in deoxyHb but not in oxyHb. As explained in detail in Chapter 4, these sites include the N-termini of the α- and β-chains and the imidazole side chains of solvent-exposed histidine residues, most of which have pK_as near pH 7.0 (Riggs, 1988, Lukin and Ho, 2004, Berenbrink, 2006). When Hb is deoxygenated, conformational changes in the Hb tetramer tend to increase the pK_a values of these Bohr groups, thereby increasing H^+ uptake at constant pH:

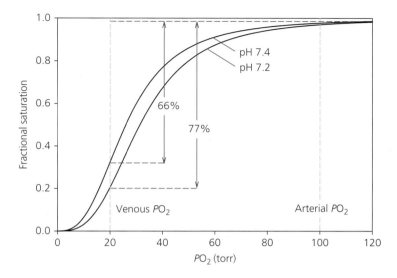

Fig. 2.17. In the physiological range, a reduction in blood pH reduces Hb-O_2 affinity (the alkaline Bohr effect). When blood pH is reduced from 7.4 to 7.2 (as would happen during exercise, when blood is acidified by metabolically produced CO_2 and lactic acid), the reduction in Hb-O_2 affinity promotes O_2 unloading to working muscle. This reduction of Hb-O_2 affinity at low pH represents a form of feedback control that stimulates the release of O_2 where it is needed most. Conversely, higher pH at the respiratory surfaces—where CO_2 is released—promotes O_2 loading in the pulmonary capillaries.

$$H\text{-}Hb + O_2 \leftrightarrow Hb\text{-}O_2 + H^+. \qquad (2.23)$$

Reciprocally, the pK_a values are decreased upon oxygenation of Hb, thereby triggering the release of bound H^+.

The second mechanism by which CO_2 reduces Hb-O_2 affinity involves a direct chemical reaction with Hb in the deoxy state. Specifically, CO_2 reversibly combines with the terminal amino groups of the α- and β-chain subunits, forming carbaminoHb (Kilmartin and Rossi-Bernardi, 1973, Jensen, 2004):

$$Hb\text{-}NH^{3+} \leftrightarrow Hb\text{-}NH_2 + H^+ \qquad (2.24)$$
$$Hb\text{-}NH_2 + CO_2 \leftrightarrow Hb\text{-}NHCOOH \leftrightarrow$$
$$Hb\text{-}NHCOO^- + H^+.$$

Similar to the effect of binding H^+ ions, carbamino formation reduces Hb-O_2 affinity by preferentially stabilizing Hb in the low-affinity T-state. In human red cells, carbamino formation accounts for ~20 percent of the Bohr effect (Severinghaus, 1979).

In addition to the importance of the Bohr effect in augmenting tissue O_2 delivery, oxygenation-linked proton binding and carbamino formation also play a critical role in CO_2 transport in the venous circulation. Most CO_2 is transported to the lungs in the form of HCO_3^- that is produced from the hydration of CO_2 inside the red blood cell. HCO_3^- is transported across the red cell membrane and enters the blood plasma; in exchange, membrane-permeable Cl^- ions are transported into the red cell, thereby maintaining charge neutrality (Fig. 2.18). This anion exchange across the red cell membrane is mediated by the same AE1 protein mentioned previously.

Due to the buffering power of deoxyHb (binding protons and aiding the formation of bicarbonate), a large concentration of CO_2 can be transported to the lungs in the form of soluble HCO_3^-. The process is reversed as blood circulates through the pulmonary capillaries. Upon oxygenation, Hb releases bound H^+ and HCO_3^- is converted into the less soluble CO_2, which is driven out of solution and then exhaled (Fig. 2.19). Oxygenation-linked carbaminoHb formation generally contributes less to CO_2 transport, accounting for ~5–15 percent of CO_2 exchange in the case of human Hb (depending on metabolic rate) (Klocke, 1988, Geers and Gros, 2000) and far less in the Hbs of some other vertebrates like teleost fishes (Jensen et al., 1998a).

Several alternative strategies for CO_2 transport have been documented in the animal kingdom.

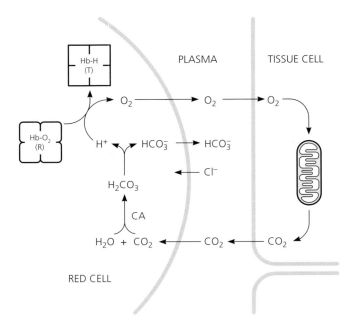

Fig. 2.18. The unloading of O_2 and the uptake of protons and CO_2 during the transit of red cells through the tissue capillaries. The enzyme carbonic anhydrase (CA) catalyzes the formation of carbonic acid (H_2CO_3) from CO_2 and water. Carbonic acid dissociates to form bicarbonate (HCO_3^-) and H^+, which results in a drop in intracellular pH. The release of O_2 from Hb and the binding of protons are associated with a shift from the R to the T quaternary structure.

Modified from Bunn and Forget (1986).

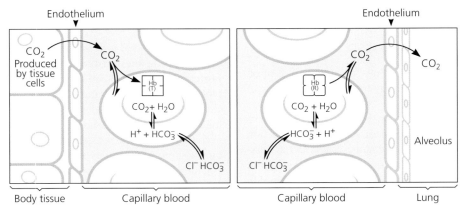

Fig. 2.19. Transport of CO_2 in the venous circulation from tissues (left panel) to the lungs (right panel). Most CO_2 is transported to the lungs in the form of bicarbonate (HCO_3^-) and a minor fraction is directly transported by Hb as a carbamino group. When Hb is oxygenated in the pulmonary capillaries, protons are released, and HCO_3^- is converted back to the less soluble CO_2, which is then exhaled.

Lampreys do not possess a means of HCO_3^-/Cl^- exchange across the red cell membrane, so CO_2 is transported as free, intracellular HCO_3^- (Tufts and Boutilier, 1989). The high HCO_3^- concentration is made possible by the maintenance of an exceedingly high red cell pH (Nikinmaa, 1986, Nikinmaa, 1997). Crocodilians also transport CO_2 in red blood cells, but not in the form of free intracellular HCO_3^-. Instead, HCO_3^- is allosterically bound to deoxyHb (Jensen et al., 1998b).

To summarize, the Bohr-Haldane effect provides the basis for a cycle of respiratory gas transport based on the reciprocal exchange of O_2 and CO_2:

(1) changes in red cell pH affect O_2 binding by Hb and, reciprocally, the oxygenation of Hb affects proton binding; and

(2) red cell PCO_2 affects O_2 binding by Hb and, reciprocally, the oxygenation of Hb affects CO_2 binding.

The Bohr effect therefore provides an important form of feedback control in O_2 and CO_2 transport: as arterial blood circulates through the tissue capillaries, CO_2 diffuses into the red blood cells, pH is decreased, and the O_2-equilibrium curve is shifted to the right, thereby promoting Hb-O_2 unloading to the cells of acidic tissue. Conversely, as venous blood circulates through the pulmonary capillaries, CO_2 is expelled, pH is increased, and the O_2-equilibrium curve is shifted to left, thereby promoting Hb-O_2 loading. This highlights how the shape and position of the *in vivo* O_2-equilibrium curve changes as blood circulates through the vascular system. Left-shifting the curve at the site of O_2 loading in the pulmonary capillaries while simultaneously right-shifting the curve at the site of O_2 unloading in the tissue capillaries increases the total amount of O_2 that is unloaded to the tissues for a given arterial-venous difference in PO_2.

2.13 Organic phosphates as allosteric modulators of Hb-O_2 affinity

In addition to H^+ and CO_2, which mediate the Bohr effect, organic phosphates are generally the most potent allosteric effectors of Hb-O_2 affinity in vertebrate red blood cells (though there are numerous

exceptions to the rule, as is often the case in biology). Whereas the roles of H^+ and CO_2 in regulating Hb-O_2 affinity have been appreciated for well over a century thanks to the pioneering work of Bohr and colleagues, the role of organic phosphates as allosteric modulators of Hb function was not fully elucidated until the late 1960s. However, as early as 1921, the Cambridge physiologist Joseph Barcroft and his students puzzled over the fact that O_2-equilibrium curves measured for purified Hbs in dilute solution were consistently different from those measured on hemolysates or whole blood (Fig. 2.20), leading them to suspect the presence of "a third substance…which forms an integral part of the oxygen hemoglobin complex" (Adair et al., 1921). The suspicions of Barcroft's group were confirmed and the enigmatic "third substance" was identified in a landmark 1967 paper by Reinhold and Ruth Benesch (Benesch and Benesch, 1967) (see also (Chanutin and Curnish, 1967)). The Benesches were a highly creative husband-and-wife team working at Columbia University (they were known to sign joint publications with the moniker "R2B2" to signify the "heterodimeric" nature of their collaboration). The Benesches noted that the organic phosphate 2,3-diphosphoglycerate (DPG) (also known as 2,3-biphosphoglycerate; BPG) was present

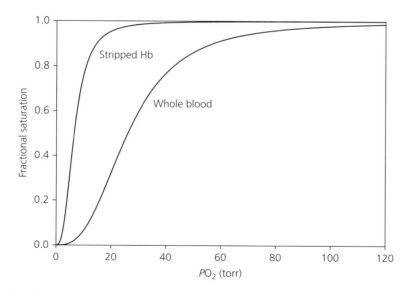

Fig. 2.20. O_2 binding by purified ("stripped") human Hb compared with whole blood. At the same temperature and pH, Hb in intact red cells has a much lower O_2 affinity and higher cooperativity than stripped Hb because it is operating in the presence of Cl^- ions and DPG.

at a surprisingly high concentration of ~5 mM in human red blood cells, equivalent to the normal erythrocytic concentration of tetrameric Hb. DPG is formed from an intermediate metabolite of glycolysis and it is present in only trace amounts in nonerythroid cells. To test whether the presence of DPG in the red blood cells played a role in regulating the oxygenation properties of Hb, the Benesches performed a series of experiments demonstrating that DPG causes a marked reduction in Hb-O_2 affinity by preferentially binding deoxyHb with 1:1 stoichiometry. These findings provided a clear explanation for Barcroft's vexing experimental results and revolutionized our understanding of the respiratory properties of red blood cells.

Under identical experimental conditions (pH 7.4, 37°C) the O_2-equilibrium curve for purified Hb approximates the hyperbolic curve of Mb, whereas the curve for whole blood is more sigmoidal (cooperativity is higher) and is shifted much further to the right (O_2-affinity is lower) (Fig. 2.20). In the former

case, the purified Hb has been stripped of Cl^- ions and DPG, so the measured curve reflects the intrinsic O_2 binding of the Hb protein. In the latter case, Hb is operating in the chemical milieu of the red blood cell with physiological concentrations of Cl^- and DPG. Adding a physiological concentration of DPG to the purified Hb solution shifts the O_2-equilibrium curve to the right, and a concomitant increase in PCO_2 yields a curve that closely approximates that for whole blood (Fig. 2.21). This highlights the importance of solvent conditions in determining the oxygenation properties of Hb.

Another example of early work that anticipated the discovery of the regulatory role of organic phosphates is provided by a classic study of the O_2-transport properties of fish blood by Krogh and Leitch (1919):

"We believe that the adaptation of the fish blood must be brought about by some substance or substances present along with the haemoglobin within the corpuscles, and we wish to point out the general significance of the haemo-

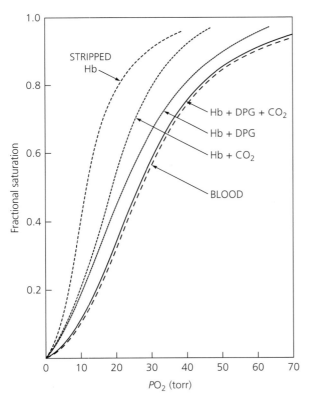

Fig. 2.21. The effect of DPG and CO_2 on the O_2-affinity of phosphate-free (stripped) Hb (37°C, pH 7.22, PCO_2 = 40 torr, DPG:tetrameric Hb = 1.2). This illustrates how the addition of physiological concentrations of DPG and acid equivalents can shift the O_2-equilibrium curve for purified Hb to match that for whole blood (PCO_2 = 40 torr, plasma pH = 7.40 [which corresponds to a red cell pH of 7.22]).

Based on data from Kilmartin and Rossi-Bernardi (1973).

globin being enclosed in corpuscles surrounded by semi-permeable membranes. By this arrangement just that chemical environment can be secured which is most suitable for the respiratory function of the haemoglobin in that particular organism, while at the same time the chemical composition of the blood plasma can be adapted...to the general requirements of the body cells."

With the benefit of a century's worth of scientific progress since Krogh and Leitch reported their insights, we now know that the "unknown substances" were organic phosphates (ATP and guanosine triphosphate [GTP] in the case of their fish). As these authors noted, the chemical microenvironment of the red blood cell plays a critical role in regulating the oxygenation properties of Hb.

2.14 The role of Hb in regulating local blood flow

In addition to Hb's familiar role as an O_2 carrier, recent discoveries have revealed that Hb also plays a role in regulating blood flow in the microcirculation. In this process Hb functions as an O_2 sensor and O_2-responsive nitric oxide (NO) signal transducer, thereby contributing to hypoxic vasodilation—a mechanism for matching perfusion to tissue O_2 demand (Singel and Stamler, 2005, Jensen, 2009a).

Research on humans and mouse models has revealed that NO functions as an important signaling molecule in the regulation of local blood flow and it also has a cytoprotective role during ischemia and reperfusion (Foster et al., 2003, Gladwin et al., 2006, Lundberg et al., 2009, van Faassen et al., 2009, Hill et al., 2010). NO typically exerts its physiological effects by binding directly to the ferrous Fe^{2+} iron of heme proteins (forming iron-nitrosyl, FeNO) or by reacting with protein thiol groups (forming S-nitrosothiols, SNO). NO produced in the vascular endothelium diffuses into the underlying vascular smooth muscle where it causes vasorelaxation via activation of heme-containing soluble guanylate cyclase (Moncada and Higgs, 1993). There are three main mechanisms by which circulating red blood cells are thought to mediate NO-dependent hypoxic vasodilation: (1) release of ATP, which stimulates NO production by activating endothelial nitric oxide synthase (eNOS)

(Ellsworth et al., 1995, Dietrich et al., 2000, González-Alonso et al., 2002, Farias et al., 2005); (2) release of vasoactive NO from S-nitrosylated Hb (SNO-Hb) upon deoxygenation (Jia et al., 1996, Singel and Stamler, 2005, Allen et al., 2009, Zhang et al., 2015); and (3) enzymatic reduction of nitrite (NO_2^-) to vasoactive NO via the nitrite reductase activity of deoxyHb (Cosby et al., 2003, Nagababu et al., 2003, Crawford et al., 2006, Gladwin and Kim-Shapiro, 2008). There is ongoing debate about the relative efficacy of these different mechanisms (Fago et al., 2003, Gladwin et al., 2003, Robinson and Lancaster, 2005, Singel and Stamler, 2005, Dalsgaard et al., 2007, Sonveaux et al., 2007, Isbell et al., 2008, Jensen, 2009a), but the common feature is that they each involve a coupling between the degree of Hb deoxygenation and local vasodilation. Hb is therefore envisioned as an O_2 sensor, as declining PO_2 in the systemic microcirculation is signaled by an increase in the relative fraction of deoxyHb, and this triggers the production and release of vasodilatory compounds such as ATP and NO. In the case of ATP release by red blood cells, there is not a clear mechanistic link with allosteric properties of Hb. By contrast, oxygenation-linked transitions in Hb quaternary structure are central to the proposed mechanisms of vasoregulation involving SNO-Hb (Fig. 2.22A) and the nitrite reductase activity of deoxyHb (Fig. 2.22B).

2.14.1 The role of SNO-Hb in hypoxic vasodilation

The SNO-Hb hypothesis was the first to postulate a direct role for Hb allostery in hypoxic vasodilation (Jia et al., 1996). According to this hypothesis, an oxidized form of NO from the small fraction of nitrosylHb (HbNO) in the arterial circulation is transferred from the β-chain heme group to a cysteine residue in the same subunit (Cys β93), forming SNO-Hb. The reactivity of this key cysteine residue depends on the quaternary state of the Hb tetramer, and the S-nitrosylation reaction is only possible in the R-state. Thus, when SNO-Hb is deoxygenated in the peripheral circulation, the allosteric transition to the T-state promotes the release of vasoactive NO from Cys β93 (Jia et al., 1996, Stamler et al., 1997, Gow and Stamler, 1998, Gow et al., 1999) (Fig. 2.23). According to this

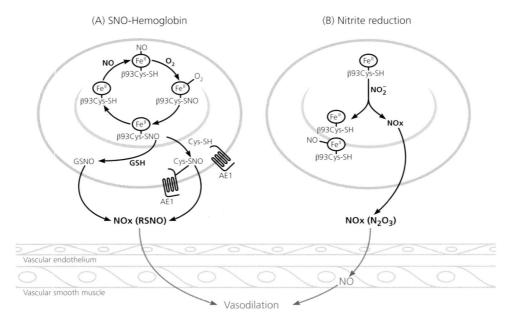

Fig. 2.22. Hypothesized mechanisms of hypoxic vasodilation mediated by allosteric properties of Hb. (A) The SNO-Hb hypothesis: In R-state Hb, *S*-nitrosylation of a cysteine residue in the β-chain (β93Cys) results in the stable formation of SNO-Hb. Upon deoxygenation in the arterial microcirculation, SNO-Hb switches to the T-state, and the NO is transferred (via transnitrosylation) to thiols such as glutathione (GSH), forming *S*-nitrosoglutathione (GSNO), and/or a cysteine in the cytoplasmic domain of the membrane-bound anion exchanger (AE1). The formation of *S*-nitrosothiols (RSNO, where *R* denotes an organic group) provides a possible means of exporting NO activity from the red cell membrane to the vessel wall because free NO is rapidly scavenged by Hb. (B) The nitrite reduction hypothesis: Nitrite (NO_2^-) entering the red cell reacts with both oxyHb (forming nitrate and metHb [Hb with oxidized Fe^{3+} hemes]) and deoxyHb (forming NO and metHb). The reaction with deoxyHb is favored at the low PO_2 that prevails in the arterial microcirculation. Some fraction of the produced NO binds to deoxy heme groups, forming HbNO, and some forms dinitrogen trioxide (N_2O_3) (via reaction of NO with a nitrite-metHb intermediate). N_2O_3 can diffuse across the cell membrane, followed by homolysis to NO and NO_2 outside the cells. The presence of reducing enzymes inside the red cell ensures that the ferric (Fe^{3+}) hemes of metHb produced in these reactions can be rapidly converted back to the ferrous state (Fe^{2+}), thereby reconstituting functional Hb that can reversibly bind O_2. "NOx" refers to nitrogen oxides such as NO and NO_2.

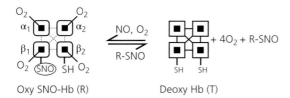

Oxy SNO-Hb (R)   Deoxy Hb (T)

Fig. 2.23. Hypothesized release of NO from SNO-Hb upon deoxygenation. According to the hypothesis of Stamler and coworkers (Jia et al., 1996, Stamler et al., 1997, Gow and Stamler, 1998, Gow et al., 1999), NO is transferred from β93Cys to thiol groups by a transnitrosylation reaction that is coupled to the allosteric R→T transition in quaternary structure.

Modified from Weber and Fago (2004).

hypothesis, NO activity is then exported from the red blood cell by transferring the NO group from SNO-Hb to the thiols on other proteins (transnitrosylation),

such as the membrane-bound anion exchanger, AE1 (Fig. 2.22A). Experiments involving transgenic mice suggest an important role for SNO-Hb in hypoxic vasodilation (Zhang et al., 2015, Zhang et al., 2016), and may help to explain the high conservation of Cys β93(F9) across all mammals. Since this same site has an amino acid other than Cys in cartilaginous fishes, ray-finned fishes, and representatives of several amphibian and reptile lineages, the same specific heme-linked SNO mechanism is not operative in all jawed vertebrates (Jensen, 2009a). Interestingly, a SNO-Mb mechanism has been described in salmonid fishes that involves oxygenation-linked *S*-nitrosylation at Cys 107(G8) (Helbo and Fago 2011; Helbo et al. 2014). Since Mb is a monomeric protein, the allosteric mechanism is thought to involve an oxygenation-linked change in tertiary conformation that promotes the targeted release of vasoactive NO to the hypoxic heart.

2.14.2 Nitrite reductase activity of deoxyHb and its role in hypoxic vasodilation

Nitrite (NO_2^-) is an endogenously produced oxidative metabolite of NO that is present at low concentrations in the blood of mammals and other vertebrates (Kleinbongard et al., 2003, Gladwin et al., 2006, Jensen, 2009b, Fago and Jensen, 2015). The reaction of nitrite with deoxyHb has attracted special attention because it suggests a mechanism of NO production that is coupled to Hb-O_2 delivery, and could therefore play a role in hypoxic vasodilation (Cosby et al., 2003, Nagababu et al., 2003, Crawford et al., 2006, Gladwin and Kim-Shapiro, 2008). Inside the red cell, nitrite reacts with the ferrous heme of deoxyHb to form NO and ferric heme:

$$Hb(Fe^{2+}) + NO_2^- + H^+ \rightarrow Hb(Fe^{3+}) \\ + NO + OH^-. \quad (2.25)$$

This mechanism requires diffusion/transport of nitrite into the red cells (either as nitrite ion [NO_2^-] or nitrous acid [HNO_2]), followed by reaction of nitrite with deoxyHb and the subsequent escape of NO activity from the red cell. This last step may occur via production of dinitrogen trioxide (N_2O_3) from the reaction of NO with a nitrite-Hb(Fe^{3+}) intermediate (Robinson and Lancaster 2005; Basu et al. 2007). Nitrite reacts with both deoxyHb and oxyHb, but only the reaction with deoxyHb results in the production of NO (Fig. 2.22B), and this is the reaction that is favored at the intermediate O_2 saturations that prevail in the arterial microcirculation. The nitrite reductase activity of deoxyHb is regulated by the heme redox potential (tendency to transfer electrons), a property that depends on the conformational state of the Hb tetramer. In mammalian Hbs, deoxy hemes have a lower redox potential (higher reductase activity) in the R-state than in the T-state, so Hb-mediated nitrite reduction occurs at the highest rate when Hb is ~50 percent saturated (Huang et al., 2005, Gladwin and Kim-Shapiro, 2008, Jensen, 2009a). In the case of ectothermic vertebrates, by contrast, the reaction rate increases in a more-or-less linear fashion with decreasing O_2 saturation (Jensen and Rohde, 2010, Jensen et al., 2017).

There are still many more questions than answers regarding mechanisms of red cell-mediated vasoregulation. The exciting idea that Hb allostery mediates the sensing of O_2 as well as the O_2-coupled transport of NO vasoactivity is sure to motivate continued research in this area for many years to come. Findings to date clearly demonstrate that vertebrate Hb has evolved physiologically important interactions with NO in addition to the more familiar interactions with O_2 and CO_2.

Now that we've explored the basic oxygenation properties of Hb and its physiological significance in respiratory gas transport, in Chapter 3 *we will conduct a brief survey of allosteric theory. This provides a framework for understanding structural mechanisms of cooperativity and intramolecular regulatory control.*

References

Adair, G. S., Barcroft, J., and Bock, A. V. (1921). The identity of haemoglobin in human beings. *Journal of Physiology-London*, **55**, 332–8.

Allen, B. W., Stamler, J. S., and Piantadosi, C. A. (2009). Hemoglobin, nitric oxide and molecular mechanisms of hypoxic vasodilation. *Trends in Molecular Medicine*, **15**, 452–60.

Arnone, A. 1972. X-ray-diffraction study of binding of 2, 3-diphosphoglycerate to human deoxyhemoglobin. *Nature*, **237**, 146–9.

Arnone, A. and Perutz, M. F. (1974). Structure of inositol hexaphosphate-human deoxyhemoglobin complex. *Nature*, **249**, 34–6.

Baldwin, J. and Chothia, C. (1979). Haemoglobin: the structural changes related to ligand binding and its allosteric mechanism. *Journal of Molecular Biology*, **129**, 175–220.

Basu, S., Grubina, R., Huang, J., et al. (2007). Catalytic generation of N_2O_3 by the concerted nitrite reductase and anhydrase activity of hemoglobin. *Nature Chemical Biology*, **3**, 785–94.

Benesch, R. and Benesch, R. E. (1967). The effect of organic phosphates from human erythrocytes on the allosteric properties of hemoglobin. *Biophysical Research Communications*, **26**, 162–7.

Berenbrink, M. (2006). Evolution of vertebrate haemoglobins: histidine side chains, specific buffer value and Bohr effect. *Respiratory Physiology and Neurobiology*, **154**, 165–84.

Berner, R. A. (2009). Phanerozoic atmospheric oxygen: new results using the GEOCARBSULF model. *American Journal of Science*, **309**, 603–6.

Berner, R. A. and Canfield, D. E. (1989). A new model for atmospheric oxygen over Phanerozoic time. *American Journal of Science*, **289**, 333–61.

Bettati, S., Mozzarelli, A., and Perutz, M. F. (1983). Allosteric mechanism of hemoglobin: rupture of salt bridges raises oxygen affinity of the T-structure. *Journal of Molecular Biology*, **281**, 581–5.

Bohr, C. (1904). Theoretische behandlung der quantitativen verhaltnis be der sauerstoffaufnahme des hamoglobins. *Abl. Physiol.*, **17**, 682–8.

Bohr, C., Hasselbalch, L., and Krogh, A. (1904). Ueber einen in biologischer beziehung wichtigen einfluss, den die kohlensaurespannung des blutes auf dessen sauerstoffbindung ubt. *Skand. Archives of Physiology and Biochemistry*, **16**, 402–12.

Briggs, D. E. G. (1985). Gigantism in Palaeozoic arthropods. *Special Papers in Paleontology*, **33**, 157.

Bunn, H. F. and Forget, B. G. (1986). *Hemoglobin: Molecular, Genetic and Clinical Aspects*, Philadelphia, PA, W. B. Saunders Company.

Burggren, W. W., Farrell, A. P., and Lillywhite, H. B. (1997). *Vertebrate Cardiovascular Systems*, Oxford, Oxford University Press.

Chanutin, A. and Curnish, R. R. (1967). Effect of organic and inorganic phosphates on the oxygen equilibrium of human erythrocytes. *Archives of Biochemistry and Biophysics*, **121**, 96.

Christiansen, J., Douglas, C. G., and Haldane, J. S. (1914). The absorption and dissociation of carbon dioxide by human blood. *Journal of Physiology*, **48**, 244–71.

Clapham, M. E. and Karr, J. A. (2012). Environmental and biotic controls on the evolutionary history of insect body size. *Proceedings of the National Academy of Sciences of the United States of America*, **109**, 10927–30.

Cosby, K., Partovi, K. S., Crawford, J. H., et al. (2003). Nitrite reduction to nitric oxide by deoxyhemoglobin vasodilates the human circulation. *Nature Medicine*, **9**, 1498–505.

Crawford, J. H., Isbell, T. S., Huang, Z., et al. (2006). Hypoxia, red blood cells, and nitrite regulate NO-dependent hypoxic vasodilation. *Blood*, **107**, 566–74.

Dalsgaard, T., Simonsen, U., and Fago, A. (2007). Nitrite-dependent vasodilation is facilitated by hypoxia and is independent of known NO-generating nitrite reductase activities. *American Journal of Physiology-Heart and Circulatory Physiology*, **292**, H3072–8.

Dejours, P. (1981). *Principles of Comparative Respiratory Physiology*, Amsterdam, Elsevier/North-Holland Biomedical Press.

Dickerson, R. E. and GEIS, I. (1983). *Hemoglobin: Structure, Function, Evolution, and Pathology*, Menlo Park, CA, Benjamin/Cummings.

Dietrich, H. H., Ellsworth, M. L., Sprague, R. S., and Dacey, R. G. (2000). Red blood cell regulation of microvascular tone through adenosine triphosphate. *American Journal of Physiology-Heart and Circulatory Physiology*, **278**, H1294–8.

Dudley, R. (1998). Atmospheric oxygen, giant Paleozoic insects and the evolution of aerial locomotor performance. *Journal of Experimental Biology*, **201**, 1043–50.

Edsall, J. T. (1986). Understanding blood and hemoglobin—an example of international relations in science. *Perspectives in Biology and Medicine*, **29**, S107–23.

Ellsworth, M. L., Forrester, T., Ellis, C. G., and Dietrich, H. H. (1995). The erythrocyte as a regulator of vascular tone. *American Journal of Physiology-Heart and Circulatory Physiology*, **269**, H2155–61.

Fago, A., Crumbliss, A. L., Peterson, J., Pearce, L. L., and Bonaventura, C. (2003). The case of the missing NO-hemoglobin: spectral changes suggestive of heme redox reactions reflect changes in NO-heme geometry. *Proceedings of the National Academy of Sciences of the United States of America*, **100**, 12087–92.

Fago, A. and Jensen, F. B. (2015). Hypoxia tolerance, nitric oxide, and nitrite: lessons from extreme animals. *Physiology*, **30**, 116–26.

Farias, M., Gorman, M. W., Savage, M. V., and Feigl, E. O. (2005). Plasma ATP during exercise: possible role in regulation of coronary blood flow. *American Journal of Physiology-Heart and Circulatory Physiology*, **288**, H1586–90.

Fitch, N. A., Johnston, I. A., and Wood, R. E. (1984). Skeletal muscle capillary supply in a fish that lacks respiratory pigments. *Respiration Physiology*, **57**, 201–11.

Foster, M. W., McMahon, T. J., and Stamler, J. S. (2003). S-nitrosylation in health and disease. *Trends in Molecular Medicine*, **9**, 160–8.

Geers, C. and Gros, G. (2000). Carbon dioxide transport and carbonic anhydrase in blood and muscle. *Physiological Reviews*, **80**, 681–715.

Gladwin, M. T. and Kim-Shapiro, D. B. (2008). The functional nitrite reductase activity of the heme-globins. *Blood*, **112**, 2636–47.

Gladwin, M. T., Lancaster, J. R., Freeman, B. A., and Schechter, A. N. (2003). Nitric oxide's reactions with hemoglobin: a view through the SNO-storm. *Nature Medicine*, **9**, 496–500.

Gladwin, M. T., Raat, N. J. H., Shiva, S., et al. (2006). Nitrite as a vascular endocrine nitric oxide reservoir that contributes to hypoxic signaling, cytoprotection, and vasodilation. *American Journal of Physiology-Heart and Circulatory Physiology*, **291**, H2026–35.

González-Alonso, J., Olsen, D. B., and Saltin, B. (2002). Erythrocyte and the regulation of human skeletal muscle blood flow and oxygen delivery—role of circulating ATP. *Circulation Research*, **91**, 1046–55.

Gow, A. J., Luchsinger, B. P., Pawloski, J. R., Singel, D. J., and Stamler, J. S. (1999). The oxyhemoglobin reaction of nitric oxide. *Proceedings of the National Academy of Sciences of the United States of America*, **96**, 9027–32.

Gow, A. J. and Stamler, J. S. (1998). Reactions between nitric oxide and haemoglobin under physiological conditions. *Nature*, **391**, 169–73.

Graham, J. B., Dudley, R., Aguilar, N. M., and Gans, C. (1995). Implications of the Late Paleozoic oxygen pulse for physiology and evolution. *Nature*, **375**, 117–20.

Gros, G., Wittenberg, B. A., and Jue, T. (2010). Myoglobin's old and new clothes: from molecular structure to function in living cells. *Journal of Experimental Biology*, **213**, 2713–25.

Haldane, J. B. S. (1927). *Possible Worlds and Other Essays*, London, UK, Chatto and Windus.

Haldane, J. S. (1922). *Respiration*, New Haven, CT, Yale University Press.

Harrison, J., Frazier, M. R., Henry, J. R., et al. (2006). Responses of terrestrial insects to hypoxia or hyperoxia. *Respiratory Physiology and Neurobiology*, **154**, 4–17.

Harrison, J. F., Kaiser, A., and Vandenbrooks, J. M. (2010). Atmospheric oxygen level and the evolution of insect body size. *Proceedings of the Royal Society B-Biological Sciences*, **277**, 1937–46.

Helbo, S. and Fago, A. (2011). Allosteric modulation by S-nitrosation in the low-O$_2$ affinity myoglobin from rainbow trout. *American Journal of Physiology-Regulatory Integrative and Comparative Physiology*, **300**, R101–8.

Helbo, S., Gow, A. J., Jamil, A., et al. (2014). Oxygen-linked S-nitrosation in fish myoglobins: a cysteine-specific tertiary allosteric effect. *PloS One*, **9**, e97012.

Helbo, S., Weber, R. E., and Fago, A. (2013). Expression patterns and adaptive functional diversity of vertebrate myoglobins. *Biochimica Et Biophysica Acta-Proteins and Proteomics*, **1834**, 1832–9.

Hemmingsen, E. A., Douglas, E. L., Johansen, K., and Millard, R. W. (1972). Aortic blood flow and cardiac output in hemoglobin-free fish *Chaenocephalus aceratus*. *Comparative Biochemistry and Physiology*, **43**, 1045–51.

Hicks, J. W. (1998). Cardiac shunting in reptiles: mechanism, regulation, and physiological function. *In:* Gans, C. and Gaunt, S. (eds.) *Biology of the Reptilia, Vol. 19*, pp. 425–83. Ithaca, NY, Society for the Study of Amphibians and Reptiles.

Hicks, J. W. and Wang, T. (2004). Hypometabolism in reptiles: behavioral and physiological mechanisms that reduce aerobic demands. *Respiratory Physiology and Neurobiology*, **141**, 261–71.

Hill, A. V. (1913). The combinations of haemoglobin with oxygen and with carbon monoxide. I. *Biochemical Journal*, **7**, 471–80.

Hill, B. G., Dranka, B. P., Bailey, S. M., Lancaster, J. R., and Darley-Usmar, V. M. (2010). What part of NO don't you understand? Some answers to the cardinal questions in nitric oxide biology. *Journal of Biological Chemistry*, **285**, 19699–704.

Hladky, S. B. and Rink, T. J. (1977). pH equilibrium across the red cell membrane. *In:* Ellory, J. C. and Lew, V. L. (eds.) *Membrane Transport in Red Cells*, pp. 115–35. London, Academic Press.

Hochachka, P. W., Buck, L. T., Doll, C. J., and Land, S. C. (1996). Unifying theory of hypoxia tolerance: molecular/metabolic defense and rescue mechanisms for surviving oxygen lack. *Proceedings of National Academy of Sciences USA*, **93**, 9493–8.

Hochachka, P. W. and Lutz, P. (2001). Mechanism, origin, and evolution of anoxia tolerance in animals. *Comparative Biochemistry and Physiology B*, **130**, 435–59.

Holeton, G. F. (1970). Oxygen uptake and circulation by a hemoglobinless Antarctic fish (*Chaenocephalus aceratus* Lonnberg) compared with three red-blooded Antarctic fish. *Comparative Biochemistry and Physiology*, **34**, 457–71.

Huang, Z., Shiva, S., Kim-Shapiro, D. B., et al. (2005). Enzymatic function of hemoglobin as a nitrite reductase that produces NO under allosteric control. *Journal of Clinical Investigation*, **115**, 2099–107.

Isbell, T. S., Sun, C. W., Wu, L. C., et al. (2008). SNO-hemoglobin is not essential for red blood cell-dependent hypoxic vasodilation. *Nature Medicine*, 14, 773–7.

Ivy, C. M. and Scott, G. R. (2015). Control of breathing and the circulation in high-altitude mammals and birds. *Comparative Biochemistry and Physiology A-Molecular & Integrative Physiology*, **186**, 66–74.

Jensen, F. B. (2004). Red blood cell pH, the Bohr effect, and other oxygenation-linked phenomena in blood O$_2$ and CO$_2$ transport. *Acta Physiologica Scandinavica*, **182**, 215–27.

Jensen, F. B. (2009a). The dual roles of red blood cells in tissue oxygen delivery: oxygen carriers and regulators of local blood flow. *Journal of Experimental Biology*, **212**, 3387–93.

Jensen, F. B. (2009b). The role of nitrite in nitric oxide homeostasis: a comparative perspective. *Biochimica et Biophysica Acta*, **1787**, 841–8.

Jensen, F. B., Fago, A., and Weber, R. E. (1998a). Hemoglobin structure and function. *In:* Perry, S. F. and Tufts, B. L. (eds.) *Fish Physiology, Vol. 17: Fish Respiration*, pp. 1–40. New York, NY, Academic Press.

Jensen, F. B., Kolind, R. A. H., Jensen, N. S., Montesanti, G., and Wang, T. (2017). Interspecific variation and plasticity in hemoglobin nitrite reductase activity and its correlation with oxygen affinity in vertebrates. *Comparative Biochemistry and Physiology A-Molecular and Integrative Physiology*, **206**, 47–53.

Jensen, F. B. and Rohde, S. (2010). Comparative analysis of nitrite uptake and hemoglobin-nitrite reactions in erythrocytes: sorting out uptake mechanisms and oxygenation dependencies. *American Journal of Physiology-Regulatory Integrative and Comparative Physiology*, **298**, R972–82.

Jensen, F. B., Wang, T., Jones, D. R., and Brahm, J. (1998b). Carbon dioxide transport in alligator blood and its erythrocyte permeability to anions and water. *American Journal of Physiology-Regulatory Integrative and Comparative Physiology*, **274**, R661–71.

Jia, L., Bonaventura, C., Bonaventura, J., and Stamler, J. S. (1996). S-nitrosohaemoglobin: a dynamic activity of blood involved in vascular control. *Nature*, **380**, 221–6.

Kilmartin, J. V. and Rossi-Bernardi, L. (1973). Interaction of hemoglobin with hydrogen ions, carbon dioxide, and organic phosphates. *Physiological Reviews*, **53**, 836–90.

Kleinbongard, P., Dejam, A., Lauer, T., et al. (2003). Plasma nitrite reflects constitutive nitric oxide synthase activity in mammals. *Free Radical Biology and Medicine*, **35**, 790–6.

Klocke, R. A. (1988). Velocity of CO_2 exchange in blood. *Annual Review of Physiology*, **50**, 625–37.

Krogh, A. and Leitch, I. (1919). The respiratory function of the blood in fishes. *Journal of Physiology*, **52**, 288–300.

Lower, R. (1671). *Tractatus de Corde: Item de motu and colore sanguinis, et chyli in eum transitu: cui accessit dissertatio de origine catarrhi, in qua ostenditur illum non provenire cerebro (3rd. ed.)*, Amstelodami, apud Danielem Elzevirium.

Lukin, J. A. and HO, C. (2004). The structure-function relationship of hemoglobin in solution at atomic resolution. *Chemical Reviews*, **104**, 1219–30.

Lundberg, J. O., Gladwin, M. T., Ahluwalia, A., et al. (2009). Nitrate and nitrite in biology, nutrition and therapeutics. *Nature Chemical Biology*, **5**, 865–9.

Mairbäurl, H. (1994). Red blood cell function in hypoxia at altitude and exercise. *International Journal of Sports Medicine*, **15**, 51–63.

Moncada, S. and Higgs, A. (1993). Mechanisms of disease—the L-arginine nitric oxide pathway. *New England Journal of Medicine*, **329**, 2002–12.

Nagababu, E., Ramasamy, S., Abernethy, D. R., and Rifkind, J. M. (2003). Active nitric oxide produced in the red cell under hypoxic conditions by deoxyhemoglobin-mediated nitrite reduction. *Journal of Biological Chemistry*, **278**, 46349–56.

Near, T. J., Dornburg, A., Kuhn, K. L., et al. (2012). Ancient climate change, antifreeze, and the evolutionary diversification of Antarctic fishes. *Proceedings of the National Academy of Sciences of the United States of America*, **109**, 3434–9.

Near, T. J., Pesavento, J. J., and Cheng, C. H. C. (2003). Mitochondrial DNA, morphology, and the phylogenetic relationships of Antarctic icefishes (Notothenioidei: Channichthyidae). *Molecular Phylogenetics and Evolution*, **28**, 87–98.

Nigen, A. M., Manning, J. M., and Alben, J. O. (1980). Oxygen-linked binding sites for inorganic anions to hemoglobin. *Journal of Biological Chemistry*, **255**, 5525–9.

Nikinmaa, M. (1986). Red cell pH of lamprey (*Lampetra fluviatilis*) is actively regulated. *Journal of Comparative Physiology B-Biochemical Systemic and Environmental Physiology*, **156**, 747–50.

Nikinmaa, M. (1997). Oxygen and carbon dioxide transport in vertebrate erythrocytes: an evolutionary change in the role of membrane transport. *Journal of Experimental Biology*, **200**, 369–80.

O'Donnell, S., Mandaro, R., Schuster, T. M., and Arnone, A. (1979). X-ray diffraction and solutions studies of specifically carbamylated human hemoglobin A—evidence for the location of a proton-linked and oxygen-linked chloride binding site at valine 1a. *Journal of Biological Chemistry*, **254**, 2204–8.

Perutz, M. F. (1970). Stereochemistry of cooperative effects in haemoglobin. *Nature*, **228**, 726–39.

Perutz, M. F. (1978). Hemoglobin structure and respiratory transport. *Scientific American*, **239**, 92–125.

Perutz, M. F. (1989). Mechanisms of cooperativity and allosteric regulation in proteins. *Quarterly Reviews of Biophysics*, **22**, 139–236.

Riggs, A. F. (1988). The Bohr effect. *Annual Review of Physiology*, **50**, 181–204.

Robinson, J. M. and Lancaster, J. R. (2005). Hemoglobin-mediated, hypoxia-induced vasodilation via nitric oxide—mechanism(s) and physiologic versus pathophysiologic relevance. *American Journal of Respiratory Cell and Molecular Biology*, **32**, 257–61.

Ruud, J. T. (1954). Vertebrates without erythrocytes and blood pigment. *Nature*, **173**, 848–50.

Severinghaus, J. W. (1979). Simple, accurate equations for human blood O_2 dissociation computations. *Journal of Applied Physiology—Respiratory, Environmental, Exercise Physiology*, **46**, 599–602.

Shear, W. A. and Kukalovareck, J. (1990). The ecology of Paleozoic terrestrial arthropods—the fossil evidence. *Canadian Journal of Zoology-Revue Canadienne De Zoologie*, **68**, 1807–34.

Sidell, B. D. and O'Brien, K. M. (2006). When bad things happen to good fish: the loss of hemoglobin and myoglobin expression in Antarctic icefishes. *Journal of Experimental Biology*, **209**, 1791–802.

Singel, D. J. and Stamler, J. S. (2005). Chemical physiology of blood flow regulation by red blood cells: the role of nitric oxide and S-nitrosohemoglobin. *Annual Review of Physiology*, **67**, 99–145.

Sonveaux, P., Lobsheva, II, Feron, O., and McMahon, T. J. (2007). Transport and peripheral bioactivities of nitrogen oxides carried by red blood cell hemoglobin: role in oxygen delivery. *Physiology*, **22**, 97–112.

Stamler, J. S., Jia, L., Eu, J. P., et al. (1997). Blood flow regulation by S-nitrosohemoglobin in the physiological oxygen gradient. *Science*, **276**, 2034–7.

Storz, J. F. (2016). Hemoglobin-oxygen affinity in high-altitude vertebrates: is there evidence for an adaptive trend? *Journal of Experimental Biology*, **219**, 3190–203.

Taylor, C. R. and Weibel, E. R. (1981). Design of the mammalian respiratory system. 1. Problem and strategy. *Respiration Physiology*, **44**, 1–10.

Tufts, B. L. and Boutilier, R. G. (1989). The absence of rapid chloride bicarbonate exchange in lamprey erythrocytes. Implications for CO_2 transport and ion distributions between plasma and erythrocytes in the blood of *Petromyzon marinus. Journal of Experimental Biology*, **144**, 565–76.

Van Faassen, E. E., Babrami, S., Feelisch, M., et al. (2009). Nitrite as regulator of hypoxic signaling in mammalian physiology. *Medicinal Research Reviews*, **29**, 683–741.

Weber, R. E. and Fago, A. (2004). Functional adaptation and its molecular basis in vertebrate hemoglobins, neuroglobins and cytoglobins. *Respiratory Physiology* and *Neurobiology*, **144**, 141–59.

West, J. B. (1969). Effect of slope and shape of dissociation curve on pulmonary gas exchange. *Respiration Physiology*, **8**, 66–85.

Wittenberg, J. B. and Wittenberg, B. A. (2003). Myoglobin function reassessed. *Journal of Experimental Biology*, **206**, 2011–20.

Wyman, J. (1964). Linked functions and reciprocal effects in hemoglobin: a second look. *Advances in Protein Chemistry*, 19, 223–86.

Yuan, Y., Tam, M. F., Simplaceanu, V., and HO, C. (2015). New look at hemoglobin allostery. *Chemical Reviews*, **115**, 1702–24.

Zhang, R. L., Hess, D. T., Qian, Z. X., et al. (2015). Hemoglobin βCys93 is essential for cardiovascular function and integrated response to hypoxia. *Proceedings of the National Academy of Sciences of the United States of America*, **112**, 6425–30.

Zhang, R. L., Hess, D. T., Reynolds, J. D., and Stamler, J. S. (2016). Hemoglobin S-nitrosylation plays an essential role in cardioprotection. *Journal of Clinical Investigation*, **126**, 4654–8.

CHAPTER 3

Allosteric theory

3.1 Homotropic and heterotropic allostery

As we saw in Chapter 2 (section 2.11), allostery refers to a modulation of protein activity that is caused by an indirect interaction between structurally remote binding sites. In this mode of intramolecular control, the binding of ligand at a protein's active site is influenced by the binding of another ligand at a different site in the same protein. This interaction at a distance is mediated by a ligation-induced transition between alternative conformational states. Hb is regarded as the "allosteric paradigm" (Cui and Karplus, 2008), and the oxygenation-linked transition between the T and R quaternary structures provides a standard textbook example of how allostery works. In many systems other than Hb, allostery is an important biochemical mechanism of regulatory control, as it provides the basis for enzyme repression/induction and feedback inhibition of metabolic pathways (Monod et al., 1963, Changeux, 1993, Changeux, 2012, Changeux, 2013). Experimental and theoretical investigations of protein allostery have helped elucidate the biological functions of regulatory molecules in diverse contexts, from hormones in the endocrine system to neurotransmitters in the nervous system.

Allosteric interactions can be positive or negative, depending on whether ligand-binding affinity is increased or decreased. Homotropic allostery refers to interactions between binding sites for identical ligands, as in the cooperative binding of O_2 by Hb. The binding of O_2 to one heme produces a positive homotropic effect by increasing the O_2 affinity of the remaining hemes (an allosteric mechanism is not compatible with negative homotropic effects). Heterotypic allostery refers to interactions between binding sites for different ligands, as in the regulation of Hb-O_2 affinity by the binding of non-heme ligands such as H^+, CO_2, Cl^- ions, and organic phosphates, all of which are chemically distinct from O_2 and other diatomic gases that bind to the ferrous heme iron. The binding of these effectors generally produces a positive heterotypic effect by inhibiting heme reactivity even though—by definition—they bind at sites remote from the heme active site.

Increasing the concentration of heterotropic effectors typically shifts the O_2-equilibrium curve to the right (O_2 affinity is reduced) and makes it more sigmoid (cooperativity is enhanced). The reduction in Hb-O_2 affinity is caused by a shift in the T↔R conformational equilibrium: allosteric effectors preferentially bind to deoxyHb, thereby stabilizing the low-affinity T-state at the expense of the high-affinity R-state. In general, allosteric effects stem from interactions between subunits of multimeric proteins. This is why changes in the concentrations of allosteric effectors do not alter the O_2-equilibrium curves of $\alpha\beta$ dimers, α- or β-chain monomers, or Mb.

3.2 Hb as a model allosteric protein

In 1938, Felix Haurowitz of the Charles University in Prague discovered that oxyHb and deoxyHb form structurally distinct crystals (Haurowitz, 1938),

Hemoglobin: Insights into Protein Structure, Function, and Evolution. Jay F. Storz, Oxford University Press (2019).
© Jay F. Storz 2019. DOI: 10.1093/oso/9780198810681.001.0001

indicating that different ligation states of the same protein have different quaternary structures. As stated by the structural biologist Max Perutz (who, incidentally, was Haurowitz's cousin), this structural difference "…implied that hemoglobin is not an oxygen tank but a molecular lung, because it changes its structure every time it takes up oxygen or releases it" (Perutz, 1978). After many years of effort, Perutz and his colleagues in Cambridge used X-ray crystallography to solve the crystal structure of unliganded (T-state) and liganded (R-state) Hb, and they demonstrated that both tetrameric structures are characterized by a symmetrical arrangement of the constituent subunits (Muirhead and Perutz, 1963, Perutz et al., 1964). These initial discoveries motivated theoretical and experimental investigations to elucidate the structural basis for the different stabilities and reactivities of T- and R-state Hb, and the nature of the oxygenation-linked transition in quaternary structure between the two states.

As stated by Dickerson and Geis (1983), the history of changing theories to explain the allosteric effects of Hb provides "…a classic example of the way that protein structures suggest mechanistic theories and models, which in turn suggest new chemical experiments, whose results provide feedback that forces the original theories to be abandoned, modified, qualified, and improved until the truth is finally approached asymptotically." The challenge of formulating models to describe the allosteric effects of Hb is to maintain compatibility with high-resolution data on Hb structure as well as experimental data on its functional properties (Eaton et al., 1999, Bellelli, 2010, Yuan et al., 2015).

Efforts to rationalize the cooperative effects of Hb inspired the development of general models to explain the allosteric regulation of enzyme activity. In fact, Hb has been described as an "honorary enzyme" with O_2 as its substrate (Wyman and Allen, 1951, Brunori, 1999). In this context, the reaction of interest is substrate binding rather than binding followed by catalysis. Hb is considered a prototype protein for the study of allosteric regulation for several reasons (Imai, 1982). First, the oxygenation of Hb is governed by both homotropic and heterotropic interactions that are experimentally well characterized. Second, our detailed understanding of structure-function relationships in Hb makes it possible to decipher the stereochemical mechanisms of

these interactions at atomic-level resolution. Third, O_2-binding by Hb is reversible and can be studied at equilibrium. By contrast, substrates of enzymatic reactions are metabolized, so binding reactions can only be observed in steady or transient states, making experimental analyses much more difficult.

In the following sections we will explore several models that were specifically formulated to explain the allosteric effects of Hb, or that were inspired by studies of Hb structure and function. We will start with an analysis of homotropic interactions that helped reveal the rules governing the cooperativity of O_2-binding by Hb.

3.3 The Adair equation

As we saw in Chapter 2 (sections 2.8–2.9), the Hill equation is a useful empirical descriptor of cooperative O_2-binding by Hb. The equation assumes all-or-none O_2 binding and therefore ignores partially liganded intermediates. In fact, Hill's original model postulated a mixture of Hb oligomers with different numbers of subunits to explain non-integral cooperativity coefficients (e.g., ~2.8 for solutions of human Hb) (Hill, 1913, Edsall, 1972, Bellelli, 2010). In 1924, Gilbert Adair, a mathematically adept student in Joseph Barcroft's Cambridge laboratory, experimentally confirmed that Hb exists as a stable tetramer (Adair, 1925b), and he therefore reasoned that non-integral cooperativity coefficients require the existence of partially liganded Hbs (i.e., Hbs with only one, two, or three bound O_2 molecules). Accordingly, Adair developed a mathematical model that described the sequential binding of ligands with unequal association constants (Adair, 1925a). The cooperative binding of O_2 by Hb can be accurately described by a parameterization of Adair's model in which successive O_2 molecules are bound with progressively increasing association constants.

If the α- and β-chain subunits are equivalent with respect to intrinsic binding affinity, and if the intra-dimer ($\alpha_1\beta_1/\alpha_2\beta_2$) and interdimer ($\alpha_1\beta_2/\alpha_2\beta_1$) contacts are equivalent with respect to intersubunit interaction, then O_2 binding to tetrameric Hb can be described by four equilibria, each governed by its own equilibrium binding constant:

$$Hb + 4O_2 \leftrightarrow Hb(O_2) + 3O_2 \leftrightarrow Hb(O_2)_2 \\ + 2O_2 \leftrightarrow Hb(O_2)_3 + O_2 \leftrightarrow Hb(O_2)_4. \quad (3.1)$$

For the sake of brevity, unliganded Hb can be written as Hb_0, $Hb(O_2)$ can be written as Hb_1, and so on, with subscripts denoting the number of bound O_2 molecules.

In order to calculate the fractional saturation of Hb at a given PO_2, Adair derived the concentration of each possible ligation intermediate relative to that of unliganded Hb:

$$[Hb_1] = 4K_1 PO_2 [Hb_0] \qquad = A_1 PO_2 [Hb_0] \quad (3.2)$$

$$[Hb_2] = \left(4K_1 PO_2\right)\left(\tfrac{3}{2} K_2 PO_2\right)[Hb_0]$$
$$= A_2 PO_2^2 [Hb_0] \quad (3.3)$$

$$[Hb_3] = \left(4K_1 PO_2\right)\left(\tfrac{3}{2} K_2 PO_2\right)\left(\tfrac{2}{3} K_3 PO_2\right)$$
$$[Hb_0] \qquad = A_3 PO_2^3 [Hb_0] \quad (3.4)$$

$$[Hb_4] = \left(4K_1 PO_2\right)\left(\tfrac{3}{2} K_2 PO_2\right)\left(\tfrac{2}{3} K_3 PO_2\right)$$
$$\left(\tfrac{1}{4} K_4 PO_2\right)[Hb_0] \qquad = A_4 PO_2^4 [Hb_0] \quad (3.5)$$

where K_i (i = 1 to 4) are the intrinsic (or "microscopic") association constants for individual binding sites, and A_i are the apparent (or "macroscopic") association constants. The apparent constants are the products of all preceding intrinsic constants in the reaction scheme and the corresponding statistical factors (4, 3/2, 2/3, 1/4). The statistical factors account for the number of binding sites in each protein at which "on" and "off" reactions can occur. For example, when the first O_2 molecule binds to deoxyHb (Hb_0), the O_2 can bind at any of the four unoccupied sites, and—once bound—there is only one site from which O_2 can be released; thus, K_1 is multiplied by 4 (there are four ways to transition from Hb_0 to Hb_1, and only one way to revert Hb_1 back to Hb_0). When the second O_2 molecule binds to Hb with one liganded site (Hb_1), the O_2 can bind at any of three unoccupied sites and there are two sites from which O_2 can be released; thus, K_2 is multiplied by 3/2 (there are three ways to transition from Hb_1 to Hb_2, and two ways to revert Hb_2 back to Hb_1). Likewise, K_3 is multiplied by 2/3 and K_4 is multiplied by 1/4. Cooperative binding requires that $K_4 > K_1$.

The sum of the concentrations of all ligation intermediates, including unliganded and fully liganded Hb (Hb_0 and Hb_4, respectively) is called the partition function or binding polynomial of Hb:

$$P = [Hb_0]\left(\begin{array}{c} 1 + A_1 PO_2 + A_2 PO_2^2 + A_3 PO_2^3 \\ + A_4 PO_2^4 \end{array}\right). \quad (3.6)$$

The fractional saturation (the concentration of liganded sites relative to the concentration of all potentially available ligand-binding sites) is expressed as

$$Y = \frac{[Hb_1] + 2[Hb_2] + 3[Hb_3] + 4[Hb_4]}{4([Hb_0] + [Hb_1] + [Hb_2] + [Hb_3] + [Hb_4])}. \quad (3.7)$$

This can be expressed in terms of apparent association constants by substituting the above-mentioned relationships and cancelling terms

$$Y = \frac{A_1 PO_2 + 2A_2 PO_2^2 + 3A_3 PO_2^3 + 4A_3 PO_2^4}{4(1 + A_1 PO_2 + A_2 PO_2^2 + A_3 PO_2^3 + A_3 PO_2^4)}. \quad (3.8)$$

This is the Adair equation (Adair, 1925a); it provides the most general description of O_2-binding equilibria for tetrameric Hb and demonstrates how non-integral cooperativity coefficients can be produced by a balance between alternative conformational states with different ligand affinities (Imai, 1982). The expression remains valid in the presence of allosteric effectors provided that their concentrations remain constant during the four-step oxygenation process. It has become common practice to refer to the K parameters as "Adair constants," although Adair himself formulated his model in terms of the A parameters.

The Adair equation can also be expressed in terms of the intrinsic association constants

$$Y = \frac{\begin{array}{c} K_1 PO_2 + 3K_1 K_2 PO_2^2 + 3K_1 K_2 K_3 PO_2^3 \\ + 4K_1 K_2 K_3 K_4 PO_2^4 \end{array}}{\begin{array}{c} 1 + 4K_1 PO_2 + 6K_1 K_2 PO_2^2 + 4K_1 K_2 K_3 PO_2^3 \\ + 4K_1 K_2 K_3 K_4 PO_2^4 \end{array}} \quad (3.9)$$

by making the following substitutions in equation 3.8: $A_1 = 4K_1$, $A_2 = 6K_1 K_2$, $A_3 = 4K_1 K_2 K_3$, and $A_4 = K_1 K_2 K_3 K_4$. If the association constants for each of the four ligation steps are equal (no cooperativity), then equation 3.9 yields a simple hyperbola. If the four association constants are not equal, that means that O_2 binding is cooperative and a plot of equation 3.9 will yield a sigmoid curve. With this model, experimental measurements of Hb-O_2 affinity over a full range of saturation levels enable us to quantify homotropic and heterotropic effects in terms of a four-step ligation scheme.

3.4 Asymptotes of the Hill plot

As we saw in section 2.9, the Hill plot for Hb (the plot of log ($Y/(1-Y)$ vs. log PO_2) describes a sigmoid

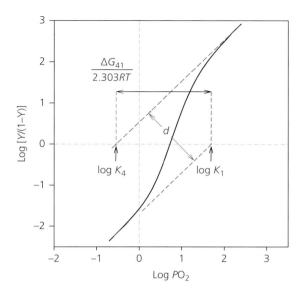

Fig. 3.1. Hill plot of O_2-equilibrium data for Hb. The X-intercepts of the lower and upper asymptotes with log $Y(1-Y) = 0$ yield the logarithms of K_1 and K_4, the equilibrium constants for binding the first and fourth O_2 molecule, respectively. The distance between these intercepts ($\Delta G_{41}/2.303RT$) and the perpendicular distance between the upper and lower asymptotes (*d*) provide alternative measures of the free energy of cooperativity.

curve with upper and lower asymptotes that approximate a slope of 1.0 (Fig. 3.1). In practice, it is difficult to obtain precise measurements at extremely low and high saturation levels (i.e., Y <0.0–0.02 and >0.98–1.00) (Roughton et al., 1955, Imai, 1982). According to Adair's scheme, the lower asymptote at low saturation reflects the initial O_2-binding to deoxyHb ($Hb_0 + O_2 \rightarrow Hb_1$), and the upper asymptote at high saturation reflects O_2-binding to

the last remaining unliganded site ($Hb_3 + O_2 \rightarrow Hb_4$). And, according to some models of allostery, O_2-binding to the first and last sites reflect binding to Hb in the T and R-states, respectively. As we saw previously, log-transformed values of the association constants for the first and fourth ligation steps, K_1 and K_4, can be estimated by extrapolating both asymptotes of the Hill plot to the horizontal axis at log $Y/(1-Y) = 0$ (Fig. 3.1). The remaining intrinsic association constants, K_2 and K_3, can be estimated by fitting equation 3.9 to a plot of Y vs. PO_2. Estimates of the Adair constants derived from O_2-equilibrium measurements of human Hb are compiled in Table 3.1. Over a range of experimental conditions, the tabulated data show that the four association constants generally increase monotonically (positive cooperativity), and $K_1 \ll K_4$. The constants K_1, K_2, and K_3 are significantly lower in the presence than in the absence of organic phosphates (DPG and inositol hexaphosphate, IHP), and this is also reflected by higher P_{50} values (Table 3.1). Whereas the binding of DPG reduces K_1, K_2, and K_3, while leaving K_4 mostly unaffected, the allosteric binding of IHP (which has a much higher association constant than DPG) reduces K_4 in addition to K_1-K_3 (Tyuma et al., 1971a, Tyuma et al., 1971b).

The distance between the X-intercepts of the upper and lower asymptotes gives $\Delta G_{41}/2.303RT$ (Fig. 3.1) where R is the molar gas constant, T is the absolute temperature, and ΔG_{41} is the free energy of cooperativity:

$$\Delta G_{41} = RT \ln\left(\frac{K_4}{K_1}\right) \qquad (3.10)$$

Table 3.1 Estimated Adair constants and P_{50} values based on O_2-binding data for human Hb. O_2-equilibria were measured at 25° in 0.05 M Bis-tris buffer (for pH 7 4 and 6.5) and in 0.05 M Tris buffer (for pH 9.1); [heme] = 0.6 mM

pH	Cl⁻	phosphates	K_1 (torr⁻¹)	K_2 (torr⁻¹)	K_3 (torr⁻¹)	K_4 (torr⁻¹)	P_{50} (torr)
9.1	0.1 M	–	0.0595	0.16	1.5	3.33	2.15
7.4	0.1 M	–	0.0218	0.062	0.30	3.45	5.32
7.4	0.1 M	2 mM DPG	0.00814	0.033	0.030	3.82	14.0
7.4	0.1 M	2 mM IHP	0.00502	0.013	0.0042	0.915	48.8
6.5	0.1 M	–	0.0119	0.011	0.069	1.24	18.8
6.5	0.1 M	2 mM IHP	0.00403	0.0044	0.0073	0.0339	136

Data from Imai (1979).

This expresses the energetic enhancement of binding affinity for the fourth ligation step (K_4) relative to the first (K_1) (Fig. 3.1). Wyman (Wyman, 1964) introduced a similar measure, termed the free energy of interaction among binding sites,

$$\Delta G = 2.303 \sqrt{2RTd} \qquad (3.11)$$

where d is the perpendicular distance between the upper and lower asymptotes of the Hill plot (Fig. 3.1). More rigorous and exact means of estimating interaction free energies were developed subsequently (Saroff and Minton, 1972, Minton and Saroff, 1974).

3.5 Models of allostery

"The discovery of an interaction among the four hemes made it obvious that they must be touching, but in science what is obvious is not necessarily true. When the structure of Hb was finally solved, the hemes were found to lie in isolated pockets on the surface of the subunits. Without contact between them how could one of them sense whether the others had combined with oxygen?"
—**Perutz (1978)**

The body of theory developed by Adair can rationalize the observed ligand-binding behavior of Hb and other multisubunit proteins. However, it provides no mechanistic insight as to why successive ligation steps do not have equal association constants. Adair's model therefore inspired the development of new models to explain how identical, symmetrically related subunits of multimeric proteins can exhibit different ligand affinities. The most influential model of protein allostery was developed in the 1960s in a collaborative effort involving two Frenchmen, Jacques Monod and Jean-Pierre Changeux, at the Pasteur Institute in Paris, and an American, Jeffries Wyman, then working at the University of Rome. This was an impressive and colorful trio. Monod was a pioneer of twentieth-century molecular biology, and is most widely known for his work on regulatory mechanisms of gene expression. Monod's research career was interrupted during World War II while he served in the French resistance, later becoming the chief of staff of the French Forces of the Interior. He was awarded the French Croix de Guerre, the American Bronze Star medal and, two decades later, the Nobel Prize in Physiology

or Medicine (Carroll, 2013). Changeux was a PhD student in Monod's lab, and he later went on to become a pioneering neurobiologist. He spent most of his career at the Pasteur Institute and, in addition to his scientific accomplishments, he became a highly acclaimed writer of popular science books that explored the nexus of neuroscience, cognitive science, and philosophy. The American member of the trio, Jeffries Wyman, was a mathematical biophysicist who spent the early part of his career at Harvard University. Like Monod and Changeux, Wyman was a true renaissance man. In addition to the lure of several highly productive collaborations with friends and colleagues in Europe, Wyman's passion for classical art and archaeology inspired him to spend the last half of his life in Egypt, Italy, and France.

In 1965, Monod, Wyman, and Changeux published a landmark paper in which they formulated the concerted model of protein allostery (also known as the "two-state" or "MWC" model) (Monod et al., 1965). This model was inspired by the idea that the cooperative effects observed in certain enzyme-catalyzed reactions might be understood by analogy with the cooperativity of O_2-binding by Hb. Monod, Wyman, and Changeux knew that deoxy and oxyHb had different quaternary structures, and they suspected that the same might be true for different ligation states of multisubunit enzymes. The MWC model has proven to be extremely useful for rationalizing the cooperative effects of Hb and other proteins. But more generally, it has provided key insights into a pervasive biochemical mechanism of regulatory control. With pardonable self-pride, Jaques Monod described allostery as the "second secret of life" (he modestly ceded primacy to the discovery of the double helix structure of DNA).

The simplest form of the MWC model assumes a multisubunit enzyme that exists in two discrete conformational states: the T-state, which has low affinity for substrate, and the R-state, which has high affinity for substrate. The two conformations are distinguished by the arrangement of subunits and the number and strength of chemical bonds between the subunits. The conformation that is less constrained by intersubunit bonds—the R-state—has a higher catalytic activity (or, in the case of Hb, a higher O_2-affinity). The multiple catalytic sites of a multisubunit enzyme are like the four O_2-binding heme groups of tetrameric Hb. In this model, an enzyme's activity toward its substrate is modulated by the binding of effector molecules (heterotropic ligands) that are chemically distinct from the enzyme's substrate and which bind at sites that are structurally remote from the enzyme's active site. Thus, effector molecules do not competitively inhibit the binding of substrate in the active site. Instead, the binding of effector molecules indirectly modulates the reactivity of the active site by inducing a global conformational change in the protein. This mode of regulation provides an especially efficient mechanism of feedback control, as the activity of an enzyme that catalyzes an early reaction in a metabolic pathway can be inhibited by the accumulation of reaction end-products further downstream in the pathway.

The MWC model is based on four postulates:

1. The allosteric protein consists of symmetrically related subunits with identical ligand affinities. This is a simplifying assumption in the case of Hb, since we know that the α- and β-chain subunits are not structurally identical and they have slightly different O_2 affinities.
2. Each multimeric assembly exists in at least two conformational states (T and R); these states are in equilibrium regardless of the ligation state of the protein.
3. Ligands can bind to individual subunits in either the T- or the R-state. The ligand affinity of a given binding site is only determined by the global conformational state of the protein.

4. The molecular symmetry of the protein is conserved during the conformational transition between the T- and R-states. This dictates that each of the constituent subunits must change conformation in concert.

The binding of ligand induces a shift in the equilibrium between the two conformational states (Fig. 3.2). Due to symmetry, the association constant for each individual site is determined by the global conformational state of the tetramer (either T or R) and—within a given conformation—is independent of the ligation state of the other three sites. For this reason, we dispense with subscripted numbers (K_1, K_2, K_3, and K_4) and we define K_T and K_R as the intrinsic association constants for O_2 in the T- and R-states, respectively. Although K_T and K_R were defined as dissociation constants in the original paper (Monod et al., 1965), we will follow the example of Imai (1982) and Bellelli (2010) by defining them as association constants to maintain consistency with the Adair equation.

In the absence of ligand, Hb is found almost exclusively in the T-state, and each O_2 that is bound by one of its subunits increases the probability that the tetramer will transition to the R-state. Once this transition occurs, the O_2 affinity of the remaining unliganded sites increases dramatically, with the result that fully oxygenated Hb is found almost exclusively in the R-state. This T→R transition rationalizes the sigmoid shape of the O_2-equilibrium curve: the curve is shallow at low PO_2 when Hb exists almost exclusively in the low-affinity T-state, it then quickly

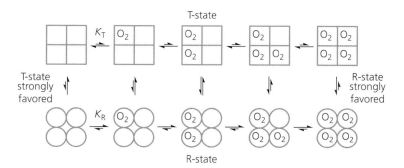

Fig. 3.2. Schematic illustration of the MWC (concerted) model of Hb allostery. The tetrameric assembly exists in one of two alternative quaternary conformations, the T-state or the R-state. When no O_2 molecules are bound, the equilibrium is shifted in the favor of the low-affinity T-state, and when Hb is fully liganded, the equilibrium is shifted in favor of the high-affinity R-state.

steepens as O_2-binding increases the relative fraction of Hb tetramers that have transitioned to the high-affinity R-state, and it flattens out again once most binding sites are liganded with O_2. According to the MWC view, the observed O_2-equilibrium curve represents a combination of curves for pure R-state and T-state Hb (Fig. 3.3). Thus, the progressive increase in O_2-affinity illustrated by Perutz's biblical parable about the rich and the poor does not stem from any direct interaction between the hemes; rather, it reflects the increasing fraction of Hb in the R-state relative to that in the T-state (Perutz, 1978).

If the number of binding sites is known for a given protein, then the MWC model can be formulated with just three parameters: the ratio of the concentrations of T- and R-state Hb in the absence of bound ligand (L) and the ligand affinities of binding sites in each of the alternative conformations (K_T and K_R). The ratio of the concentrations of T- and R-state Hb with n bound ligands is defined as the allosteric equilibrium constant:

$$L_n = \left[{}^T Hb_n / {}^R Hb_n \right] \tag{3.12}$$

where $n = 0$ to 4. There are two requirements for cooperativity: the T-state must predominate in the absence of O_2 ($L_0 \gg 1$) and, conversely, the R-state must predominate in the presence of saturating concentrations of O_2 ($L_4 \ll 1$). Because of the differ-

ent ligand affinities of the alternative conformational states, O_2-binding progressively shifts the allosteric equilibrium in favor of the R-state. The relationship between the allosteric constant and O_2 saturation can be generalized as

$$L_n = L_0 \left(K_T / K_R \right)^n . \tag{3.13}$$

This equation indicates that L decreases monotonically as the number of bound ligands (n) increases from zero to four.

To calculate the fractional saturation at any given PO_2, we need to define the concentration of all ligation intermediates of R- and T-state Hb. The expected concentrations for the ligation intermediates of R-state Hb are defined relative to unliganded R-state Hb (${}^R Hb_0$):

$$\left[{}^R Hb_1 \right] = 4 \left[{}^R Hb_0 \right] PO_2 K_R \tag{3.14}$$

$$\left[{}^R Hb_2 \right] = \frac{3}{2} \left[{}^R Hb_1 \right] PO_2 K_R = 6 \left[{}^R Hb_0 \right] PO_2^2 K_R^2 \tag{3.15}$$

$$\left[{}^R Hb_3 \right] = \frac{3}{2} \left[{}^R Hb_2 \right] PO_2 K_R = 4 \left[{}^R Hb_0 \right] PO_2^3 K_R^3 \tag{3.16}$$

$$\left[{}^R Hb_4 \right] = \frac{3}{2} \left[{}^R Hb_3 \right] PO_2 K_R = \left[{}^R Hb_0 \right] PO_2^4 K_R^4 . \tag{3.17}$$

These equations make use of the same statistical factors (4, 3/2, 2/3, and 1/4) that we saw in the derivation of the Adair constants.

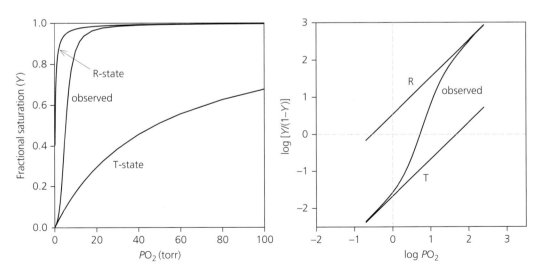

Fig. 3.3. According to the MWC model, the observed O_2-equilibrium curve (left) and Hill plot (right) for Hb reflect weighted averages of curves for the alternative T- and R-states.

The ligation intermediates of ^THb are calculated in the same way by using $^THb_0$ as a reference ($=L_0[^RHb_0]$) and by substituting K_T for K_R. This yields the following partition function:

$$P = \left[^RHb_0\right]\left[\left(1 + PO_2K_R\right)^4 + L_0\left(1 + PO_2K_T\right)^4\right]. \quad (3.18)$$

We can then calculate the fractional saturation at any PO_2 by dividing the sum of the ligation intermediates by the partition function:

$$Y = \frac{PO_2K_R\left(1 + PO_2K_R\right)^3 + L_0PO_2K_T\left(1 + PO_2K_T\right)^3}{\left(1 + PO_2K_R\right)^4 + L_0\left(1 + PO_2K_T\right)^4} \quad (3.19)$$

This equation describes the MWC model for homotropic interactions. Comparison of equation 3.19 with equation 3.9 reveals the relationship between the Adair constants and the MWC parameters (Imai, 1973, Imai, 1982) (Table 3.2). The MWC model rationalizes Adair's four O_2-binding constants as the weighted averages of the O_2 affinities of mixtures of the two allosteric states, ^RHb and ^THb. The MWC model predicts that O_2 binding is cooperative at each step, such that that the association constants

Table 3.2 Relationship between the Adair constants and MWC parameters

Adair scheme	K_1	K_2	K_3	K_4
MWC (two-state)	$\dfrac{K_R + LK_T}{1 + L}$	$\dfrac{K_R^2 + LK_T^2}{K_R + LK_T}$	$\dfrac{K_R^3 + LK_T^3}{K_R^2 + LK_T^2}$	$\dfrac{K_R^4 + LK_T^4}{K_R^3 + LK_T^3}$

increase montonically ($K_1 < K_2 < K_3 < K_4$), reflecting the increase in the fraction of ^RHb over ^THb as ligand-binding progresses. Thus, cooperativity becomes greater when L is large and $K_R \gg K_T$. Under such conditions, the relations in Table 3.2 yield the approximations

$$K_1 \approx K_T \text{ and } K_4 \approx K_R. \quad (3.20)$$

By quantifying the expected $[^THb_n/^RHb_n]$ ratio at each stage of binding, the MWC model predicts the conformational shift in the population of Hb molecules as a function of PO_2. Figure 3.4 shows PO_2-dependent changes in the concentrations of unliganded (deoxy) $^THb_0$, along with partially liganded T-state Hbs ($^THb_1$ and $^THb_2$), and fully oxygenated $^RHb_4$, along with two

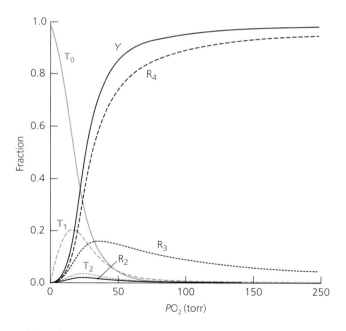

Fig. 3.4. Whereas curve-fitting with the Hill equation yields fractional saturations as a function of PO_2, the MWC model yields concentrations of ^THb and ^RHb in all possible ligation states. The fractional saturation (Y) as a function of PO_2. The fraction of Hb in the T-state with zero, one, and two liganded sites (T_0, T_1, and T_2, respectively) and the fraction of Hb in the R-state with two, three, and four liganded sites (R_2, R_3, and R_4). The fractions of Hb in the other possible ligation states are too small to be visible in the graph.

Modified from Berg et al. (2015) with permission from W. H. Freeman.

partially liganded R-state Hbs (^{R}Hb$_2$ and ^{R}Hb$_3$). As one might guess from a consideration of the schematic shown in Figure 3.2, the other partially liganded states are thermodynamically unstable and are therefore present at exceedingly low concentrations. These unstable states include ^{T}Hb with three or four bound O$_2$ molecules (^{T}Hb$_3$ and ^{T}Hb$_4$, respectively) and R-state Hbs with zero or one bound O$_2$ molecules (^{R}Hb$_0$ and ^{R}Hb$_1$, respectively).

Successive binding of O$_2$ molecules to ^{T}Hb produces a progressively increasing strain, until the protein finally transitions to the R-state. Once that transition occurs, any remaining unliganded sites will bind O$_2$ with the high affinity characteristic of ^{R}Hb (we will explore the structural basis for this transition in Chapter 4). Under physiological conditions, normal human Hb typically undergoes the T- to R-state transition after binding two or three O$_2$ molecules (Gibson and Parkhurst, 1968, MacQuarrie and Gibson, 1971), which is why T-state Hbs with two or more liganded sites are present at such low concentrations (Fig. 3.4). Conversely, fully oxygenated R-state Hb will typically transition to the T-state after releasing the first one or two O$_2$ molecules, and once the transition occurs, the remaining liganded sites will jettison their cargo due to their suddenly reduced O$_2$-affinity (this is why R-state Hbs with two or more unoccupied sites are present at such low concentrations; Fig. 3.4). This is the essence of a positive homotropic interaction: binding O$_2$ at one site increases the heme reactivity of other subunits by promoting the T→R transition; reciprocally, releasing O$_2$ at one site decreases the heme reactivity of other subunits by promoting the R→T transition.

As shown by Bellelli (2018), the two-state MWC model can be extended to include heterotropic interactions by coupling the original formulation with Wyman's linked functions (Wyman, 1948, Wyman, 1964, Wyman and Gill, 1990). This new formulation describes the O$_2$-affinity of ^{T}Hb with and without bound effector, yielding the reaction scheme shown in Fig. 3.5 (using DPG as an example). For simplicity, this three-state reaction scheme ignores effector-binding to ^{R}Hb. Whereas ^{T}Hb and ^{R}Hb are freely interconverting structural isomers, ^{T}Hb and ^{T}Hb$_{DPG}$ interconvert via binding and release of the effector.

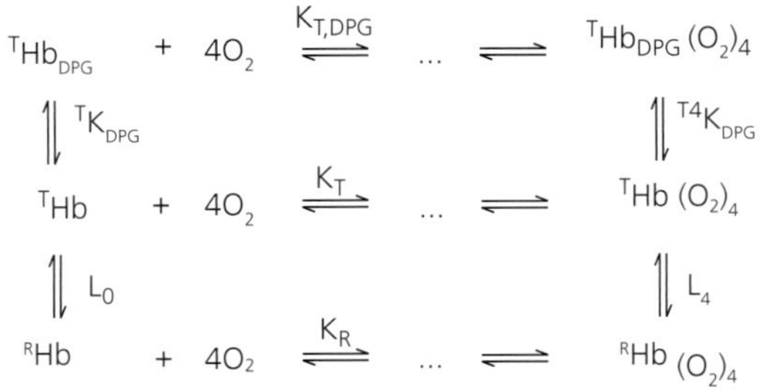

Fig. 3.5. Reaction scheme for human Hb in the presence of DPG, a representative heterotypic effector with 1:1 binding stoichiometry (Bellelli, 2018). For simplicity, this scheme ignores DPG binding to ^{R}Hb, and it also ignores the presence of other heterotypic effectors that may be present in solution (e.g., H$^+$ and Cl$^-$).

Reproduced from Bellelli (2018) with permission from Bentham Science.

The reaction scheme shown in Fig. 3.5 extends the two-state MWC model by introducing two additional equilibrium constants, ^{T}K$_{DPG}$ (the DPG binding constant of ^{T}Hb) and $K_{T,DPG}$ (the O$_2$ binding constant for ^{T}Hb with bound DPG). We can therefore expand the partition function considered previously, once again using ^{R}Hb$_0$ as the reference state:

$$P = \begin{bmatrix} ^R Hb_0 \end{bmatrix} \begin{bmatrix} \left(1+PO_2 K_R\right)^4 + L_0\left(1+PO_2 K_T\right)^4 \\ + L_0[DPG]^T K_{DPG}\left(1+PO_2 K_{T,DPG}\right)^4 \end{bmatrix} \quad (3.21)$$

This leads to the following expression for calculating saturation values as a function of PO_2:

$$Y = \frac{\begin{array}{l} PO_2 K_R\left(1+PO_2 K_R\right)^3 + L_0 PO_2 K_T\left(1+PO_2 K_T\right)^3 \\ + L_0 PO_2[DPG]K_{T,DPG}{}^T K_{DPG}\left(1+PO_2 K_{T,DPG}\right)^3 \end{array}}{\begin{array}{l} \left(1+PO_2 K_R\right)^4 + L_0\left(1+PO_2 K_T\right)^4 \\ + L_0[DPG]^T K_{DPG}\left(1+PO_2 K_{T,DPG}\right)^4 \end{array}} \quad (3.22)$$

In this formulation, the relationship between [^{T}Hb] and [^{R}Hb] is described by L_0, as before, and the relationship between [^{T}Hb] and [^{T}Hb$_{DPG}$] is described by the product of the effector concentration and its association constant, [DPG] $K_{T,DPG}$.

The MWC model assumes that allosteric effectors reduce Hb-O$_2$ affinity by shifting the allosteric equilibrium in favor of ^{T}Hb relative to ^{R}Hb but without affecting the intrinsic binding constants of either structure. In other words, allosteric effectors reduce Hb-O$_2$ affinity by increasing L without affecting K_T or K_R. If the intrinsic binding constants are not affected, then the free energy of cooperativity remains constant. Contrary to this expectation, studies of vertebrate Hb have demonstrated that the binding of organic phosphates and H^+ typically decrease K_T in addition to shifting the allosteric equilibrium (Imai, 1973, Imai and Yonetani, 1975, Goodford et al., 1978, Kilmartin et al., 1978, Imai, 1983, Weber et al., 1987, Tsuneshige et al., 2002, Yonetani et al., 2002, Grispo et al., 2012). In experiments on human Hb under standardized conditions (25°, pH 7.4, 0.1 M Cl⁻, [heme] = 0.6 mM), the addition of 2 mM DPG increases L as expected, but it also produces a 3.0-fold reduction in K_T (Table 3.3). This effect is evident in a Hill plot of the data: the addition of DPG produces a rightward shift in the lower asymptote (K_T is reduced) but the upper asymptote (K_R) is not significantly affected (Fig. 3.6). At low pH, high concentrations of DPG and IHP have the effect of reducing both K_T and K_R (Imai, 1982, Yonetani et al., 2002) (Table 3.3). These effects are also evident in the Hill plots, as the addition of IHP produces a rightward shift in both the lower and the upper asymptotes of the curve (Fig. 3.6). The observation that the binding of allosteric effectors alters the association constants in addition to the allosteric constant motivated refinements and elaborations of the canonical MWC model while still retaining the basic two-state framework (Szabo and Karplus, 1972, Szabo and Karplus, 1975, Bellelli, 2018).

In summary, the MWC model demonstrates that both homotropic and heterotropic effects of Hb can be explained by a conformational switch between low-affinity ^{T}Hb and high-affinity ^{R}Hb. A key requirement of the model is that the molecular symmetry of the tetramer is maintained through the allosteric transition in quaternary structure. An alternative model of protein allostery was formu-lated by Daniel Koshland, George Némethy, and David Filmer in 1966 (Koshland et al., 1966). This model, known as the sequential model (or KNF model), is premised on the "induced-fit hypothesis" of ligand binding, which postulates that a flexible interaction between the ligand and protein induces a conformational change that increases ligand-binding affinity (Fig. 3.7). The KNF model refined and extended a much earlier model of Hb coopera-tivity that was developed by the American chemist, Linus Pauling (Pauling, 1935). According to the KNF model, the localized conformational change that is induced by ligand binding to one subunit is transmitted to neighboring subunits. This mechanical coupling between binding sites gives rise to subunit–subunit interaction—the *sine qua non* of cooperativity. Importantly, the KNF model postulates that the binding of ligand to one subunit in the multimeric assembly increases the binding affinity of structur-ally adjacent subunits, but without inducing a T→R conversion of the assembly as a whole. Instead, con-formational changes occur sequentially as more and more ligand is bound (Fig. 3.8). The essence of the KNF model is that a protein's ligand affinity depends on the number of bound ligands, whereas the essence of the MWC model is that ligand affin-ity is solely determined by the global conform-ational state of the multimeric assembly. In the limiting case of the KNF model when the coupling between subunits is very strong, conformational changes of individual subunits may be nearly concerted and the multimeric assembly maintains symmetry through the shift in quaternary structure, in accordance with the MWC model.

Table 3.3 Estimated values of the two-state MWC parameters based on O_2-binding data for human Hb. O_2 equilibria were measured at 25° in 0.05 M Bis-tris buffer (for pH 7.4 and 6.5) and in 0.05 M Tris buffer (for pH 9.1); [heme] = 0.6 mM

pH	Cl⁻	phosphates	K_T (torr⁻¹)	K_R (torr⁻¹)	L_0 (=T_0/R_0)	L_4 (=T_4/R_4)
9.1	0.1 M	–	0.0604	3.34	2.7×10^3	2.9×10^{-4}
7.4	0.1 M	–	0.0241	3.31	8.7×10^4	2.5×10^{-4}
7.4	0.1 M	2 mM DPG	0.0080	3.00	3.0×10^6	1.5×10^{-4}
7.4	0.1 M	2 mM IHP	0.0060	0.91	2.8×10^6	5.4×10^{-3}
6.5	0.1 M	–	0.0117	1.60	6.2×10^5	1.8×10^{-3}
6.5	0.1 M	2 mM IHP	0.0040	0.06	4.0×10^3	6.5×10^{-2}

Data from Imai (1982).

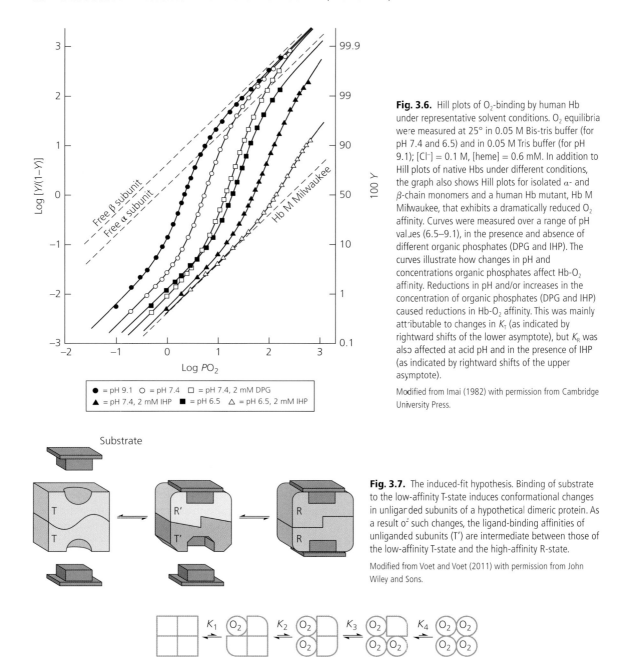

Fig. 3.6. Hill plots of O_2-binding by human Hb under representative solvent conditions. O_2 equilibria were measured at 25° in 0.05 M Bis-tris buffer (for pH 7.4 and 6.5) and in 0.05 M Tris buffer (for pH 9.1); [Cl^-] = 0.1 M, [heme] = 0.6 mM. In addition to Hill plots of native Hbs under different conditions, the graph also shows Hill plots for isolated α- and β-chain monomers and a human Hb mutant, Hb M Milwaukee, that exhibits a dramatically reduced O_2 affinity. Curves were measured over a range of pH values (6.5–9.1), in the presence and absence of different organic phosphates (DPG and IHP). The curves illustrate how changes in pH and concentrations organic phosphates affect Hb-O_2 affinity. Reductions in pH and/or increases in the concentration of organic phosphates (DPG and IHP) caused reductions in Hb-O_2 affinity. This was mainly attributable to changes in K_T (as indicated by rightward shifts of the lower asymptote), but K_R was also affected at acid pH and in the presence of IHP (as indicated by rightward shifts of the upper asymptote).

Modified from Imai (1982) with permission from Cambridge University Press.

Fig. 3.7. The induced-fit hypothesis. Binding of substrate to the low-affinity T-state induces conformational changes in unligarded subunits of a hypothetical dimeric protein. As a result of such changes, the ligand-binding affinities of unliganded subunits (T') are intermediate between those of the low-affinity T-state and the high-affinity R-state.

Modified from Voet and Voet (2011) with permission from John Wiley and Sons.

Fig. 3.8. Schematic illustration of the KNF (sequential) model of Hb allostery. The binding of ligand changes the tertiary structure of the individual subunit to which it binds. This conformational change induces changes in adjacent subunits that increase their ligand-binding affinity.

The MWC and KNF models represent idealized, limiting cases, and neither of them perfectly accounts for the O_2-binding properties of Hb. Nonetheless, the MWC model has provided the most influential paradigm for understanding Hb allostery, not only because of its elegant formulation and its success in rationalizing observed structural and functional properties, but also because the model makes clear predictions that can be decisively tested with experimental data (Eaton et al., 1999, Eaton et al., 2007, Bettati et al., 2009, Bellelli, 2010).

3.6 Experimental tests of theoretical predictions

One clear prediction of the MWC model is that tetrameric Hb exists in at least two discrete structural conformations with different O_2 affinities. This prediction was supported by early X-ray crystallographic studies and subsequent analyses that investigated the structural basis of oxygenation-linked transitions in quaternary structure (Perutz, 1970, Baldwin and Chothia, 1979, Gelin et al., 1983, Perutz et al., 1998). In the decades since Perutz's pioneering studies, increasingly refined crystallographic analyses have characterized atomic level structural differences between unliganded human Hb and fully liganded human Hb ($Hb(O_2)_4$ or $Hb(CO)_4$), the standard reference structures for ^THb and ^RHb, respectively (Park et al., 2006). Consistent with inferences drawn from crystallographic data (Paoli et al., 1996, Tame, 1999, Park et al., 2006), nuclear magnetic resonance studies of Hb structure in solution indicate that alternative ensembles of ^THb and ^RHb structures exist in a dynamic equilibrium that is affected by heme ligation and the binding of allosteric effectors (Ho, 1992, Lukin and Ho, 2004, Yuan et al., 2015). Importantly, structural studies generally indicate that partially liganded Hbs are referable to either the T- or the R-state, as predicted by the MWC model, and do not represent intermediate conformational states, as predicted by the KNF model (Arnone et al., 1986, Luisi and Shibayama, 1989, Luisi et al., 1990). A variety of experimental results has also demonstrated that the Hb tetramer transitions between alternative conformational states with different ligand affinities (Gibson, 1959, Giardina et al., 1975, Ferrone and Hopfield, 1976,

Hofrichter et al., 1983, Bellelli and Brunori, 1994, Bellelli, 2010, Bellelli, 2018).

A second prediction of the MWC model is that both allosteric states of Hb coexist over the full range of saturation levels. This prediction is difficult to test experimentally since $^RHb_0$ is expected to exist at exceedingly low concentration (because $L_0>>1$) and, likewise, $^THb_4$ is expected to exist at low concentration (because $L_4<<1$). Nonetheless, kinetic measurements of O_2-binding and dissociation have recorded spectroscopic features that indicate the presence of R- and T-state Hbs in these rare ligation states (Giardina et al., 1975, Sawicki and Gibson, 1976, Sawicki and Gibson, 1977, Bellelli and Brunori, 1994).

A third prediction of the MWC model is that ligand binding to each individual conformation is non-cooperative if the allosteric transition in quaternary structure is prevented. This prediction has been tested in a set of clever and technically sophisticated experiments by A. Mozzarelli, W. A. Eaton, and colleagues. Using a technique pioneered by Mozzarelli's group at the University of Parma in Italy, this team recorded O_2-binding equilibria of deoxy ^THb immobilized in a crystal lattice that prevented the oxygenation-linked T→R transition in quaternary structure. As predicted by the MWC model, O_2-equilibrium curves of the immobilized ^THbs were hyperbolic (Hill coefficient $n = 1.0$), demonstrating that O_2-binding was non-cooperative when the liganded T-state molecule was prevented from transitioning to the R-state (Mozzarelli et al., 1991, Rivetti et al., 1993). The immobilized ^THb was also unresponsive to changes in pH (no Bohr effect). Subsequent experiments in which ^THb and ^RHb were immobilized in a silica gel matrix also demonstrated that O_2-binding is non-cooperative when the allosteric transition in quaternary structure is inhibited (Shibayama and Saigo, 1995, Bettati and Mozzarelli, 1997, Viappiani et al., 2004, Henry et al., 2015). These results are consistent with kinetic studies of ligand binding and dissociation, which indicate that the O_2 affinity of human Hb is principally determined by quaternary structural state (as predicted by the MWC model), not the number of bound ligands (as predicted by the KNF model) (Parkhurst and Gibson, 1967, Antonini and Brunori, 1970, Antonini and Brunori, 1971, Hopfield et al.,

1971, Edelstein, 1975, Shulman et al., 1975, Sawicki and Gibson, 1976, Shulman, 2001).

The two-state MWC model has generally held up well to experimental tests and most O_2-binding data for Hb can be rationalized in terms of a conformational equilibrium between the T- and R-states (Edelstein, 1975, Shulman et al., 1975, Eaton et al., 1999, Shulman, 2001, Eaton et al., 2007, Bettati et al., 2009, Bellelli, 2010, Bellelli and Brunori, 2011). As stated by Bellelli (2010): "Only in the case of the changes of K_T and K_R induced by allosteric effectors were the experimental results found to be grossly incompatible with the model's predictions, but this defect proved the easiest to amend." Several extensions and refinements of the MWC model have been formulated over the years, most of which concern the interplay between tertiary and quaternary effects in the stepwise oxygenation of Hb. However, these extensions are still essentially two-state models. For example, the tertiary two-state model (Henry et al., 2002) can be considered an extension of the MWC model that postulates an equilibrium between high- and low-affinity tertiary conformations of individual subunits (t and r, respectively) in both the T and R quaternary structures (Lee and Karplus, 1983). In the tertiary two-state model, cooperative O_2-binding stems from an all-or-none T↔R transition in quaternary structure, as in the canonical MWC model, but with representation of both t and r tertiary conformations in liganded ^{T}Hb and unliganded ^{R}Hb quaternary structures. Overall, the two-state MWC model and the extensions that it has inspired provide a highly useful theoretical framework for understanding and interpreting the allosteric effects of Hb.

The foregoing account provides a brief overview of a vast literature. Having reviewed some of the basic theory to explain the allosteric effects of Hb, in Chapter 4 we will explore the structural basis of homotropic and heterotropic interactions.

References

Adair, G. S. (1925a). The hemoglobin system. VI. The oxygen dissociation curve of hemoglobin. *Journal of Biological Chemistry*, **63**, 529–45.

Adair, G. S. (1925b). The osmotic pressure of haemoglobin in the absence of salts. *Proceedings of the Royal Society of London Series A-Containing Papers of a Mathematical and Physical Character*, **A109**, 292–300.

Antonini, E. and Brunori, M. (1970). Hemoglobin. *Annual Review of Biochemistry*, **39**, 977–1042.

Antonini, E. and Brunori, M. (1971). *Hemoglobin and Myoglobin in their Reactions with Ligands*, Amsterdam, North-Holland Publishing Company.

Arnone, A., Rogers, P., Blough, N. V., McGourty, J. L., and Hoffman, B. M. (1986). X-ray diffraction studies of a partially liganded hemoglobin [a^{FEII}-COb^{MNII}]2. *Journal of Molecular Biology*, **188**, 693–706.

Baldwin, J. and Chothia, C. (1979). Haemoglobin: the structural changes related to ligand binding and its allosteric mechanism. *Journal of Molecular Biology*, **129**, 175–220.

Bellelli, A. (2010). Hemoglobin and cooperativity: experiments and theories. *Current Protein and Peptide Science*, **11**, 2–36.

Bellelli, A. (2018). Non-allosteric cooperativity in hemoglobin. *Current Protein and Peptide Science*, **19**, 1–16.

Bellelli, A. and Brunori, M. (1994). Optical measurements of quaternary structural changes in hemoglobin. *Methods in Enzymology*, **232**, 56–71.

Bellelli, A. and Brunori, M. (2011). Hemoglobin allostery: variations on the theme. *Biochimica Et Biophysica Acta-Bioenergetics*, **1807**, 1262–72.

Bettati, S. and Mozzarelli, A. (1997). T state hemoglobin binds oxygen noncooperatively with allosteric effects of protons, inositol hexaphosphate, and chloride. *Journal of Biological Chemistry*, **272**, 32050–5.

Bettati, S., Viappiani, C., and Mozzarelli, A. (2009). Hemoglobin, an "evergreen" red protein. *Biochimica Et Biophysica Acta-Proteins and Proteomics*, **1794**(9), 1317–24.

Brunori, M. (1999). Hemoglobin is an honorary enzyme. *Trends in Biochemical Sciences*, **24**, 158–61.

Carroll, S. B. (2013). *Brave Genius*, New York, NY, Crown Publishers.

Changeux, J. P. (1993). Allosteric proteins: from regulatory enzymes to receptors—personal recollections. *Bioessays*, **15**, 625–34.

Changeux, J. P. (2012). Allostery and the Monod-Wyman-Changeux model after 50 years. *Annual Review of Biophysics*, **41**, 103–33.

Changeux, J. P. (2013). 50 years of allosteric interactions: the twists and turns of the models. *Nature Reviews Molecular Cell Biology*, **14**, 819–29.

Cui, Q. and Karplus, M. (2008). Allostery and cooperativity revisited. *Protein Science*, **17**, 1295–307.

Dickerson, R. E. and Geis, I. 1983. *Hemoglobin: Structure, Function, Evolution, and Pathology*, Menlo Park, CA, Benjamin/Cummings.

Eaton, W. A., Henry, E. R., Hofrichter, J., et al. (2007). Evolution of allosteric models for hemoglobin. *IUBMB Life*, **59**, 586–99.

Eaton, W. A., Henry, E. R., Hofrichter, J., and Mozzarelli, A. (1999). Is cooperative oxygen binding by hemoglobin really understood? *Nature Structural Biology*, **6**, 351–8.

Edelstein, S. J. (1975). Cooperative interactions of hemoglobin. *Annual Review of Biochemistry*, **44**, 209–32.

Edsall, J. T. (1972). Blood and hemoglobin: the evolution of knowledge of functional adaptation in a biochemical system. Part I: the adaptation of chemical structure to function in hemoglobin. *Journal of the History of Biology*, **5**, 205–57.

Ferrone, F. A. and Hopfield, J. J. (1976). Rate of quaternary structure change in hemoglobin measured by modulated excitation. *Proceedings of the National Academy of Sciences of the United States of America*, **73**, 4497–501.

Gelin, B. R., Lee, A. W. M., and Karplus, M. (1983). Hemoglobin tertiary structural change on ligand-binding. Its role in the co-operative mechanism. *Journal of Molecular Biology*, **171**, 489–559.

Giardina, B., Ascoli, F., and Brunori, M. (1975). Spectral changes and allosteric transition in trout hemoglobin. *Nature*, **256**, 761–2.

Gibson, Q. H. (1959). Photochemical formation of a quickly reacting form of haemoglobin. *Biochemical Journal*, **71**, 293–303.

Gibson, Q. H. and Parkhurst, L. J. (1968). Kinetic evidence for a tetrameric functional unit in hemoglobin. *Journal of Biological Chemistry*, **243**, 5521–4.

Goodford, P. J., Stlouis, J., and Wootton, R. (1978). Quantitative analysis of effects of 2,3-diphosphoglycerate, adenosine triphosphate and inositol hexaphosphate on oxygen dissociation curve of human hemoglobin. *Journal of Physiology-London*, **283**, 397–407.

Grispo, M. T., Natarajan, C., Projecto-Garcia, J., et al. (2012). Gene duplication and the evolution of hemoglobin isoform differentiation in birds. *Journal of Biological Chemistry*, 287, 37647–58.

Haurowitz, F. (1938). The balance between hemoglobin and oxygen. *Hoppe-Seylers Zeitschrift Für Physiologische Chemie*, **254**, 266–74.

Henry, E. R., Bettati, S., Hofrichter, J., and Eaton, W. A. (2002). A tertiary two-state allosteric model for hemoglobin. *Biophysical Chemistry*, **98**, 149–64.

Henry, E. R., Mozzarelli, A., Viappiani, C., et al. (2015). Experiments on hemoglobin in single crystals and silica gels distinguish among allosteric models. *Biophysical Journal*, **109**, 1264–72.

Hill, A. V. (1913). The combinations of haemoglobin with oxygen and with carbon monoxide. I. *Biochemical Journal*, **7**, 471–80.

Ho, C. (1992). Proton nuclear magnetic resonance studies on hemoglobin: cooperative interactions and partially liganded intermediates. *Advances in Protein Chemistry*, **43**, 153–312.

Hofrichter, J., Sommer, J. H., Henry, E. R., and Eaton, W. A. (1983). Nanosecond absorption-spectroscopy of hemoglobin—Elementary processes in kinetic cooperativity. *Proceedings of the National Academy of Sciences of the United States of America-Biological Sciences*, **80**, 2235–9.

Hopfield, J. J., Shulman, R. G., and Ogawa, S. (1971). Allosteric model of hemoglobin. 1. Kinetics. *Journal of Molecular Biology*, **61**, 425–43.

Imai, K. (1973). Analyses of oxygen equilibria of native and chemically modified human adult hemoglobins on basis of Adair's stepwise oxygenation theory and allosteric model of Monod, Wyman, and Changeux. *Biochemistry*, **12**, 798–808.

Imai, K. (1979). Thermodynamic aspects of the co-operativity in 4-step oxygenation equilibria of hemoglobin. *Journal of Molecular Biology*, **133**, 233–47.

Imai, K. (1982). *Allosteric Effects in Haemoglobin*, Cambridge, UK, Cambridge University Press.

Imai, K. (1983). The Monod-Wyman-Changeux allosteric model describes hemoglobin oxygenation with only one adjustable parameter. *Journal of Molecular Biology*, **167**, 741–9.

Imai, K. and Yonetani, T. (1975). pH-dependence of Adair constants of human hemoglobin—nonuniform contribution of successive oxygen bindings to alkaline Bohr effect. *Journal of Biological Chemistry*, **250**, 2227–31.

Kilmartin, J. V., Imai, K., Jones, R. T., et al. (1978). Role of Bohr group salt bridges in cooperativity in hemoglobin. *Biochimica Et Biophysica Acta*, **534**, 15–25.

Koshland, D. E., Némethy, G., and Filmer, D. (1966). Comparison of experimental binding data and theoretical models in proteins containing subunits. *Biochemistry*, **5**, 365–85.

Lee, A. W. M. and Karplus, M. (1983). Structure-specific model of hemoglobin cooperativity. *Proceedings of the National Academy of Sciences of the United States of America-Biological Sciences*, **80**, 7055–9.

Luisi, B., Liddington, B., Fermi, G., and Shibayama, N. (1990). Structure of deoxy-quaternary hemoglobin with liganded beta subunits. *Journal of Molecular Biology*, **214**, 7–14.

Luisi, B. and Shibayama, N. (1989). Structure of hemoglobin in the deoxy-quaternary state with ligand bound at the alpha hemes. *Journal of Molecular Biology*, **206**, 723–36.

Lukin, J. A. and Ho, C. (2004). The structure-function relationship of hemoglobin in solution at atomic resolution. *Chemical Reviews*, **104**, 1219–30.

Macquarrie, R. and Gibson, Q. H. (1971). Use of a fluorescent analogue of 2,3-diphosphoglycerate as a probe of human hemoglobin conformation during carbon monoxide poisoning. *Journal of Biological Chemistry*, **246**, 5832–5.

Minton, A. P. and Saroff, H. A. (1974). General formulation of free energy of interaction in cooperative ligand-binding—application to hemoglobin. *Biophysical Chemistry*, **2**, 296–9.

Monod, J., Changeux, J. P., and Jacob, F. (1963). Allosteric proteins and cellular control systems. *Journal of Molecular Biology*, **6**, 306–29.

Monod, J., Wyman, J., and Changeux, J. P. (1965). On the nature of allosteric transitions—a plausible model. *Journal of Molecular Biology*, **12**, 88–118.

Mozzarelli, A., Rivetti, C., Rossi, G. L., Henry, E. R., and Eaton, W. A. (1991). Crystals of hemoglobin with the T-quaternary structure bind oxygen noncooperatively with no Bohr effect. *Nature*, **351**, 416–19.

Muirhead, H. and Perutz, M. F. (1963). Structure of haemoglobin—a 3-dimensional Fourier synthesis of reduced human haemoglobin at 5.5 Å resolution. *Nature*, **199**, 633–8.

Paoli, M., Liddington, R., Tame, J., Wilkinson, A., and Dodson, G. (1996). Crystal structure of T state haemoglobin with oxygen bound at all four haems. *Journal of Molecular Biology*, **256**, 775–92.

Park, S. Y., Yokoyama, T., Shibayama, N., Shiro, Y., and Tame, J. R. H. (2006). 1.25 angstrom resolution crystal structures of human haemoglobin in the oxy, deoxy and carbonmonoxy forms. *Journal of Molecular Biology*, **360**, 690–701.

Parkhurst, L. J. and Gibson, Q. H. (1967). Reaction of carbon monoxide with horse hemoglobin in solution in erythrocytes and in crystals. *Journal of Biological Chemistry*, **242**, 5762–70.

Pauling, L. (1935). The oxygen equilibrium of hemoglobin and its structural interpretation. *Proceedings of the National Academy of Sciences of the United States of America*, **21**, 186–91.

Perutz, M. F. (1970). Stereochemistry of cooperative effects in haemoglobin. *Nature*, **228**, 726–39.

Perutz, M. F. (1978). Hemoglobin structure and respiratory transport. *Scientific American*, **239**, 92–125.

Perutz, M. F., Watson, H. C., Muirhead, H., Diamond, R., and Bolton, W. (1964). Structure of haemoglobin—X-ray examination of reduced horse haemoglobin. *Nature*, **203**, 687–90.

Perutz, M. F., Wilkinson, A. J., Paoli, M., and Dodson, G. G. (1998). The stereochemical mechanism of the cooperative effects in hemoglobin revisited. *Annual Review of Biophysics and Biomolecular Structure*, **27**, 1–34.

Rivetti, C., Mozzarelli, A., Rossi, G. L., Henry, E. R., and Eaton, W. A. (1993). Oxygen binding by single crystals of hemoglobin. *Biochemistry*, **32**, 2888–906.

Roughton, F. J. W., Otis, A. B., and Lyster, R. L. J. (1955). The determination of the individual equilibrium constants of the 4 intermediate reactions between oxygen

and sheep haemoglobin. *Proceedings of the Royal Society Series B-Biological Sciences*, **144**, 29–54.

Saroff, H. A. and Minton, A. P. (1972). Hill plot and energy of interaction in hemoglobin. *Science*, **175**, 1253–5.

Sawicki, C. A. and Gibson, Q. H. (1976). Quaternary conformational changes in human hemoglobin studied by laser photolysis of carboxyhemoglobin. *Journal of Biological Chemistry*, **251**, 1533–42.

Sawicki, C. A. and Gibson, Q. H. (1977). Quaternary conformational changes in human oxyhemoglobin studied by laser photolysis. *Journal of Biological Chemistry*, **252**, 5783–8.

Shibayama, N. and Saigo, S. (1995). Fixation of the quaternary structures of human adult hemoglobin by encapsulation in transparent porous silica gels. *Journal of Molecular Biology*, **251**, 203–9.

Shulman, R. G. (2001). Spectroscopic contributions to the understanding of hemoglobin function: implications for structural biology. *IUBMB Life*, **51**, 351–7.

Shulman, R. G., Hopfield, J. J., and Ogawa, S. (1975). Allosteric interpretation of hemoglobin properties. *Quarterly Reviews of Biophysics*, **8**, 325–420.

Szabo, A. and Karplus, M. (1972). Mathematical model for structure-function relations in hemoglobin. *Journal of Molecular Biology*, **72**, 163–97.

Szabo, A. and Karplus, M. (1975). Analysis of cooperativity in hemoglobin. Valency hybrids, oxidation, and methemoglobin replacement reactions. *Biochemistry*, **14**, 931–40.

Tame, J. R. H. (1999). What is the true structure of liganded haemoglobin? *Trends in Biochemical Sciences*, **24**, 372–7.

Tsuneshige, A., Park, S., and Yonetani, T. (2002). Heterotropic effectors control the hemoglobin function by interacting with its T and R states—a new view on the principle of allostery. *Biophysical Chemistry*, **98**, 49–63.

Tyuma, I., Imai, K., and Shimizu, K. (1971a). Effect of inositol hexaphosphate and other organic phosphates on cooperativity in oxygen binding of human hemoglobin. *Biochemical and Biophysical Research Communications*, **44**, 682–6.

Tyuma, I., Shimizu, K., and Imai, K. (1971b). Effect of 2,3-diphosphoglycerate on cooperativity in oxygen binding of human adult hemoglobin. *Biochemical and Biophysical Research Communications*, **43**, 423–8.

Viappiani, C., Bettati, S., Bruno, S., et al. (2004). New insights into allosteric mechanisms from trapping unstable protein conformations in silica gels. *Proceedings of the National Academy of Sciences of the United States of America*, **101**, 14414–19.

Voet, D. and Voet, J. G. (2011). *Biochemistry*, Hoboken, NJ, John Wiley and Sons.

Weber, R. E., Jensen, F. B., and Cox, R. P. (1987). Analysis of teleost hemoglobin by Adair and Monod-Wyman-Changeux models—effects of nucleoside triphosphates and pH on oxygenation of tench hemoglobin. *Journal* of Comparative Physiology B-Biochemical Systemic and Environmental Physiology, **157**, 145–52.

Wyman, J. (1948). Heme proteins. *Advances in Protein Chemistry*, **4**, 407–531.

Wyman, J. (1964). Linked functions and reciprocal effects in hemoglobin: a second look. *Advances in Protein Chemistry*, **19**, 223–86.

Wyman, J. and Allen, D. W. (1951). The problem of the heme interactions in hemoglobin and the basis of the Bohr effect. *Journal of Polymer Science*, **7**, 499–518.

Wyman, J. and Gill, S. J. (1990). *Binding and Linkage*, Mill Valley, CA, University Science Books.

Yonetani, T., Park, S., Tsuneshige, A., Imai, K., and Kanaori, K. (2002). Global allostery model of hemoglobin—modulation of O_2 affinity, cooperativity, and Bohr effect by heterotropic allosteric effectors. *Journal of Biological Chemistry*, **277**, 34508–20.

Yuan, Y., Tam, M. F., Simplaceanu, V., and Ho, C. (2015). New look at hemoglobin allostery. *Chemical Reviews*, **115**, 1702–24.

CHAPTER 4

Hemoglobin structure and allosteric mechanism

"When the allosteric proteins display their functions in biological systems they behave just like a precision machine. Thus all living cells, tissues, organs, and consequently our bodies, owe their normal physiological functions solely to the splendid actions of the 'molecular machines'."

—Imai (1982)

"All these various machines, and even their most minute parts, are adjusted to each other with an accuracy which ravishes into admiration all men who have ever contemplated them."

—Cleanthes, in *Dialogues Concerning Natural Religion* (Hume, 1779)

4.1 Max Perutz, X-ray crystallography, and the origins of structural biology

Insight into the mechanistic basis of a protein's functional properties requires a detailed understanding of its three-dimensional structure. In principle, X-ray diffraction patterns of a crystallized protein can reveal the position of every constituent atom in three-dimensional space. In 1958, John Kendrew and colleagues at the Medical Research Council (MRC) Laboratory in Cambridge, England, used X-ray crystallography to solve the complete molecular structure of sperm whale Mb (Kendrew et al., 1958, Kendrew et al., 1960). This work was conducted several years after the discovery of the double helix structure of DNA, and it is clear that Kendrew and his colleagues were expecting to discover similarly deterministic rules of protein structure. As stated in their 1958 *Nature* paper that reported the crystal structure of Mb: "Perhaps the most remarkable features of the molecule are its complexity and its lack of symmetry. The arrangement seems to be almost totally lacking in the kind of regularities which one instinctively anticipates, and it is

more complicated than has been predicted by any theory of protein structure."

One year after Kendrew's team solved the crystal structure of sperm whale Mb, Max Perutz (also at the MRC) solved the structure of horse "met-hemoglobin" (metHb) (Perutz et al., 1960), a Hb derivative in which the ferrous (Fe^{2+}) heme iron of each subunit is oxidized to the ferric (Fe^{3+}) state. Mb and Hb were therefore the first proteins to have their atomic level structures characterized by means of X-ray crystallography. In recognition of these pioneering achievements and their role in establishing the foundation for the nascent field of structural biology, Kendrew and Perutz shared the Nobel Prize in Chemistry in 1962. In Perutz's case, solving the initial crystal structure of Hb was the culmination of over twenty years of dogged, obsessive work, involving numerous setbacks and blind alleys that would have been more than enough to discourage less tenacious personalities (Ferry, 2007).

In the years that followed, crystal structures of horse and human Hbs were solved at increasingly high resolution (Perutz, 1965, Perutz et al., 1968,

Hemoglobin: Insights into Protein Structure, Function, and Evolution. Jay F. Storz, Oxford University Press (2019).
© Jay F. Storz 2019. DOI: 10.1093/oso/9780198810681.001.0001

Arnone, 1972, Fermi, 1975, Heidner et al., 1976, Ladner et al., 1977, Baldwin, 1980, Shaanan, 1983, Fermi et al., 1984, Liddington et al., 1988), and Perutz and his associates at the MRC remained at the vanguard of these efforts. At present, the structures of liganded and unliganded human Hb have been solved by X-ray crystallography with such a high degree of resolution that almost all of the ~4,600 non-hydrogen atoms of this complex molecule can be positioned with an accuracy of 0.5 Å or better (Park et al., 2006). In the last few decades, X-ray diffraction analyses of Hb crystal structures have been complemented by nuclear magnetic resonance (NMR) spectroscopy and X-ray wide-angle scattering, which provide information on the dynamic ensemble of structures of Hb that occur in solution (Yuan et al., 2015).

Perutz (1914–2002), a native of Austria, was an avid alpinist in his youth. In an interview he once compared making scientific discoveries to summiting a peak: "When you get to the top after a hard climb, a view of a new landscape opens before you." This is certainly an apt metaphor to describe the impact of his research on the structural basis of Hb allostery. The fact that it is possible to explain allosteric mechanisms of Hb function in terms of atomic level interactions is one of the great triumphs of twentieth-century molecular biology and protein science. As stated by Bunn and Forget (1986): "If the hemoglobin story has a hero, it is Max Perutz."

4.2 Tertiary structure: the globin fold

The α- and β-chain subunits of human Hb are 141 and 146 amino acids in length, respectively, and the fully assembled Hb tetramer has a molecular weight of 64,650 Daltons including the four prosthetic heme groups. Among different vertebrate taxa, there is some variation in the lengths of the subunit polypeptides due to amino acid insertions or deletions. Whereas the Hb β-chain polypeptides of humans and almost all other amniote vertebrates are 146 amino acids in length, those of cartilaginous fishes, teleost fishes, and amphibians range from 140 to 147 in length. However, in all cases the same secondary and tertiary structures of the globin chains are conserved. The α-chain polypeptide folds into seven α-helices and the β-chain folds into eight. These α-helices are labeled A-H (the D-helix is missing from the α-chain) and are linked together by short interhelical segments, labeled AB, BC, and so on. The N- and C- terminal extensions of each chain are labeled NA and HC, respectively. In each polypeptide chain, individual residues are labeled according to their helical position and their sequential number from the N-terminus. For example, His α87(F8) refers to the histidine that occupies the eighty-seventh residue position of the α-chain and the eighth position of the F-helix.

The tertiary structures of Mb and the individual Hb subunits are defined by the characteristic "globin

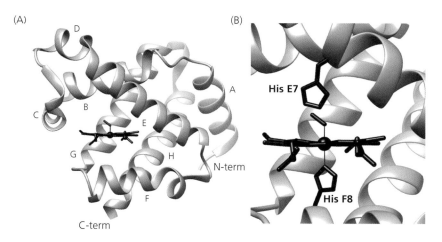

Fig. 4.1. The globin fo d. (A) Tertiary structure of the Hb β-chain in the oxygenated state. (B) A zoomed-in view of the heme pocket showing the arrangement of the proximal and distal histidines (His F8 and His E7, respectively).

fold," a three-dimensional folding of α-helices that enclose the non-covalently bound heme group in a hydrophobic pocket (Fig. 4.1A). Within this hydrophobic pocket, the heme is held in place by a coordination bond between the iron atom and the imidazole side chain of His F8 (α87, β92), the "proximal histidine" (Fig. 4.1B). The residue Phe CD1(α43, β42) assists by wedging the heme into a stable position. His F8 and Phe CD1 are among the few residues that are completely invariant among all vertebrate Hb chains. One other nearly invariant position in vertebrate Hb is Leu F4(α83, β88), which

prevents hydrolysis of the Fe-His F8 bond by restricting solvent access to the proximal heme pocket (Liong et al., 2001). The reversible binding of O_2 is further facilitated by the His E7(α58, β63) residue, the "distal histidine," which lies opposite of His F8 on the other side of the heme plane. His E7 stabilizes the Fe-O bond by donating a hydrogen bond to the bound O_2 (Phillips and Schoenborn, 1981, Olson et al., 1988, Lukin et al., 2000, Park et al., 2006, Birukou et al., 2010, Yuan et al., 2010) (Fig. 4.2). The remaining hydrophobic residues lining the distal heme pocket help stabilize the heme within the

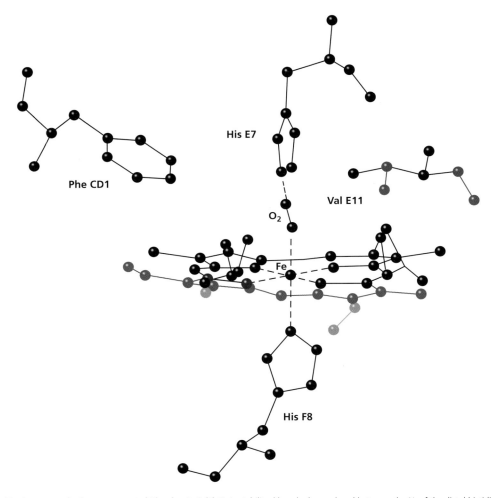

Fig. 4.2. The heme complex in an oxygenated Hb subunit. Fe(II)-O_2 is stabilized by a hydrogen bond between the N_ϵ of the distal histidine (His E7) side chain and the free atom of the bound O_2 molecule (Phillips, 1980). The hydrogen bond donated by His E7 also prevents protonation of the bound O_2, thereby protecting against oxidation of the heme iron, in combination with the surrounding apolar amino acid side chains (e.g., Phe CD1, Val E11, and Leu G8) which exclude water from the distal pocket (Brantley et al., 1993, Aranda et al., 2009).

folded globin chain and inhibit autoxidation by excluding water from the vicinity of the bound ligand (Brantley et al., 1993, Aranda et al., 2009).

The tertiary structures of the α- and β-chain Hb subunits are highly similar to one another and to that of Mb, even though only 18 percent of homologous amino acid residues are identical among the three polypeptides (Fig. 4.3). In each of these globins, the α-helices are folded back and forth on top of one another to form the heme pocket (Fig. 4.1A). As

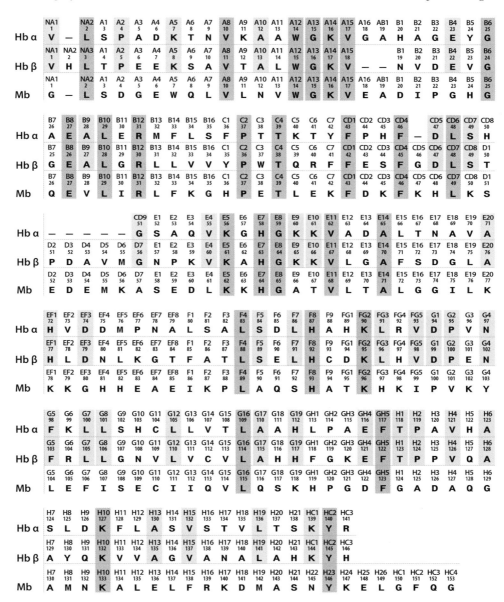

Fig. 4.3. Alignment of amino acid sequences from human α-globin, β-globin, and Mb. Light gray bars denote amino acids that are conserved in both Hb chains but not in Mb. Dark gray bars that span all three rows of the alignment denote amino acids that are conserved in both Hb subunits and in Mb. The β-chain of human Hb, with 146 residues, has a shorter H-helix than the Mb chain, with 153 residues. The α-chain, with 141 residues, is shorter than both because the entire D-helix has been deleted. In the helical notation shown above the sequences, note that the conserved, penultimate tyrosine is shown as HC2 in the α- and β-chain subunits of Hb, this same residue position is H23(146) in the longer final H-helix of Mb.

already mentioned, the D-helix is only present in Mb and in the Hb β-chain, not in the Hb α-chain.

The tertiary structures of the monomeric Mb and the α/β-globin chains of jawed vertebrate Hb are highly similar to those of O_2-transport Hbs from cyclostomes (lampreys and hagfish), and to the far more distantly related globins from nonvertebrate deuterostomes, protostomes, and even leguminous plants, reflecting the rigid specifications for preserving the geometry of the heme pocket. Nonetheless, these diverse globins vary in subtle structural details that belie their distant evolutionary relationships (Lesk and Chothia, 1980, Lecomte et al., 2005). As stated by (Perutz, 1983): "The angles between helical segments differ by up to 30° and the points of contact between them by up to 7 Å. Many different combinations of side chains are found to produce helix interfaces that are comparably well packed, as if the tertiary structure had been conserved by a patchwork of improvisations . . . " The tertiary structures of these diverse globins are primarily stabilized by hydrophobic interactions between non-polar residues and by hydrogen bonds between polar residues (Perutz et al., 1965, Perutz, 1983).

4.2.1 The heme-ligand complex

The heme group consists of an iron atom held at the center of a tetrapyrrole ring (Fig. 4.2). In globins such as vertebrate Hb and Mb, the iron atom is coordinated by four nitrogen atoms at the center of the porphyrin ring and the imidazole ϵ_2 nitrogen of the proximal histidine (His F8). In this pentacoordinate deoxy state, the sixth coordination site of the iron atom is open and accessible to the binding of gaseous ligands. Reversible binding of O_2 to the heme is only possible if the iron atom remains in the ferrous state. The distal histidine (His E7) and the hydrophobic nature of the distal heme pocket protects the ferrous iron atom, Fe(II), from being oxidized to the ferric state (forming metHb), which does not bind O_2 because the sixth coordination site is occupied by either a water molecule or hydroxide ion, depending on pH. Red blood cells contain reducing systems including cytochrome b_5 and metHb reductase, which rapidly convert the ferric heme iron of metHb back to the ferrous form using NADPH (nicotinamide adenine dinucleotide phosphate-oxidase) generated from glucose metabolism.

The globin moiety also modulates the relative binding preferences for different diatomic ligands. Carbon monoxide binds Fe(II) with a higher affinity than O_2 and at high concentrations can severely reduce Hb-O_2 transport capacity, which accounts for its high toxicity. Free heme in solution has a ~10,000-fold higher binding affinity for CO than for O_2, whereas heme groups that are incorporated into the globin moiety have a dramatically increased affinity for O_2 (Springer et al., 1994). In Mb and Hb, the increase in the relative preference for O_2 binding is attributable to electrostatic interactions between the partial negative charge on bound O_2 and polar side chains in the distal pocket. His E7 is typically the key hydrogen bond donor and reduces the ratio of CO/O_2 affinities to ~20 in the case of Mb and ~200 in the case of human Hb (Springer et al., 1989, Perutz, 1990, Springer et al., 1994, Olson and Phillips, 1997, Lukin et al., 2000, Birukou et al., 2010). This preferential enhancement of O_2 binding relative to CO binding is physiologically important because CO is endogenously produced by the catabolic breakdown of heme (for example, when senescent red cells are cleared from the circulation). If Mb or Hb had the same CO/O_2 affinities as free heme, the endogenous production of CO would be fatally toxic (Collman et al., 1976). Thus, enclosing the heme group inside the hydrophobic environment of the folded globin polypeptide protects the heme iron against autoxidation and also enhances ligand discrimination in favor of O_2 binding.

4.3 Quaternary structure

The original X-ray crystallographic studies by Perutz and associates revealed that Hb assumes two alternative quaternary structures, one characteristic of the non-liganded, deoxy conformation (T-state, ^THb) and the other characteristic of the fully liganded conformation (R-state, ^RHb). Results of more recent crystallographic and NMR studies indicate that ^THb and ^RHb should be viewed not as discrete, static structures, but as dynamic ensembles of structures (Paoli et al., 1996, Tame, 1999, Lukin et al., 2003, Lukin and Ho, 2004, Gong et al., 2006, Park et al., 2006, Sahu et al., 2006, Sahu et al., 2007, Yuan et al., 2015) (Plate 4). Park et al. (2006) expressed the view that "each quaternary structure, R and T, represents a significant volume (rather than a point) in conformational space." This view is supported by evidence for ligation-induced conformational changes within a single crystal form (Shibayama et al., 2014, Shibayama et al., 2017). Nonetheless, it is convenient shorthand to refer to canonical "R" and "T" conformations when describing the allosteric transition in quaternary structure.

The four subunits of Hb form a tetrahedron with a twofold (dyad) axis of symmetry that runs through the central water-filled cavity. The twofold symmetry means that if one dimeric half-molecule is rotated 180° about the axis it will superimpose on the other dimer. If the individual subunits are designated as α_1, α_2, β_1, and β_2, then—

due to the dyad symmetry—we can define two different interfaces between unlike subunits: intradimer interfaces ($\alpha_1\beta_1/\alpha_2\beta_2$) and interdimer interfaces ($\alpha_1\beta_2/\alpha_2\beta_1$) (Fig. 4.4). The quaternary structure is mainly stabilized by hydrophobic interactions and hydrogen bonds between unlike subunits. In ^{T}Hb, the tetrameric assembly is further stabilized by salt bridges between unlike subunits and between the α_1 and α_2 subunits (Fig. 4.5). The symmetrically identical $\alpha_1\beta_1$ and $\alpha_2\beta_2$ interfaces involve mostly hydrophobic interactions between the B and H helices and between the G helices of opposing subunits. In total, the intradimeric contacts between helices α_1B-β_1H, α_1G-β_1G, and α_1H-

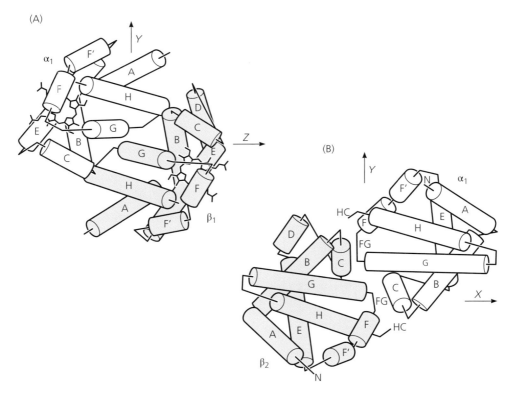

Fig. 4.4. Schematic diagram of $\alpha\beta$ dimers showing intra- and interdimeric interfaces between unlike subunits (α shown in white, β in gray). The α-helices are depicted as cylindrical rods labelled A through H, and non-helical connecting loops and chain termini are depicted as lines. (A) Hb half-molecule, $\alpha_1\beta_1$, viewed from the interface it forms with the opposing $\alpha_2\beta_2$ dimer, which would be in the foreground. The intradimer $\alpha_1\beta_1/\alpha_2\beta_2$ interfaces mostly involve hydrophobic interactions between amino acids in the B, G, and H helices (α_1B-β_1H, α_1G-β_1G, and α_1H-β_1B). (B) Subunits α_1 and β_2 viewed from the interface they form with the opposing α_2 and β_1 subunits, which would be in the foreground. Heme groups are not pictured. The interdimer interface is formed by interactions between the C and G helices and FG corner. The major interdimeric contacts involve α_1C-β_2FG and α_1FG-β_2C, and the minor contacts involve α_1C-β_2C and α_1FG-β_2FG. There is also a comparatively weak interface between the N- and C-termini of opposing α_1 and α_2 chains. In the T↔R transition in quaternary structure, the interdimer interfaces ($\alpha_1\beta_2/\alpha_2\beta_1$) are shifted and the network of atomic contacts is reconfigured.

Modified from Baldwin (1980) with permission from Elsevier.

deoxyHb (T)

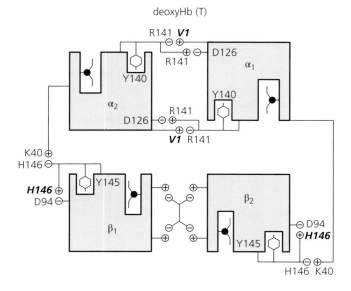

oxyHb (R)

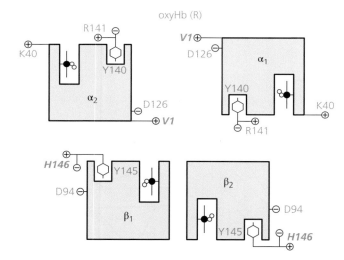

Fig. 4.5. Diagrammatic representation of electrostatic interactions within and between subunits of human Hb. Salt bridges within and between subunits are present in deoxy ᵀHb (top) but are broken in the oxygenation-linked transition to ᴿHb (bottom). A DPG molecule with four negatively charged phosphate groups binds in a cleft between the β-chain subunits of ᵀHb, but is excluded in ᴿHb. The two residues shown in bold, Val $\alpha1$(NA1) and His β146(HC3), experience an increased pK_a upon breakage of key salt bridges, resulting in the dissociation of Bohr protons in the T→R transition.

Diagram modified from Perutz (1970) and Eaton et al. (2007).

β_1B bury a surface of 3,300 Å² in ᵀHb and 3,500 Å² in ᴿHb (Chothia et al., 1976). The $\alpha_1\beta_2$ and $\alpha_2\beta_1$ interfaces involve interactions among amino acids in the C and G helices and FG corner (Fig. 4.6, Plate 5, Tables 4.1–4.2).

The major interdimeric contacts involve α_1C-β_2FG and α_1FG-β_2C, and the minor contacts involve α_1C-β_2C and α_1FG-β_2FG. Together, the $\alpha_1\beta_2$/$\alpha_2\beta_1$ interfaces bury a surface of 2,620 Å² in ᵀHb and only 1920 Å² in ᴿHb (Chothia et al., 1976, Lesk et al., 1985). The specific amino acids involved in the intra- and interdimeric contacts are annotated in Figure 4.7. The

comparatively weak interface between the α-chains of opposing dimers involves residues 1–2 and 127–141 at the N- and C-termini, respectively (burying only 470 Å² in ᵀHb and 710 Å² in ᴿHb); there are no direct interactions between the β-chains of opposing dimers (Lesk et al., 1985).

4.3.1 Subunit motion and the allosteric T↔R transition

Due to asymmetry in the strength of intersubunit contacts, tetrameric Hb is best described as a dimer of

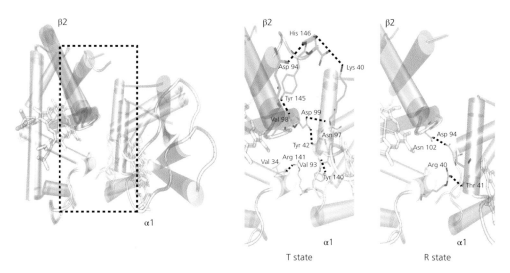

Fig. 4.6. Network of atomic contacts at interdimer ($\alpha_1\beta_2/\alpha_2\beta_1$) interfaces involving the C and G helices and FG corner. These interdimer "sliding" contacts involve α_1C-β_2G and α_1FG-β_2C, and the minor contacts involve α_1C-β_2C and α_1FG-β_2FG. The complete set of subunit contacts are detailed in Table 4.1.

Modified from Baldwin and Chothi (1979) with permission from Elsevier.

Table 4.1 Atomic contacts at the interdimer $\alpha_1\beta_2/\alpha_2\beta_1$ interface of human Hb

	deoxyHb (T)	oxyHb (R)
α_1FG/β_2C (joint)	Arg $\beta_2$40(C6) against Arg $\alpha_1$92(FG4)	Same as ^{T}Hb
	Trp $\beta_2$37(C3) against Asp $\alpha_1$94(G1) and Pro $\alpha_1$95(G2)	Same as ^{T}Hb
α_1C/β_2FG (switch)	His $\beta_2$97(FG4) between Pro $\alpha_1$44(CD2) and Thr $\alpha_1$41(C6)	His $\beta_2$97(FG4) between Thr $\alpha_1$41(C6) and Thr $\alpha_1$38(C3)
	H-bond between Asp $\beta_2$99(G1) and Tyr $\alpha_1$42(C7)	H-bond between Asp $\beta_2$102(G4) and Asp $\alpha_1$94(G1)
FG corners	H-bond between Asp $\beta_2$99(G1) and Asn $\alpha_1$97(G4)	
C helices	Arg $\beta_2$40(C6) against Tyr $\alpha_1$42(C7)	Same as ^{T}Hb, plus H-bond between Arg $\beta_2$40(C6) and CO of Thr $\alpha_1$41(C6)
Carboxy termini of α-chains	Arg $\alpha_1$141(HC3) side chain forms H-bonds with Asp $\alpha_2$126(H9) side chain, and to CO of Val $\beta_2$34(B16); it forms salt bridge via Cl$^-$ ion with Val $\alpha_2$1(NA1), the N-terminus of the opposing α-chain. Packed against Tyr $\beta_2$35(C1), Pro $\beta_2$36(C2), and Trp $\beta_2$37(C3)	All bonds described for ^{T}Hb are broken; both Arg $\alpha_1$141(HC3) and Tyr $\alpha_1$140(HC2) are freed
	Arg $\alpha_1$141(HC3) C-terminus forms salt bridge with Lys $\alpha_2$127(H10) side chain	
	Tyr $\alpha_1$140(HC2) side chain forms H-bond with CO of Val $\alpha_1$93(FG5). Ring is tucked in $\alpha_1\beta_2$ interface, in contact with Pro $\beta_2$36(C2) and Trp $\beta_2$37(C3)	
	Tyr $\beta_2$35(C1) side chain forms H-bond with Asp $\alpha_2$126(H9)	Same as ^{T}Hb, since α_2 and β_2 move as a unit
Carboxy termini of β-chains	His $\beta_2$146(HC3) side chain forms H-bond with Asp $\beta_2$94 (FG1) side chain	All bonds described for ^{T}Hb are broken (including His $\beta_2$146$\cdots$Asp $\beta_2$94); both His $\beta_2$146(HC3) and Tyr $\beta_2$145(HC2) are freed
	His $\beta_2$146(HC3) C-terminus forms salt bridge or H-bond to Lys $\alpha_1$40(C5) side chain	
	Tyr $\beta_2$145(HC2) side chain forms H-bond to CO of Val $\beta_2$98(FG5). Ring is tucked in $\alpha_1\beta_1$ interface, in contact with Thr $\alpha_1$41(C6)	

Modified from Dickerson and Geis (1983).

Table 4.2 Side chain hydrogen bonds characteristic of T- and R-states of human Hb. Distances are given in Å. The expected error is of the order 0.1 Å. In the case of hydrogen bonds in the deoxy Hb structure, the distance for the non-crystallographic symmetry equivalent bond is shown in parentheses

Residues	deoxyHb (T)	oxyHb (R)
Lys $40\alpha_1$-His $146\beta_2$	2.83 (2.59)	–
Tyr $42\alpha_1$-Asp $99\beta_2$	2.51 (2.52)	–
Asp $94\alpha_1$-Trp $37\beta_2$	2.85 (2.84)	3.68
Asp $94\alpha_1$-Asn $102\beta_2$	– (–)	2.83
Arg $141\alpha_1$-Asp $126\alpha_2$	2.78 (2.66)	–
Asp $94\beta_1$-His $146\beta_1$	2.60 (2.84)	–
Trp $37\beta_1$-Asn $102\beta_1$	– (–)	2.90

From Park et al. (2006).

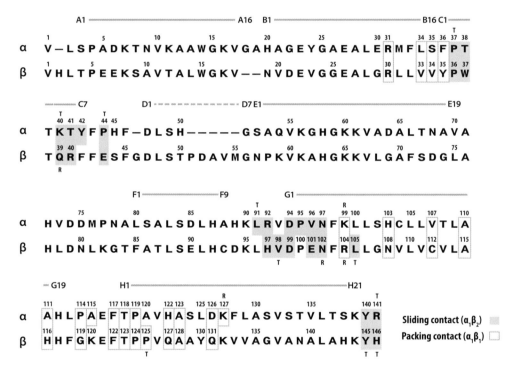

Fig. 4.7. Alignment of amino acid sequences from the α- and β-chains of human Hb. Light gray bars denote interdimeric ($\alpha_1\beta_2/\alpha_2\beta_1$) sliding contacts, and those with gray outlines denote intradimeric ($\alpha_1\beta_1/\alpha_2\beta_2$) packing contacts. The T and R shown above or below select residue positions indicate intersubunit contacts that are only present in ^{T}Hb or ^{R}Hb, respectively. The boundaries of helices A–H are indicated above each row of the alignment. The D-helix is indicated by a dashed line because it is present in the β-chain but not in the α-chain.

semi-rigid $\alpha_1\beta_1$ and $\alpha_2\beta_2$ dimers. During the oxygenation-linked transition in quaternary structure between ^{T}Hb and ^{R}Hb, the $\alpha_1\beta_1$ and $\alpha_2\beta_2$ dimers undergo a 15° relative rotation around the twofold (dyad) axis (Perutz, 1972, Baldwin and Chothia, 1979, Shaanan,

1983, Fermi et al., 1984) (Fig. 4.8). This mutual rotation of the dimers involves no appreciable change in the $\alpha_1\beta_1/\alpha_2\beta_2$ intradimer interfaces (the "packing contacts") but causes substantial changes in the nature of atomic contacts at the $\alpha_1\beta_2/\alpha_2\beta_1$ interdimer contact

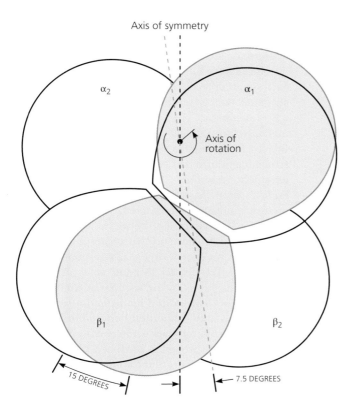

Axis of symmetry

α_2

α_1

Axis of
rotation

β_1

β_2

15 DEGREES

7.5 DEGREES

Fig. 4.8. Schematic diagram illustrating the change in quaternary structure that accompanies the ligation of Hb. In the oxygenation-linked T→R transition, two semi-rigid $\alpha\beta$ dimers ($\alpha_1\beta_1$ and $\alpha_2\beta_2$) rotate relative to one another. The $\alpha_1\beta_1$ dimer is depicted in liganded (R) and unliganded (T) states (subunits in gray and white, respectively); the R- and T-states of the $\alpha_2\beta_2$ dimer are superimposed. If the $\alpha_2\beta_2$ dimer is held fixed, the $\alpha_1\beta_1$ dimer turns by 15° about an off-center axis. The two-fold symmetry of the tetrameric protein is preserved during the transition in quaternary structure, but the axis of symmetry is rotated by 7.5°.

Modified from Baldwin and Chothia (1979).

surfaces (the "sliding contacts") (Baldwin, 1980, Shaanan, 1983, Fermi et al., 1984, Silva et al., 1992, Paoli et al., 1996, Park et al., 2006) (Tables 4.1, 4.2). Although analyses based on X-ray crystallography had suggested that the $\alpha_1\beta_1/\alpha_2\beta_2$ intradimer interfaces are not appreciably altered during the T↔R transition (Perutz et al., 1998), solution NMR studies have suggested that subtle changes in the strength of hydrogen bonds between unlike subunits in the same dimer may play a role in mediating the allosteric transition in quaternary structure (Mihailescu and Russu, 2001, Chang et al., 2002). These NMR results are consistent with the hypothesis that intradimer ($\alpha_1\beta_1/\alpha_2\beta_2$) interactions are coupled to ligation-induced changes in the interdimer ($\alpha_1\beta_2/\alpha_2\beta_1$) interface (Ackers et al., 2002). Most of the free energy difference between the T- and R-states is concentrated in the $\alpha_1\beta_2/\alpha_2\beta_1$ sliding contacts (Pettigrew et al., 1982). In some cases, this free energy difference can be abolished by a single amino acid substitution (Dickerson and Geis, 1983). It is therefore not surprising that these intersubunit contacts are among the most highly conserved sites in vertebrate Hb.

When the heme groups of the Hb tetramer are not liganded, the T-state is thermodynamically more stable than the R-state owing to the additional salt bridges and other non-covalent bonds within and between subunits (Perutz, 1970, Baldwin and Chothia, 1979, Perutz, 1989a) (Fig. 4.5). In dilute solution under otherwise physiological conditions, ^{R}Hb can reversibly dissociate into symmetrical $\alpha_1\beta_1$ and $\alpha_2\beta_2$ dimers. This dissociation occurs at the $\alpha_1\beta_2/\alpha_2\beta_1$ interdimer interface due to the fact that the $\alpha_1\beta_1/\alpha_2\beta_2$ intradimer interface is held together by stronger atomic contacts (Rosemeyer and Huehns, 1967, Park, 1970). As a result of oxygenation-linked changes in the overall strength of the $\alpha_1\beta_2/\alpha_2\beta_1$ intersubunit contacts, tetramer-dimer dissociation occurs 10^5 to 10^6 times more readily in ^{R}Hb than in ^{T}Hb.

4.4 The structural basis of homotropic effects: Perutz's stereochemical mechanism

The O_2-transport Hbs of ancestral vertebrates likely existed in a monomer-oligomer equilibrium as in

Table 4.3 Distances in Å between heme iron atoms in human Hb

	deoxyHb (T)	oxyHb (R)
$\alpha_1\text{-}\alpha_2$	34.9	36.0
$\alpha_1\text{-}\beta_2/\alpha_2\text{-}\beta_1$	24.6	25.0
$\alpha_1\text{-}\beta_1/\alpha_2\text{-}\beta_2$	36.9	35.0
$\beta_1\text{-}\beta_2$	39.9	33.4

the Hbs of modern-day lampreys and hagfish, where cooperative O_2-binding stems from association-dissociation dynamics. In jawed vertebrates, by contrast, the efficiency of Hb as a specialized O_2-carrier molecule is chiefly attributable to its multisubunit quaternary structure, as the cooperativity of O_2-binding stems from an interaction between subunits of the intact tetramer. This interaction is indirect, as the heme groups are well separated in the tetramer by iron–iron distances ranging from 24 to 40 Å (Table 4.3). In the absence of direct interactions between the heme groups, how does a change in the ligation state of one heme-bearing subunit affect the O_2 affinity of the other subunits? Specifically, what is the trigger that produces the T→R transition after two or three O_2 molecules have bound? This question occupied Max Perutz for much of his career. With characteristic fixity of purpose, he leveraged his detailed understanding of Hb structure to decipher the causal determinants of homotropic and heterotropic allostery. As stated by Eaton et al. (Eaton et al., 1999): "Perutz took a bold approach that went far beyond simply identifying the R and T quaternary structures of [oxyHb and deoxyHb] and describing them in detail. . . . In a *tour de force* he saw through the complexity of his atomic models and developed a structural explanation for exactly how hemoglobin worked."

Perutz contrasted his empirically minded approach to the problem with the theoretical approach adopted by Jacques Monod and colleagues in their formulation of the MWC model (Perutz, 1978): "There are two ways out of an impasse in science: to experiment or to think. By temperament, perhaps, I experimented, whereas Jacques Monod thought. In the end our paths converged." Perutz's investigations into the molecular mechanisms of Hb cooperativity revealed that subunit–subunit interactions are attributable to an indirect coupling between

ligation-induced changes in the tertiary structure of individual subunits and global changes in the quaternary structure of the tetrameric assembly. Specifically, the binding of O_2 to the heme Fe(II) of each subunit produces a localized change in tertiary structure that is mechanically transmitted to the $\alpha_1\beta_2/\alpha_2\beta_1$ interface. This perturbation of the interdimer interface induces the global T→R transition, thereby increasing the individual O_2-affinities of the remaining unliganded heme-bearing subunits in the tetrameric assembly. The convergence between Perutz's crystallographic results and Monod's theoretical results helped to define the role of the conformational equilibrium between liganded and unliganded Hbs (^{R}Hb and ^{T}Hb, respectively) in governing subunit cooperativity, as envisioned by the two-state MWC model (Monod et al., 1965). As described in the following sections, a combination of experimental work and mathematical modeling revealed how the connection between the ligation state of each subunit's heme iron and the quaternary structure of the protein is mediated by rearrangements of atomic contacts at the $\alpha_1\beta_2/\alpha_2\beta_1$ interdimer interfaces, and how differences in the relative stabilities of ^{R}Hb and ^{T}Hb stem from ligation-induced changes in the strength of intra- and intersubunit salt bridges.

4.4.1 The movement of Fe(II) into the heme plane triggers the T→R conformational shift

"To be guided by the atomic models towards the molecular mechanism of respiratory transport seemed like a dream. But was it true? Would the mechanism stand the cold scrutiny of experiment?"   **—Perutz (1978)**

In unliganded subunits of ^{T}Hb, the pentacoordinate Fe(II) is situated ~0.4–0.6 Å out of the heme plane on the side of the proximal histidine due to its larger diameter in the high-spin state. This causes pyramidal doming of the porphyrin skeleton (Fermi et al., 1984, Park et al., 2006) (Fig. 4.9, Table 4.4). When Fe(II) is not coplanar with the porphyrin ring, the individual Fe-Nporphyrin bonds are elongated, and on the distal side of the heme, the out-of-plane Fe(II) is not readily accessible for O_2 binding (Fig. 4.10). However, the binding of O_2 to Fe(II) causes a change to a lower spin state with a smaller diameter.

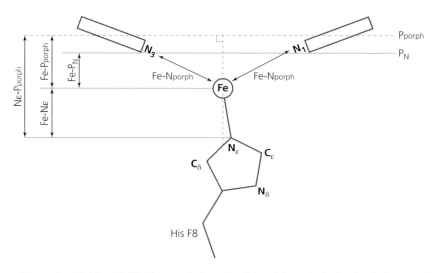

Fig. 4.9. Geometry of the proximal histidine (His F8) side chain, the heme iron (Fe), and the porphyrin ring. P_{porph} is the mean plane of the porphyrin ring, and P_N is the plane of the porphyrin nitrogen atoms. See Table 4.4 for estimates of the indicated distances and bond lengths derived from X-ray crystallography. When unliganded, the heme is domed slightly toward His F8, such that the distance from Fe to the mean plane of the four porphyrin nitrogens (Fe-P_N) is shorter than the distance to mean porphyrin plane (Fe-P_{porph}).

Table 4.4 Heme geometry in liganded and unliganded Mb and subunits of Hb. Distances to mean planes and bond lengths are given in Å, along with estimates of measurement error. Pporph is the mean plane of the porphyrin together with the first side chain atoms. PN is the plane of the porphyrin nitrogen atoms. See Fig. 4.9 for a graphical depiction of measured distances and bond lengths

	$N\epsilon$-Pporph (Å)	Fe-Pporph (Å)	Fe-P_N (Å)	Fe-Nporph (mean, Å)	Fe-$N\epsilon$ (Å)	Reference
deoxyHb (T)						
α-chain	2.72 ± 0.06	0.58 ± 0.03	0.40 ± 0.05	2.08 ± 0.03	2.16 ± 0.06	Fermi et al. (1984)
		0.50, 0.60[a]				Park et al. (2006)
β-chain	2.58 ± 0.06	0.50 ± 0.03	0.36 ± 0.05	2.05 ± 0.03	2.09 ± 0.06	Fermi et al. (1984)
		0.45, 0.40[a]				Park et al. (2006)
deoxyMb	2.67	0.47	0.42	2.03 ± 0.10	2.22	Takano (1977)
oxyHb (R)						
α-chain	2.1 ± 0.01	0.16 ± 0.08	0.12 ± 0.08	1.99 ± 0.05	1.94 ± 0.09	Shaanan (1983)
		0.09				Park et al. (2006)
β-chain	2.1 ± 0.01	0.00 ± 0.08	-0.11 ± 0.08	1.96 ± 0.06	2.07 ± 0.09	Shaanan (1983)
		0.06				Park et al. (2006)
oxyMb	2.28 ± 0.06	0.22 ± 0.03	0.18 ± 0.03	1.95 ± 0.06	2.07 ± 0.06	Phillips (1980)

[a] Separate distances are reported for each hydrogen bond in the asymmetric unit.

Consequently, Fe(II) moves into the plane of the porphyrin ring, causing the Fe-Nporphyrin bonds to contract by ~0.1 Å (Table 4.4). As a result of this ligation-induced change in Fe(II) diameter, the distance from Fe(II) to the plane of the porphyrin ring is reduced by 0.51–0.41 Å in the α-chain and by 0.39–0.34 Å in the β-chain (Park et al., 2006). The movement of Fe(II) into the heme plane causes the pyramidal doming of the porphyrin to subside (Fig. 4.11). In T-state Hb, the proximal movement of Fe(II) into the heme plane is inhibited by interactions between the F-helix and the opposing subunit at

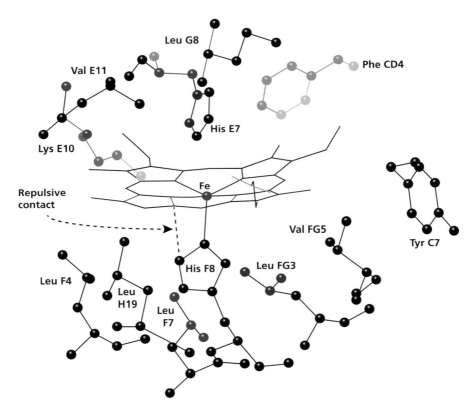

Fig. 4.10. The heme group and its structural environment in the unliganded α-chain of human Hb. Only select side chains are shown and the heme D propionate is omitted for clarity. When unliganded, the high-spin Fe(II) atom is situated outside the heme plane, which causes a pyramidal doming of the porphyrin skeleton. Fe(II) is inhibited from moving into the heme plane by repulsive contact between the imidazole ε hydrogen atom of F8 His and the N_1 nitrogen atom of the porphyrin ring and by contacts at the $\alpha_1\beta_2/\alpha_1\beta_2$ interface that restrict the upward movement of the F-helix.

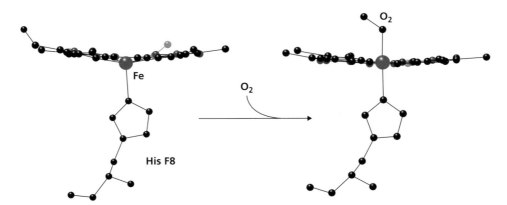

Fig. 4.11. O_2 binding changes the position of the heme iron. The unliganded, high-spin iron atom lies just below the plane of the porphyrin ring (left); upon ligation, the iron atom becomes low-spin and moves into the porphyrin plane, thereby forming a stronger Fe-O bond (right).

the $\alpha_1\beta_2/\alpha_2\beta_1$ interface. When those interactions are weakened by the conformational switch to the R-state, movements of the Fe-His F8 complex and the F-helices are far less restricted in the remaining unliganded subunits, resulting in greatly enhanced O_2 affinities. The role of interactions at the $\alpha_1\beta_2/\alpha_2\beta_1$ interface in impeding the in-plane movement of Fe-His F8 is indicated by the fact that O_2-affinities of $\alpha\beta$ dimers and isolated α- and β-chain monomers are nearly identical to those of R-state Hb tetramers (Antonini and Brunori, 1971, Mathews and Olson, 1994). Interestingly, the α- and β-subunits of human Hb have very similar O_2-affinities (Unzai et al., 1998) even though ligand binding is governed by somewhat different stereochemical mechanisms. Diverse lines of evidence suggest that proximal constraints on Fe(II) reactivity play a

key role in regulating the ligand affinity of the α-subunits, whereas steric hindrance by distal heme pocket residues is more important in the β-subunits (Perutz, 1970, Nagai et al., 1987, Perutz, 1989b, Mathews et al., 1991, Tame et al., 1991, Balakrishnan et al., 2009, Jones et al., 2014, Bringas et al., 2017).

When the liganded Fe(II) moves to the center of the heme plane, it pulls the covalently attached proximal His with it, causing a subtle reorientation of the F8 His imidazole ring, and a ~1 Å lateral translation of the entire F-helix across the heme plane (Fig. 4.12). The C-terminal end of the F-helix forms part of the $\alpha_1\beta_2/\alpha_2\beta_1$ contact surface, and the ligation-induced movement of the helix produces strain on this intersubunit interface—strain that can only be alleviated by a global conformational switch

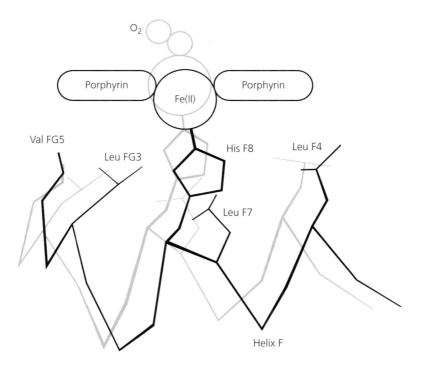

Fig. 4.12. The oxygenation-linked T→R transition in Hb quaternary structure is triggered by the movement of the Fe(II) heme iron into the plane of the porphyrin ring. In the T-structure (black lines), the center of the Fe(II) atom is ~0.6 Å below the plane. When Hb transitions to the R-state (gray lines), the Fe(II) atom moves into the plane of the porphyrin ring, pulling with it the covalently attached proximal histidine (His F8), which causes a shift in the orientation of the F-helix. Fe(II) binds O_2 far more readily when it is coplanar with the porphyrin ring. In the reverse movement, which triggers the deoxygenation linked R→T transition, Fe(II) is pulled out of the plane and the Fe(II)-O bond is broken due to disruption of interactions with the iron d-orbitals and steric repulsion from the porphyrin nitrogen atoms. This concerted movement of the Fe(II) and His F8 is propagated to the $\alpha_1\beta_2/\alpha_2\beta_1$ intersubunit sliding contacts such that changes in the ligation state of the heme trigger the allosteric transition in quaternary structure.

to the R-state. Thus, the ligation-induced movement of Fe(II) into the heme plane is transduced by the proximal histidine (F8) to other residues in the F-helix and FG corner that comprise the so-called allosteric core at the $\alpha_1\beta_2/\alpha_2\beta_1$ interface (Gelin and Karplus, 1977, Gelin et al., 1983, Janin and Wodak, 1985, Abraham et al., 1992). Since the gears and levers also operate in reverse, the same steric interactions explain why the R→T transition causes O_2 to be released when the reconfiguration of atomic contacts at the $\alpha_1\beta_1/\alpha_2\beta_2$ interfaces pulls Fe(II) into its out-of-plane position, thereby doming the porphyrin and markedly decreasing O_2 affinity.

4.4.2 The $\alpha_1\beta_2$ and $\alpha_2\beta_1$ contacts have two stable positions

As mentioned in section 4.3.1, the main structural differences between ^{T}Hb and ^{R}Hb are concentrated in the $\alpha_1\beta_2/\alpha_2\beta_1$ interdimer interface, which is formed by the network of atomic contacts between the C-helix of α_1 and the FG segment of β_2 (due to symmetry, the same points of interdimer contact exist at the $\alpha_2\beta_1$ interface) (Fig. 4.6, Tables 4.1–4.2). At the α_1C-β_2FG interface, the sidechains of residues in β_2FG fit into grooves formed by successive turns of the α_1 C-helix. The oxygenation-linked T→R

transition in quaternary structure produces a ~6 Å relative shift at the α_1C-β_2FG interface, which is why it is called the "switch region." This shift involves a ratchet-like movement in which the side chains of the β_2FG residues disengage from their (T-state) positions between grooves of the adjoining α_1 C-helix, and then slot into new (R-state) positions formed by a new set of grooves one turn away in the helix. For example, in the T→R transition, the imidazole side chain of His β_2FG4(97) ratchets into a new groove between the sidechains of Thr α_1C3(38) and Thr α_1C6(41) (Fig. 4.13). In this turn of the ratchet, one set of interdimer hydrogen bonds is broken and a new set of R-state bonds is formed. Site-directed mutagenesis experiments have demonstrated that the hydrogen bond between Tyr α_1(C7)42 and Asp β_2(G1)99 plays a key role in stabilizing the T-state (Imai et al., 1991). As stated by Perutz (1978), this α_1C-β_2FG interface "acts as a snap-action switch, with two alternative stable positions, each braced by a different set of hydrogen bonds." Intermediate positions of the interface are sterically prevented (Paoli et al., 1996), consistent with the all-or-nothing conformational transition envisioned by the MWC model.

In contrast to the ~6 Å shift in the α_1C-β_2FG/α_2C-β_1FG interface (the "switch"), the same T→R transition

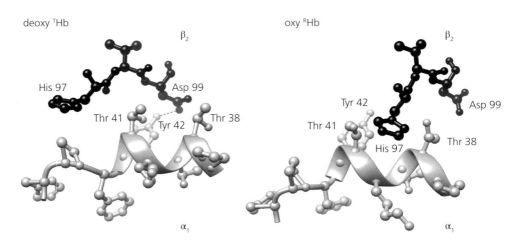

Fig. 4.13. Snap-action switch at the interdimer $\alpha_1\beta_2/\alpha_2\beta_1$ interface. At the α_1C-β_2FG interface, the sidechains of residues in β_2FG fit into grooves formed by successive turns of the α_1 C-helix. The T→R transition in quaternary structure produces a shift at the α_1C-β_2FG interface, causing the side chains of β_2FG residues to ratchet into a new set of grooves one turn away in the helix. For example, in the oxygenation-linked transition to the R-state, the interdimer hydrogen bond between Tyr α_1(C7)42 and Asp β_2(G1)99 is broken and the imidazole side chain of His β_2FG4(97) ratchets into a new groove between the sidechains of Thr α_1C3(38) and Thr α_1C6(41).

causes only a ~1 Å shift at the α_1FG-β_2C/α_2FG-β_1C interface and the interdimer contacts remain intact. This intersubunit contact therefore acts as a hinge for the pivoting of the α_1 and β_2 subunits and is referred to as the "flexible joint" (Baldwin and Chothia, 1979) (Fig. 4.6).

4.4.3 The T-state is stabilized by a network of salt bridges that break in the T→R transition

The atomic contacts between opposing $\alpha\beta$ dimers of deoxyHb include both α_1-α_2 and α_1-β_2 interactions. With regard to the former, Figure 4.14 shows that the C-terminal Arg α141(HC3) of one α-chain is involved in three separate interactions with amino acids in the opposing α-chain: (1) the carboxyl group of Arg $\alpha_1$141(HC3) interacts with the ϵ-amino group of Lys $\alpha_2$127(H10); (2) the guanidinium group of Arg $\alpha_1$141(HC3) forms an electrostatic bond with the carboxyl group of Asp $\alpha_2$126(H9); and (3) a Cl$^-$ ion forms ionic contacts with both the guanidinium group of Arg $\alpha_1$141(HC3) and the NH$_2$-terminal amino group of $\alpha_2$1Val(NA1). None of these three α_1-α_2 salt bridges are present in ^{R}Hb.

In addition to these interdimeric α_1-α_2 interactions in deoxy ^{T}Hb, the C-terminal histidine of each β-chain, His β146(HC3), is involved in two electrostatic interactions: (1) the positively charged imidazole forms a salt bridge with Asp β94(FG1) in the same

subunit; and (2) the negatively charged carboxyl group forms a $\alpha_1\beta_2$ salt bridge with Lys α40(C5) (Fig. 4.15A). In the oxygenation-linked T→R transition, both salt bridges are ruptured because the interacting residues are shifted too far apart to maintain the electrostatic attraction (Fig. 4.15B). Finally, in ^{T}Hb, the penultimate N-terminal Tyr residues of each subunit, α140(HC2) and β145(HC2), are anchored into a cleft between the F and H helices (Fig. 4.5). The Tyr residues are ejected during the oxygenation-linked transition to the R-state, thereby freeing the C-terminal ends of each globin chain.

4.5 Reconciling Perutz's stereochemical model with allosteric theory

To summarize, Perutz's stereochemical model of Hb cooperativity is based on the following structural changes that occur during the oxygenation-linked T→R transition: (1) the $\alpha_1\beta_1$ and $\alpha_2\beta_2$ dimers rotate relative to one another; (2) Fe(II) moves into the porphyrin plane, causing a shift in the F-helix relative to the heme group; (3) six intersubunit salt bridges between α_1 and α_2 (Arg α141(HC3)-Val α1(NA1) and Arg α141(HC3)-Asp α126(H9)) and between the α- and β-chain subunits (Lys α40(C5)-His β146(HC3)) are broken; and (4) two intra β-chain salt bridges (His β146(HC3)-Asp β94(FG1)) are broken (Fig. 4.5). Additionally, in the $\alpha_1\beta_2$/$\alpha_2\beta_1$ interdimer interface, the T→R transition results in the elimination of some atomic contacts (Tyr α42(C7)-Asp β99(G1), Asn α97(G4)-Asp β99(G1), Leu α91(FG3)-Arg β40(C6), and Arg α92(FG4)-Arg β40(C6)) and the formation of new contacts (Thr α38(C3)-His β97(FG4), Thr α41(C6)-Arg β40(C6), and Asp α94(G1)-Asn β102(G4)) (Perutz, 1970, Perutz, 1989a, Perutz et al., 1998, Baldwin and Chothia, 1979, Park et al., 2006, Yuan et al., 2015). The oxygenation-linked T→R transition involves a gears-and-levers coupling of tertiary and quaternary structural changes, with the interdimeric interfaces ($\alpha_1\beta_2$/$\alpha_2\beta_1$) serving as a binary switch. Upon heme oxygenation in ^{T}Hb, the movement of the F-helix relieves strain in the allosteric core while simultaneously introducing new strain in the FG corner that can only be relieved via transition to the

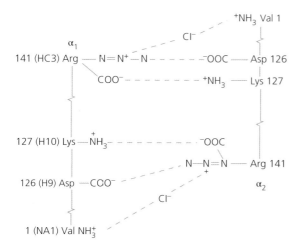

Fig. 4.14. Electrostatic interactions between α-chains of human Hb in the deoxy T-state.

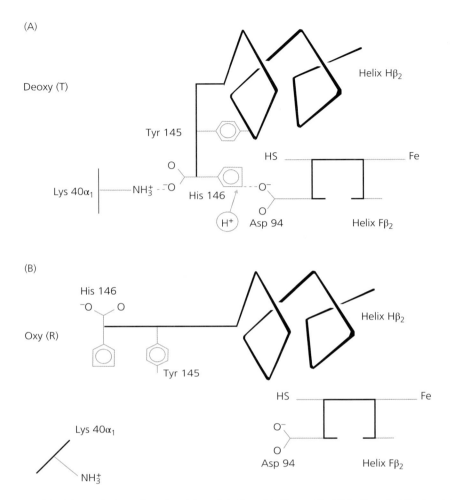

Fig. 4.15. Oxygenation-linked changes in the conformation of the β-chain C-termini of human Hb. (A) The C-terminal His of each β-chain subunit (β146His) participates in two electrostatic interactions in the deoxy (T) state. The positively charged imidazole side chain of β146His forms an intrasubunit salt bridge with β94Asp (which increases its pK_a) and its negatively charged carboxyl group forms an intradimer salt bridge with α40Lys. (B) When Hb is oxygenated, the allosteric transition in quaternary structure shifts the triad of residues apart from one another, outside the range of electrostatic interaction. The consequent rupturing of the β146His-β94Asp salt bridge results in the deprotonation of the His side chain (two protons are released per tetramer), which makes a major contribution to the Bohr effect.

Modified from Perutz (1970).

R-state. This strain weakens and—as additional hemes are liganded—eventually ruptures the salt bridges at the chain termini, thereby permitting the T→R transition to occur. The stability of ^{R}Hb relative to ^{T}Hb increases with the degree of oxygenation due to the strain on the Fe(II)-His F8 bond induced by its movement into the heme plane in deoxy Hb. This strain has been experimentally confirmed using chemical manipulations that force liganded Hb into the T-state (Maxwell and Caughey, 1976,

Perutz et al., 1976). Similarly, crystallographic analyses (Paoli et al., 1997) and experiments involving chemically modified Hbs or recombinant Hb mutants (Barrick et al., 1997, Miyazaki et al., 1999, Barrick et al., 2001) have confirmed Perutz's prediction that the proximal histidine (F8) is the main transducer of the ligation state of the heme.

Whereas Monod, Wyman, and Changeux developed a very general model with little structural or mechanistic detail, Perutz proposed a detailed structural model which appeared to be consistent with the basic tenets of the MWC model "but contained no prescription for making it quantitative" (Eaton et al., 1999). Szabo and Karplus (1972) formulated a thermodynamic model to provide a quantitative description of the role of the intra- and intersubunit salt bridges in stabilizing ^{T}Hb relative to ^{R}Hb; this model was later refined and generalized by Lee and Karplus (1983, 1988). The model derived a partition function that describes Hb allostery in terms of coupled transitions in both tertiary and quaternary structure. In a similar fashion, Gelin and Karplus (1977) and Gelin et al. (1983) formulated a thermodynamic model that describes the reaction path linking the ligation-induced movement of Fe(II)-His F8 to the reconfiguration of atomic contacts at the interdimer $\alpha_1\beta_2/\alpha_2\beta_1$ interface. Their calculations of conformational energies suggested that heme oxygenation of ^{T}Hb subunits induces strain on the bond between Fe(II) and the His F8 imidazole nitrogen as the Fe(II) moves into the plane of the porphyrin ring. The forced tilting of the liganded heme propagates strain in the allosteric core to the residues at the $\alpha_1\beta_2/\alpha_2\beta_1$ interdimer interface. Comparison of high-resolution structures of ^{T}Hb and ^{R}Hb confirm the proposed reaction path, and indicates that the accumulated strain in the FG corners of partially liganded ^{T}Hb can only be relieved by a transition in quaternary structure to the R-state (Perutz, 1970, Perutz, 1979, Baldwin and Chothia, 1979, Gelin et al., 1983, Perutz et al., 1987, Liddington et al., 1988). The conformational switch dramatically increases the O_2 affinity of the remaining unliganded heme groups because of the increased flexibility of Hs F8 and the F-helix in unliganded subunits, giving rise to subunit cooperativity that is qualitatively in accord with the two-state MWC model.

Perutz's stereochemical model involves an "induced fit" change in the tertiary structure of liganded subunits, which is seemingly consistent with the KNF sequential model of cooperativity (Koshland et al., 1966). However, these tertiary structural changes do not alter the functional symmetry of the tetramer and O_2 affinity is still uniquely determined by the quaternary state, in keeping with the MWC concerted model (Monod et al., 1965). Since the stepwise ligation of the hemes in the tetrameric assembly is coupled to the all-or-none transition in quaternary structure, the change is more concerted than sequential.

4.6 The structural basis of heterotropic effects: the binding of allosteric ligands

"[H]ow could as heterogeneous a collection of chemical agents as protons, chloride ions, carbon dioxide, and diphosphoglycerate influence the oxygen equilibrium curve in a similar way? It did not seem plausible that any of them could bind directly to the hemes, that all of them could bind at any other common site... To add to the mystery, none of these agents affected the oxygen equilibrium of Mb or of isolated subunits of Hb."

—Perutz (1978)

As discussed in Chapter 2, the binding of allosteric effectors usually stabilizes deoxyHb through the formation of additional salt bridges within and between subunits (Perutz, 1970, Bettati et al., 1983, Perutz, 1989a), thereby shifting the conformational equilibrium in favor of the low-affinity T-state. The resultant reduction in Hb-O_2 affinity promotes O_2 unloading to the cells of metabolizing tissues. In the absence of allosteric regulation by heterotypic effectors, the intrinsic O_2 affinity of Hb would be too high to sustain adequate tissue O_2 delivery *in vivo*. Allosteric effectors primarily bind to deoxyHb at sites located at the N- and C-termini of the globin chains and in the positively charged central cavity of the Hb tetramer (Arnone, 1972, Arnone and Perutz, 1974, O'Donnell et al., 1979, Nigen et al., 1980) (Fig. 4.16).

In the following sections we will explore the structural basis for the oxygenation-linked binding of H^+, CO_2, Cl^-, and organic phosphates—the allosteric effectors that are generally most important in regulating the O_2 affinity of vertebrate Hb (Bohr et al., 1904, Margaria and Green, 1933, Benesch and Benesch, 1967, Kilmartin and Rossi-Bernardi, 1973, Rollema et al., 1975, Jensen, 2004, Berenbrink, 2006, Mairbaurl and Weber, 2012). Other cofactors such as lactate anions (Guesnon et al., 1979) and nicotinamide dinucleotides (Cashon et al., 1986) may also have physiologically relevant allosteric effects. There are other cofactors that are important modulators of Hb function in particular vertebrate taxa, for example, bicarbonate ions (HCO_3^-) in the blood of crocodilians (Bauer et al., 1981) and urea in the blood of sharks and other elasmobranchs (Weber et al., 1983).

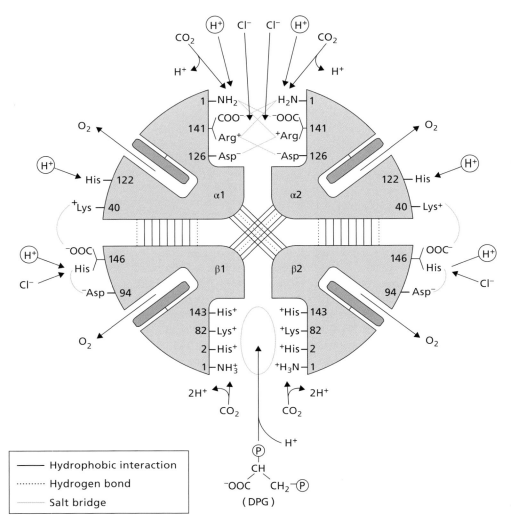

Fig. 4.16. Stylized depiction of tetrameric human Hb showing allosteric binding sites as well as intra- and intersubunit bonds. Arrows depict the allosteric binding of non-heme ligands (Cl^-, DPG, CO_2, H^+) coincident with the release of O_2 from each of the four heme groups. Specific amino acid residues are indicated by sequence numbers. The α-amino (N-terminal) and α-carboxyl (C-terminal) groups are shown as $-NH_3^+$ and $-NH_2$, respectively. Circled protons are those that bind Bohr groups in deoxy ^{T}Hb. The indicated salt bridges are present in deoxy ^{T}Hb, but not in oxy ^{R}Hb.

Modified from Imai (1982) with permission from Cambridge University Press.

4.6.1 Proton binding

"The reason why certain acid groups are affected by oxygenation is simply the alteration of their position and environment which results from the change of configuration of the hemoglobin molecule as a whole accompanying oxygenation." —**Wyman and Allen (1951)**

Since deoxyHb has a higher affinity for protons than does oxyHb, the metabolic acidosis of working muscles reduces Hb-O_2 affinity and promotes O_2 unloading in the tissue capillaries (Chapter 2, section 2.12). The magnitude of the Bohr effect depends on the range of pH values under consideration, and it also varies as a function of temperature, PCO_2, and the intracellular concentration of Cl^- ions and organic phosphates. In the absence of the organic phosphate DPG but under otherwise standard physiological conditions, deoxygenated human Hb binds two additional protons relative to fully oxygenated Hb. Under the same experimental conditions, Hbs of teleost fish often bind four or more Bohr protons (Jensen and Weber, 1985, Jensen, 1989, Jensen, 2001). There is considerable variation in the magnitude of the Bohr effect among the Hbs of different vertebrate taxa (Bailey et al., 1970, Berenbrink et al., 2005, Berenbrink, 2006).

Preferential binding of protons to deoxyHb requires that the pK_as of proton-binding sites increase upon deoxygenation in the R→T transition. In human Hb, the free NH_3^+ groups of N-terminal residues have pK_a values that vary between 7.25 and 8.0 (van Beek and de Bruin, 1980) and the imidazole side chains of solvent-exposed histidines have pK_a values that span the physiologically relevant range between pH 6.0 and 8.0. However, the pK_a values of histidine side chains are highly dependent on temperature, solvent composition, and the structural context in which they occur in the native protein (Busch and Ho, 1990, Sun et al., 1997). Consequently, estimating the relative contributions of different amino acids to the Bohr effect is experimentally challenging, and hard-won insights have been attained through a variety of approaches including X-ray crystallography and NMR imaging studies of native, recombinant, and chemically modified Hbs (Imai et al., 1989, Fang et al., 1999, Lukin and Ho, 2004, Zheng et al., 2013, Yuan et al., 2015).

At physiological pH, and in the presence of DPG, the longstanding view has been that the Bohr effect of human Hb is primarily attributable to proton binding at the following residues: Val $α1$(NA1), His $α122$(H5), His $β2$(NA2), Lys $β82$(EF6), His $β143$(H21), and His $β146$(HC3) (Kilmartin and Rossi-Bernardi, 1969, Perutz et al., 1969, Kilmartin and Rossi-Bernardi, 1973, Kilmartin, 1977, Kilmartin et al., 1978, Ho and Russu, 1987, Riggs, 1988). At physiological pH, several workers have suggested that the Bohr effect of human Hb is mainly attributable to the oxygenation-linked deprotonation of two key residues: the N-terminus of the $α$-chain, Val $α1$(NA1), and the C-terminus of the $β$-chain, His $β146$(HC3) (Kilmartin and Wootton, 1970, Perutz and Ten Eyck, 1971, Kilmartin et al., 1978, Kilmartin et al., 1980, Perutz et al., 1980). According to Perutz's stereochemical mechanism, separate intrachain salt bridges, Val $α1$(NA1)-Arg $α141$(HC3) and His

$β146$(HC3)-Asp $β94$(FG1), stabilize ^{T}Hb relative to ^{R}Hb and they reduce the heme reactivities in ^{T}Hb because of the energy required to break them (Fig. 4.14, 4.15). These same salt bridges also release protons upon breakage, thereby contributing to the Bohr effect. In ^{T}Hb, positive charges on Val $α1$(NA1) and His $β146$(HC3) are stabilized by specific electrostatic interactions, ensuring that they remain fully protonated (Fig. 4.17).

The positive charge on the amino terminus of the $α$-chain may be stabilized by a Cl^- ion bound between Val $α1$(NA1) and the C-terminus of the other $α$-chain, Arg 141(HC3) (Fig. 4.17A). The positive charge on His $β146$(HC3) is stabilized by the negative charge of the spatially adjacent Asp $β94$(FG1) in the same subunit (Fig. 4.17B). In the oxygenation-linked T→R transition, a shift in the relative positions of the $α_1$ and $α_2$ C-termini ruptures each of the bonds shown in Fig. 4.17A. Likewise, the same conformational shift moves Lys $α40$(C5) out of the range of the $β$-chain N-terminus, and the intrasubunit H-bond between Tyr $β145$(HC2) and the main-chain carbonyl group of Tyr $β98$(FG5) is weakened by a ligation-induced change in $β$-chain tertiary structure (Fig. 4.17B). The resultant loss of charge stabilization causes both Val $α1$(NA1) and His $β146$(HC3) to release their Bohr protons in the R→T transition.

Consistent with this biophysical explanation, human Hb mutants with replacements of His $β146$(HC3) exhibit a severely diminished Bohr effect (Perutz et al., 1984, Shih et al., 1984, Busch et al., 1991, Shih et al., 1993, Ivaldi et al., 1999) (Table 4.5). Likewise, naturally occurring Hb isoforms (isoHbs) of some teleost fish have Phe substituted for His at $β146$(HC3) and, as expected, the O_2-affinities of these isoHbs are largely unresponsive to changes in pH (Fago et al., 1995). Experiments in which His $β146$(HC3) was enzymatically cleaved from carp Hb demonstrated that this residue accounts for roughly 50 percent of the Bohr effect (Parkhurst et al., 1983). In the major larval isoHb of American bullfrog (*Rana catesbeiana*) tadpoles, the N-terminal $α$-chain residue is acetylated (so it does not bind H^+) and asparagine is present at $β94$(FG1) so no salt bridge is formed with His $β146$(HC3) in ^{T}Hb. As expected, this larval isoHb is pH-insensitive (Watt and Riggs, 1975, Maruyama et al., 1980, Watt et al., 1980).

(A)   (B)

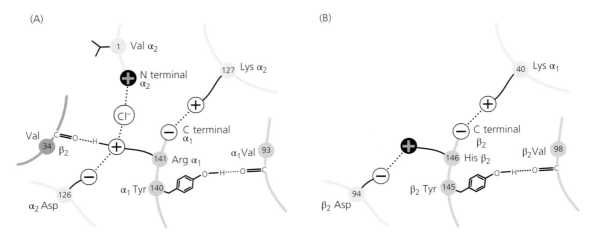

Fig. 4.17. The oxygenation-linked breakage of salt bridges in ^{T}Hb results in the partial deprotonation of the N-terminal Val α(NA1) and the C-terminal His β146(HC3), thereby contributing to the Bohr effect. In ^{T}Hb, positive charges on both residues (indicated by plus signs) are stabilized by specific electrostatic interactions. In the transition to ^{R}Hb, all of the indicated bonds are broken and the loss of charge stabilization causes both residues to release their Bohr protons. (A) In ^{T}Hb, the positive charge on the α_2 N-terminus is stabilized by a bound Cl$^-$ ion between Val $\alpha_2$1(NA1) and Arg $\alpha_1$141(HC3). The T→R transition shifts the α_1 and α_2 C-termini apart, thereby breaking the intersubunit interactions. (B) In ^{T}Hb, the positive charge on His β146(HC3) is stabilized by the negatively charged Asp β94(FG1) in the same subunit. The T→R transition ruptures the intersubunit bond between Lys α40(C5) and His β146(HC3), which untethers the β-chain N-terminus and results in the breakage of intrasubunit interactions with Asp β94 (FG1) and the main-chain carbonyl group of Tyr β98(FG5).

Modified from Dickerson and Geis (1983).

Table 4.5 Human Hb mutants associated with a greatly diminished Bohr effect, caused by replacements of His β146(HC3). Mutated nucleotides are underlined

	β146(HC3)	Codon
Wildtype	His	CAC
Hb Hiroshima	Asp	<u>G</u>AC
Hb Bologna St. Orsola	Tyr	<u>T</u>AC
Hb Cochin-Port Royal	Arg	C<u>G</u>C
Hb Cowtown	Leu	C<u>T</u>C
Hb York	Leu	C<u>C</u>C
Hb Kodaira	Gln	CA<u>A</u>/CA<u>G</u>

The original view, articulated by Kilmartin and Rossi-Bernardi (1973) and Perutz (1983), was that the pH sensitivity of vertebrate Hbs is attributable to the oxygenation-linked deprotonation of a small number of key residues with major effects. More recent experimental and computational analyses of human Hb have confirmed that the deprotonation of His β146(HC3) accounts for ~60 percent of the Bohr effect in 0.1 M chloride solution, but the

remainder appears to be attributable to Val α1(NA1), Val β1(NA1), and numerous solvent-exposed histidine residues that have individually small effects (Shih et al., 1993, Fang et al., 1999, Lukin and Ho, 2004, Berenbrink, 2006). NMR-based studies of human Hb have provided a means of estimating pK_a values for each solvent-exposed histidine in the liganded, R-state Hb (carbonmonoxyHb) and the unliganded T-state (Fig. 4.18). The contribution of each individual histidine to the Bohr effect is indicated by the magnitude of the increase in pK_a upon deoxygenation, and the degree to which the oxy- and deoxyHb pK_a values match the solvent pH. For tetrameric Hb, the contribution of each individual Bohr group, ΔH_i^+, is two times the difference in fractional occupation of the proton-binding site between ^{T}Hb and ^{R}Hb:

$$\Delta H_i^+ = 2 \left(\frac{\left[H^+ \right]}{\left[H^+ \right] + K_{ai}^T} - \frac{\left[H^+ \right]}{\left[H^+ \right] + K_{ai}^R} \right) \quad (4.1)$$

or

$$\Delta H_i^+ = 2 \left(\frac{10^{-pH}}{10^{-pH} + 10^{-pK_{ai}^T}} - \frac{10^{-pH}}{10^{-pH} + 10^{-pK_{ai}^R}} \right), \quad (4.2)$$

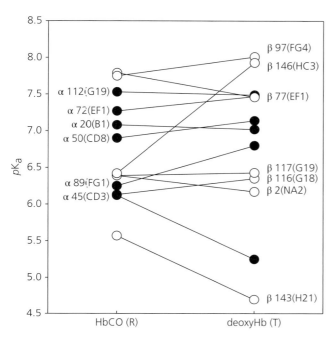

Fig. 4.18. Changes in the acid dissociation constants, pK_a, of solvent-exposed histidine side chains in the transition from liganded to unliganded human Hb (HbCO and deoxyHb, respectively). Filled symbols denote paired values for α-chain histidines (labeled on the left) and open symbols refer to β-chain histidines (labeled on the right). At physiological pH, a given histidine makes a contribution to the alkaline Bohr effect if its pK_a increases in the transition from the liganded R-state (COHb) to the unliganded T-state (deoxyHb). Conversely, a given histidine makes a negative contribution if its pK_a decreases in the same R→T transition. In such cases, the histidine contributes to the acid Bohr effect, whereby Hb-O_2 affinity is increased by a reduction in pH. His β146(HC3) makes the largest contribution to the alkaline Bohr effect, whereas β143(H21) and α45(CD3) make the largest contributions to the acid Bohr effect (Berenbrink, 2006). The remaining histidine residues in human Hb (such as the heme-associated proximal and distal histidines) do not buffer at physiological pH, and are therefore not shown in this figure because they do not make a direct contribution to the Bohr effect. All values were estimated using NMR imaging at 29° C in 0.1 M HEPES and 0.1 M NaCl in D_2O (Fang et al., 1999, Lukin and Ho, 2004).

where K_{ai}^T, and K_{ai}^R are the acid dissociation constants of an individual site i in ^THb and ^RHb, respectively. Using equation 4.2 with experimentally estimated pK_a values for all surface histidines indicates that the summed contribution of all such residues accounts for ~90 percent of the alkaline Bohr effect of human Hb in 0.1 M chloride (Berenbrink, 2006). Likewise, results of NMR studies (Fang et al., 1999) and computational titration analyses based on high-resolution crystal structures (Zheng et al., 2013) indicate that His β146(HC3) makes the largest contribution to the alkaline Bohr effect, and that the remainder is accounted for by the other twenty-four surface histidines and the N-terminal valines of both chains, Val α1(NA1) and β1(NA1). Other non-histidine residues such as Lys β82(EF6) appear to make smaller direct contributions than previously supposed.

Several lines of evidence suggest that the Bohr effect has somewhat different mechanistic underpinnings in different vertebrate taxa (Berenbrink et al., 2005, Berenbrink, 2006). As just mentioned, the oxygenation-linked deprotonation of His β146(HC3) accounts for most of the Bohr effect of human Hb in the presence of Cl^-, but this same residue makes no appreciable contribution to the large Bohr effect that is characteristic of some teleost fish Hbs.

Crystallographic studies have revealed that this is because the salt bridge that raises the pK_a of His β146(HC3) in deoxy human Hb does not form in the T-state Hbs of some teleost species (Camardella et al., 1992, Ito et al., 1995, Yokoyama et al., 2004). In teleost Hbs, moreover, the N-termini of the α-chains are typically acetylated (*Ac*-Ser or *Ac*-Thr) and are therefore not titratable, and the total number of surface histidines is generally much lower relative to the Hbs of humans and other mammals (Jensen, 1989, Berenbrink et al., 2005, Berenbrink, 2006). It therefore appears that the Bohr effect of teleost fish Hbs has a different molecular basis than that of human Hb, with no role for the α-chain N-termini and contributions from different combinations of surface histidines (Yokoyama et al., 2004).

4.6.2 CO_2 binding

As mentioned in Chapter 2, CO_2 contributes to the Bohr effect because protons are released when it is hydrated in the red cell to form bicarbonate ions. CO_2 also produces a pH-independent allosteric effect by directly reacting with the α- and β-chain N-terminal residues of deoxyHb to form carbamino compounds (Rossi-Bernardi and Roughton, 1967,

Arnone, 1974). In human Hb, the N-terminal amino groups of the β-chain subunits have a much higher affinity for CO_2 than those of the α-chains (Kilmartin and Rossi-Bernardi, 1973, Bauer et al., 1975, Perrella et al., 1975). Under physiological conditions, it is difficult to quantify the effect of carbamino formation on Hb-O_2 affinity due to interactions between CO_2 and other allosteric effectors that bind to deoxyHb. For example, direct binding of CO_2 to deoxyHb promotes the release of protons, thereby reducing the Bohr effect (Rossi-Bernardi and Roughton, 1967), but CO_2 also competes with organic phosphates, Cl⁻ ions, and lactate ions for oxygenation-linked binding of the α- and β-chain N-termini, respectively (Bauer, 1969, Bauer, 1970, Chiancone et al., 1972, Imaizumi et al., 1979, Nielsen and Weber, 2007). Consequently, carbamino formation can be expected to make a larger contribution to tissue O_2 unloading in species whose Hbs do not strongly bind organic phosphates (Baumann and Haller, 1975, Baumann et al., 1975, Campbell et al., 2010).

4.6.3 Chloride binding

Cl⁻ ions bind to the N-terminal residues of the α- and β-chain N-termini, and are jointly coordinated with Ser α131(H14) and Lys β82(EF6), respectively. However, whereas Cl⁻ binding to the α-chain between Val α1(NA1) and Ser α131(H14) produces a reduction in Hb-O_2 affinity, Cl⁻ binding to the β-chain between Val β1(NA1)/His β2(NA2) and Lys β82(EF6) is not oxygenation-linked, and therefore does not have a direct allosteric effect (Chiancone et al., 1972, Imaizumi et al., 1979, O'Donnell et al., 1979, Bonaventura and Bonaventura, 1980, Nigen et al., 1980). Cl⁻ binding to deoxyHb via either of the above-mentioned site pairs can have significant indirect effects by inhibiting the oxygenation-linked binding of protons, CO_2, and/or organic phosphates at the N-terminal residues and Lys β82(EF6) (Imai, 1982, Mairbaurl and Weber, 2012).

There is experimental evidence to suggest that Cl⁻ chiefly modulates O_2 affinity through delocalized electrostatic effects that do not involve binding to specific residues (Perutz et al., 1993, Shih et al., 1993, Bonaventura et al., 1994, Perutz et al., 1994). According to this view, Cl⁻ partially neutralizes the excess of positive charges in the water-filled central cavity of deoxy Hb, thereby stabilizing the T-state conformation. If this view is correct, then it is possible that Hb-O_2 affinity could be increased not just by substitutions at specific Cl⁻ binding sites (Weber et al., 2002), but also by substitutions at any residue positions that increase the net electropositivity of the central cavity. In human Hb, the amino acids that confer the excess positive charge include Val α1 (NA1), Lys α99(G6), His α103(G10), Val β1(NA1), His β2 (NA2), Lys β82(EF6), Arg β104(G6), and His β143(H21).

4.6.4 Phosphate binding

Different organic phosphates serve as the main allosteric effectors in the red blood cells of different groups of vertebrates: DPG in mammals, inositol pentaphosphate (IPP) in birds, ATP in turtles, lizards, and snakes, ATP and DPG in amphibians, and ATP and GTP in lobe-finned fishes, ray-finned fishes, and cartilaginous fishes (Rapoport and Guest, 1941, Bartlett, 1980, Isaacks and Harkness, 1980, Hazard and Hutchison, 1982, Weber and Jensen, 1988) (Fig. 4.19). It is probably no accident that a metabolite of anaerobic glycolysis (DPG) was recruited as a chief modulator of Hb-O_2 affinity in the enucleated red cells of mammals, whereas metabolites of aerobic metabolism (ATP, GTP, and IPP) were recruited as modulators of Hb-O_2 affinity in the nucleated, mitochondria-rich red cells of other vertebrates.

In the red cells of almost all vertebrates, these various polyphosphate molecules bind to a constellation of cationic residues lining the cleft between the β_1 and β_1 chains of deoxyHb, thereby stabilizing the T-state via electrostatic interactions (Arnone, 1972, Arnone and Perutz, 1974, Richard et al., 1993). For example, DPG carries four negative charges which allows it to bind between the β-chains of deoxy human Hb via charge–charge interactions with the Val β1(NA1) residue of one chain, and with His β2(NA2), Lys β82(EF6), and His β143(H21) of both chains (Fig. 4.20). Preferential binding of DPG to deoxy Hb stabilizes the low-affinity T-state, thereby promoting the release of O_2. Conversely, the mutual rotation of the $\alpha_1\beta_1$ and $\alpha_2\beta_2$ dimers during the T→R transition shifts the carboxyl ends of the

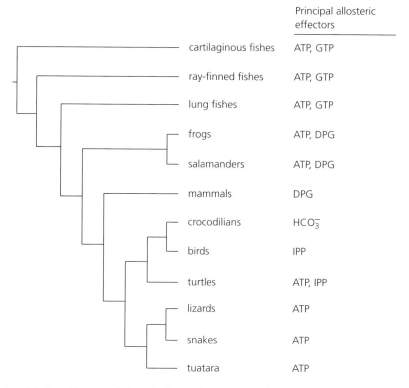

	Principal allosteric effectors
cartilaginous fishes	ATP, GTP
ray-finned fishes	ATP, GTP
lung fishes	ATP, GTP
frogs	ATP, DPG
salamanders	ATP, DPG
mammals	DPG
crocodilians	HCO_3^-
birds	IPP
turtles	ATP, IPP
lizards	ATP
snakes	ATP
tuatara	ATP

Fig. 4.19. Phylogenetic distribution of the principal allosteric effectors of Hb function in the adult red blood cells of jawed vertebrates. Information from Bartlett (1980).

β-chains 7 Å closer together in ^{R}Hb than in ^{T}Hb. This shift in the relative positions of the opposing β-chains ejects DPG from the binding cleft. DPG binds tetrameric ^{T}Hb with a ~1:1 stoichiometry (Fig. 4.16), and the same is generally true for other organic phosphates. However, evidence from several vertebrate Hbs suggests that one or more additional phosphate-binding sites in the α- and or β-chains also exert small but detectable contributions to the allosteric regulation of Hb-O_2 affinity (Amiconi et al., 1985, Tamburrini et al., 2000, Riccio et al., 2001, Laberge et al., 2005, Olianas et al., 2005, Weber et al., 2010, Grispo et al., 2012, Fan et al., 2013, Weber et al., 2013).

The importance of the above-mentioned sites for DPG binding is demonstrated by the functional effects of mutations in human Hb (Wajcman and Galacteros, 2005, Steinberg and Nagel, 2009). In particular, mutations at β82(EF6) and β143(H21) are

typically associated with dramatically increased O_2 affinities under physiological conditions (Table 4.6), and they are often associated with pathologically elevated hematocrits (erythrocytosis). This is because the increase in Hb-O_2 affinity reduces the O_2 partial pressure at which O_2 is released in the systemic circulation, resulting in a diminished gradient for O_2 diffusion to the cells of respiring tissues. The reduced level of tissue oxygenation triggers a compensatory increase in red blood cell production to increase the O_2-carrying capacity of the blood. The importance of sites β82(EF6) and β143(H21) for phosphate binding is also indicated by the allosteric properties of Hb Trout I, one of four main isoHbs in rainbow trout (*Oncorhynchus mykiss*). Hb Trout I is unusual in that its O_2 affinity is completely unaffected by the presence of organic phosphates (Brunori et al., 1975). The unresponsiveness of this isoHb is explained by the fact that positively

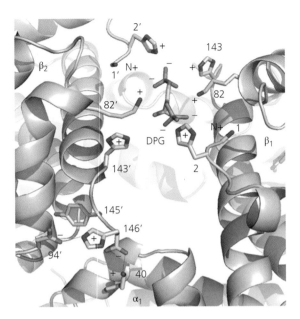

Fig. 4.20. Binding of 2,3-diphosphoglycerate (DPG) in the central cavity between the β_1 and β_2 chains of deoxy ^{T}Hb. Also shown (lower left-hand corner) is an intrachain salt bridge that is formed in ^{T}Hb between the imidazole ring of the N-terminal His β146(HC3) and the negatively charged Asp β94(FG1).

Reproduced from Storz and Moriyama (2008).

Table 4.6 Human Hb mutants with reduced DPG sensitivity and increased O_2 affinity, caused by replacements of Lys β82(EF6) or His β143(H21). Mutated nucleotides are underlined

	β82(EF6)	Codon
Wildtype	Lys	AAG
Hb Tsurumai	Gln	C̲AG
Hb Gámbara	Glu	G̲AG
Hb Helsinki	Met	AT̲G
Hb Rahere	Thr	AC̲G
Hb Providence	Asn	AAT̲/AAC̲

	β143(H21)	Codon
Wildtype	His	CAC
Hb Rancho Mirage	Asp	G̲AC
Hb Old Dominion/Burton-upon-Trent	Tyr	T̲AC
Hb Sapporo	Asn	A̲AC
Hb Vancleave	Leu	CT̲C
Hb Syracuse	Pro	CC̲C
Hb Abruzzo	Arg	CG̲C
Hb Little Rock	Gln	CAA̲/CAG̲

charged residues at β82(EF6) and β143(H21) have been substituted by uncharged residues, Leu and Ser, respectively (Barra et al., 1983, Tame et al., 1996). Interestingly, substitutions at other key DPG-binding sites have occurred in the adult Hbs of some mammals, such as lemurs, ruminant artiodactyls (e.g., cattle, goats, sheep, antelopes, and deer), and feliform carnivores (e.g., cats, hyenas, and civets), all of which exhibit low DPG sensitivities relative to adult Hbs of other mammals (Bunn, 1971, Taketa et al., 1971, Bonaventura et al., 1974, Hamilton and Edelstein, 1974, Scott et al., 1977, Bunn, 1980, Janecka et al., 2015) (Fig. 4.21). These species have invariably evolved low intrinsic Hb-O_2 affinities, presumably to compensate for the fact that intrinsic affinities in the normal range would be too high for effective tissue-O_2 delivery in the absence of regulation by DPG. The low intrinsic Hb-O_2-affinities are attributable to reductions of K_T and higher values of L relative to the DPG-sensitive Hbs of other mammals, while K_R remains unchanged (Perutz and Imai, 1980, Perutz, 1983).

The low-affinity, DPG-insensitive Hbs of lemurs, ruminants, and feliform carnivores are consistently distinguished from the high-affinity, DPG-sensitive Hbs of other mammals by substitutions at β2(NA2) (Fig. 4.22). The DPG-insensitive Hbs have large, hydrophobic residues at this position (Leu, Met, or Phe), whereas the adult Hbs of all other mammals have hydrophilic residues (His, Gln, or Asn). Crystallographic analysis of human deoxyHb complexed with DPG reveals that charged phosphate groups of the DPG molecule form salt bridges with Val β1(NA1) and His β2(NA2) from both β chains, and these electrostatic interactions pull the β_1 and β_2 A helices toward one another, locking them firmly in place (Arnone, 1972). In the ruminant artiodactyls, the N-terminal Val β1(NA1) is deleted and His β2(NA2) is replaced by Met, so the β-chain N-terminus plays no role in DPG binding. In the Hbs of these taxa, Perutz (1983) suggested that the hydrophobic side chain of Met β2(NA2) plays the same role as DPG in locking the β_1 and β_2 A helices together at the entrance of the central cavity, thereby reducing O_2 affinity by stabilizing the T-state. Similarly, in the Hbs of lemurs and feliform carnivores, hydrophobic Phe or Leu at β2(NA2) would produce the same effect.

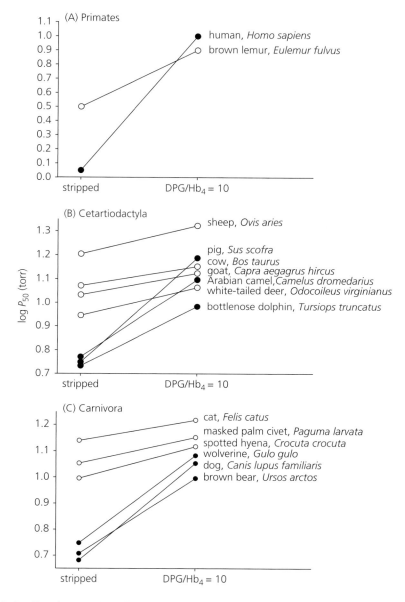

Fig. 4.21. Variation in the effect of DPG on Hb-O_2 affinity in mammalian species that are representative of three different eutherian groups: Primates, Cetartiodactyla, and Carnivora. Within each of the three groups there are species that have adult Hbs with suppressed sensitivities to DPG (data points denoted by open circles) and other species with normal, DPG-sensitive Hbs (denoted by filled circles). O_2-equilibria were measured for purified Hb in the absence of DPG ("stripped") and in the presence of DPG at 10-fold molar excess over tetrameric Hb. (A) Among primates, lemurs have adult Hbs with low intrinsic O_2 affinity and reduced sensitivity to DPG relative to humans and all other primates that have been examined. Measurements were conducted in 0.05 M bis-Tris buffer at pH 7.3, 20°C [tetrameric Hb] = 0.1 mM, in the absence and presence of 1 mM DPG. Data are from Bonaventura et al. (1974). (B) Among cetartiodactyls, ruminants such as cows, goats, sheep, and deer have adult Hbs with low intrinsic O_2 affinity and reduced sensitivity to DPG compared to representatives of other artiodactyl families (Suidae and Camelidae) and cetaceans. Hbs from sheep are representative of the "BB" genotype. Measurements on purified Hbs were conducted in 0.05 M bis-Tris HCl buffer, 0.1 NaCl, at pH 7.2, 20°C, [tetrameric Hb] = 0.1 mM, in the absence and presence of 1 mM DPG. Data are from Scott et al. (1977). (C) Among carnivores, members of the suborder Feliformia (such as cats, hyenas, and civets) have Hbs with low-affinity/reduced DPG sensitivity compared with representatives of three other carnivore families (Mustelidae, Canidae, and Ursidae). Experimental conditions are the same as in B.

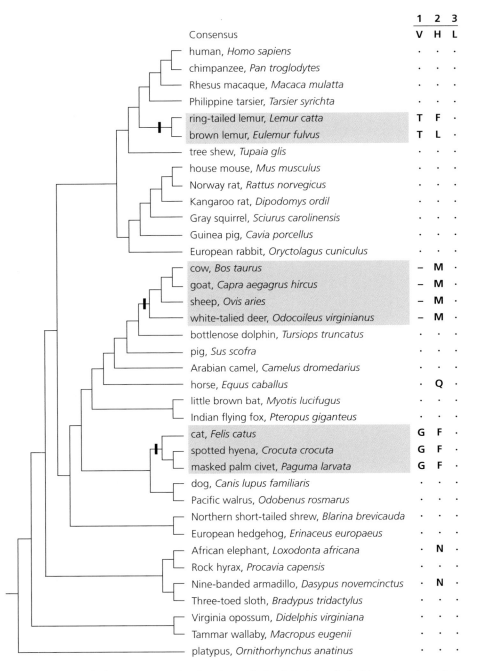

	1	2	3
Consensus	V	H	L
human, *Homo sapiens*	·	·	·
chimpanzee, *Pan troglodytes*	·	·	·
Rhesus macaque, *Macaca mulatta*	·	·	·
Philippine tarsier, *Tarsier syrichta*	·	·	·
ring-tailed lemur, *Lemur catta*	T	F	·
brown lemur, *Eulemur fulvus*	T	L	·
tree shew, *Tupaia glis*	·	·	·
house mouse, *Mus musculus*	·	·	·
Norway rat, *Rattus norvegicus*	·	·	·
Kangaroo rat, *Dipodomys ordil*	·	·	·
Gray squirrel, *Sciurus carolinensis*	·	·	·
Guinea pig, *Cavia porcellus*	·	·	·
European rabbit, *Oryctolagus cuniculus*	·	·	·
cow, *Bos taurus*	–	M	·
goat, *Capra aegagrus hircus*	–	M	·
sheep, *Ovis aries*	–	M	·
white-talied deer, *Odocoileus virginianus*	–	M	·
bottlenose dolphin, *Tursiops truncatus*	·	·	·
pig, *Sus scofra*	·	·	·
Arabian camel, *Camelus dromedarius*	·	·	·
horse, *Equus caballus*	·	Q	·
little brown bat, *Myotis lucifugus*	·	·	·
Indian flying fox, *Pteropus giganteus*	·	·	·
cat, *Felis catus*	G	F	·
spotted hyena, *Crocuta crocuta*	G	F	·
masked palm civet, *Paguma larvata*	G	F	·
dog, *Canis lupus familiaris*	·	·	·
Pacific walrus, *Odobenus rosmarus*	·	·	·
Northern short-tailed shrew, *Blarina brevicauda*	·	·	·
European hedgehog, *Erinaceus europaeus*	·	·	·
African elephant, *Loxodonta africana*	·	N	·
Rock hyrax, *Procavia capensis*	·	·	·
Nine-banded armadillo, *Dasypus novemcinctus*	·	N	·
Three-toed sloth, *Bradypus tridactylus*	·	·	·
Virginia opossum, *Didelphis virginiana*	·	·	·
Tammar wallaby, *Macropus eugenii*	·	·	·
platypus, *Ornithorhynchus anatinus*	·	·	·

Fig. 4.22. Phylogeny of mammals showing representatives from three eutherian clades, lemurs (order Primates), ruminants (order Cetartiodactyla), and feliform carnivores (order Carnivora), that have adult Hbs with unusually low sensitivities to DPG (see *Fig. 4.20*). Amino acid states at the first three N-terminal β-chain sites are shown for each species. The suppressed DPG sensitivities are invariably associated with substitutions or deletions at β1(NA1) and β2(NA2). In all three clades that possess Hbs with suppressed DPG sensitivities, the ancestral His β2(NA2) is replaced by large, hydrophobic residues at this position (Leu, Met, or Phe). The adult Hbs of all other mammals have hydrophilic residues at this site (His, Gln, or Asn). Hashmarks on the stem lineage of each DPG-insensitive clade indicate when the causative substitutions occurred. In the case of lemurs, it is clear that the His→Thr substitution at β1(NA1) occurred in the common ancestor of the group, but separate lineage-specific substitutions must have occurred at β1(NA2) as all examined species either have Phe or Leu at this site. For species that express two or more adult isoHbs with different β-chains, amino acid states are shown for one representative subunit isoform.

(A)

(B)

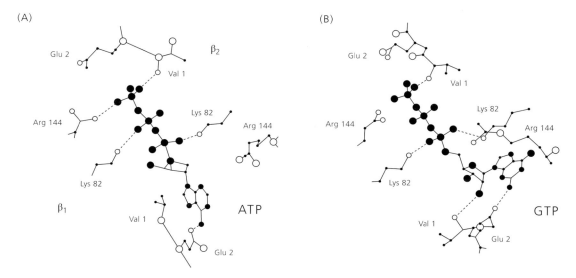

Fig. 4.23. Binding sites of ATP (A) and GTP (B) to adult Hbs of carp (*Cyprinus carpio*), a representative teleost fish.
Modified from Gronenborn et al. (1984).

In avian Hbs, IPP appears to bind at the same sites that DPG binds in human Hb, except that Hisβ141(H21) is replaced by Arg. Additionally, two basic residues in the central cavity of avian Hb, Arg β135(H13) and His β139(H17), do not directly contact bound IPP, but they clearly help to neutralize the polyphosphate's negative charges, thereby contributing indirectly to its favorable free energy of electrostatic binding (Perutz, 1983). In fish Hbs, the amino acids involved in the binding of ATP include the Val β1(NA1) of one β-chain, Glu/Asp β2(NA2) and Arg β144(H22) of the opposing β-chain, and Lys β82(EF6) of both chains (Fig. 4.23A). GTP binds to each of these five residues and forms an additional sixth H-bond with the other N-terminal Val (Gronenborn et al., 1984) (Fig. 4.23B), which explains why GTP generally exerts a more potent allosteric effect on the O_2 affinity of fish Hbs than ATP (Weber et al., 1976, Weber and Lykkeboe, 1978).

4.6.5 Interactions among allosteric effectors

In summary, H^+, CO_2, Cl^-, and organic phosphates reduce Hb-O_2 affinity and promote O_2 unloading by preferentially binding and stabilizing the T-state. H^+ ions protonate the free N-terminal residues and

surface histidines to a greater extent in ^{T}Hb than in ^{R}Hb, and Cl^- ions assist the protonation of key Bohr groups. Anions bind to ^{R}Hb far less readily because the positive charges are dispersed, thereby ensuring that Bohr protons are released when Hb binds O_2. Different allosteric ligands may also bind the same residues: H^+, CO_2, and Cl^- all bind the free NH_3^+ of the α-chain termini, H^+, CO_2, and organic phosphates bind the free NH_3^+ of the β-chain termini, and H^+, Cl^-, and organic phosphates bind the same charged residues in the central cavity. Table 4.7 summarizes the sign of interaction effects between different non-heme ligands on Hb-O_2 affinity. The Bohr effect is enhanced by the oxygenation-linked binding of Cl^- and organic phosphates, which favors the coupled uptake of protons by deoxyHb (Benesch et al., 1969, Bailey et al., 1970, Rollema et al., 1975, van Beek et al., 1979, van Beek and de Bruin, 1979, Imai, 1982), but is reduced by CO_2 binding, which causes protons to be released (Rossi-Bernardi and Roughton, 1967, Bauer, 1969). As stated by Lukin and Ho (2004): "[T]he Bohr effect depends on the intricate arrangement and interactions of all hydrogen and anion binding sites in the Hb molecule. It is an example of a global network of electrostatic interactions, rather than a few specific amino acid residues, playing a dominant role in an important physiological function."

The O_2-equilibrium curve of purified human Hb approximates the *in vivo* situation in the presence of 0.1 M Cl⁻, 1.2 moles DPG per mole Hb tetramer, 40 torr PCO_2, and an intracellular pH of 7.2 (Horvath et al., 1977, Imai, 1982, Mairbaurl and Weber, 2012). However, it is important to note that allosteric ligands do not have additive, independent effects on Hb-O_2 affinity *in vivo*. In addition to the antagonistic and synergistic binding interactions summarized in Table 4.7, the various allosteric ligands also affect Hb-O_2 affinity by indirectly modulating each other's relative concentrations in the red blood cell. For example, the reciprocal exchange of Cl⁻ and H^+ ions across the red cell membrane varies as a function of plasma pH, PCO_2, and bicarbonate concentration. Moreover, since organic phosphates are non-diffusible/non-transportable anions, changes in their intracellular concentration modulates Cl⁻/H^+ exchange (and, hence, intracellular pH), which in turn modulates Hb-O_2 affinity via the Bohr effect (Duhm, 1971, Duhm, 1976, Jensen, 2004). Thus, the oxygenation properties of blood reflect a complex and dynamic interplay between the intrinsic O_2-binding properties of Hb and its operating conditions in the chemical milieu of the red blood cell.

Having now completed our survey of Hb structure and function in Chapters 1–4, we will explore the evolution of oxygenation properties of Hb in the remaining chapters. We will start with the origin story of the vertebrate globin gene family.

Table 4.7 Interaction effects on Hb-O_2 affinity involving heme ligands (O_2) and non-heme ligands. Synergistic effects are denoted by a plus sign, antagonistic effects are denoted by a minus sign, and the absence of effect is denoted by a zero

	O_2	H^+	CO_2	DPG	Cl⁻
O_2	+	−	−	−	−
H^+		0	−	+	+
CO_2			0	−	−
DPG				0	−
Cl⁻					0

Modified from Imai (1982).

References

Abraham, D. J., Peascoe, R. A., Randad, R. S., and Panikker, J. (1992). X-ray diffraction study of di-ligated and tetra-ligated T-state hemoglobin from high salt crystals. *Journal of Molecular Biology*, **227**, 480–92.

Ackers, G. K., Dalessio, P. M., Lew, G. H., Daugherty, M. A., and Holt, J. M. (2002). Single residue modification of only one dimer within the hemoglobin tetramer reveals autonomous dimer function. *Proceedings of the National Academy of Sciences of the United States of America*, **99**, 9777–82.

Amiconi, G., Bertollini, A., Bellelli, A., et al. (1985). Evidence for two oxygen-linked binding sites for polyanions in dromedary hemoglobin. *European Journal of Biochemistry*, **150**, 387–93.

Antonini, E. and Brunori, M. (1971). *Hemoglobin and Myoglobin in their Reactions with Ligands*, Amsterdam, North-Holland Publishing Company.

Aranda, R., Cai, H., Worley, C. E., et al. (2009). Structural analysis of fish versus mammalian hemoglobins: effect of the heme pocket environment on autooxidation and hemin loss. *Proteins-Structure Function and Bioinformatics*, **75**, 217–30.

Arnone, A. (1972). X-ray-diffraction study of binding of 2,3-diphosphoglycerate to human deoxyhemoglobin. *Nature*, **237**, 146–9.

Arnone, A. (1974). X-ray studies of interaction of CO_2 with human deoxyhemoglobin. *Nature*, **247**, 143–5.

Arnone, A. and Perutz, M. F. (1974). Structure of inositol hexaphosphate-human deoxyhemoglobin complex. *Nature*, **249**, 34–6.

Bailey, J. E., Beetlestone, J. G., and Irvine, D. H. (1970). Reactivity differences between haemoglobins. XVII. The variability of the Bohr effect between species and the effect of 2,3-diphosphoglyceric acid on the Bohr effect. *Journal of the Chemical Society A*, **0**, 756–62.

Balakrishnan, G., Zhao, X., Podstawska, E., et al. (2009). Subunit-selective interrogation of CO recombination in carbonmonoxy hemoglobin by isotope-edited time-resolved resonance Raman spectroscopy. *Biochemistry*, **48**, 3120–6.

Baldwin, J. and Chothia, C. (1979). Haemoglobin: the structural changes related to ligand binding and its allosteric mechanism. *Journal of Molecular Biology*, **129**, 175–220.

Baldwin, J. M. (1980). Structure of human carbonmonoxy hemoglobin at 2.7 A resolution. *Journal of Molecular Biology*, **136**, 103–28.

Barra, D., Petruzzelli, R., Bossa, F., and Brunori, M. (1983). Primary structure of hemoglobin from trout (*Salmo irideus*). Amino acid sequence of the β-chain of trout Hb I. *Biochimica et Biophysica Acta*, **742**, 72–7.

Barrick, D., Ho, N. T., Simplaceanu, V., Dahlquist, F. W., and Ho, C. (1997). A test of the role of the proximal histidines in the Perutz model for cooperativity in haemoglobin. *Nature Structural Biology*, **4**, 78–83.

Barrick, D., Ho, N. T., Simplaceanu, V., and Ho, C. (2001). Distal ligand reactivity and quaternary structure studies of proximally detached hemoglobins. *Biochemistry*, **40**, 3780–95.

Bartlett, G. R. (1980). Phosphate compounds in vertebrate red blood cells. *American Zoologist*, **20,** 103–14.

Bauer, C. (1969). Antagonistic influence of CO_2 and 2,3-diphosphoglycerate on the Bohr effect of human haemoglobin. *Life Sciences*, **8**, 1041–6.

Bauer, C. (1970). Reduction of carbon dioxide affinity of human haemoglobin solutions by 2,3-diphosphoglycerate. *Respiration Physiology*, **10**, 10–19.

Bauer, C., Baumann, R., Engels, U., and Pacyna, B. (1975). Carbon dioxide affinity of various human hemoglobins. *Journal of Biological Chemistry*, **250**, 2173–6.

Bauer, C., Forster, M., Gros, G., and Vogel, D. (1981). Analysis of bicarbonate binding to crocodilian hemoglobin. *Journal of Biological Chemistry*, **256**, 8429–35.

Baumann, R., Bauer, C., and Haller, E. A. (1975). Oxygen-linked CO_2 transport in sheep blood. *American Journal of Physiology*, **229**, 334–9.

Baumann, R. and Haller, E. A. (1975). Cat hemoglobin A and hemoglobin B. Differences in interaction with Cl^-, phosphate and CO_2. *Biochemical and Biophysical Research Communications,* **65**, 220–7.

Benesch, R. and Benesch, R. E. (1967). The effect of organic phosphates from human erythrocytes on the allosteric properties of hemoglobin. *Biophysical Research Communications*, **26**, 162–7.

Benesch, R. E., Benesch, R., and Yu, C. I. (1969). Oxygenation of hemoglobin in the presence of 2,3-diphosphoglycerate. Effect of temperature, pH, ionic strength, and hemoglobin concentration. *Biochemistry*, **8**, 2567–71.

Berenbrink, M. (2006). Evolution of vertebrate haemoglobins: histidine side chains, specific buffer value and Bohr effect. *Respiratory Physiology and Neurobiology*, **154**, 165–84.

Berenbrink, M., Koldkjaer, P., Kepp, O., and Cossins, A. R. (2005). Evolution of oxygen secretion in fishes and the emergence of a complex physiological system. *Science*, **307**, 1752–7.

Bettati, S., Mozzarelli, A., and Perutz, M. F. (1983). Allosteric mechanism of hemoglobin: rupture of salt bridges raises oxygen affinity of the T-structure. *Journal of Molecular Biology*, **281**, 581–5.

Birukou, I., Schweers, R. L., and Olson, J. S. (2010). Distal histidine stabilizes bound O_2 and acts as a gate for ligand entry in both subunits of adult human hemoglobin. *Journal of Biological Chemistry*, **285**, 8840–54.

Bohr, C., Hasselbalch, L., and Krogh, A. (1904). Ueber einen in biologischer beziehung wichtigen einfluss, den die kohlensaurespannung des blutes auf dessen sauerstoffbindung ubt. *Skand Arch Physiol*, **16**, 402–12.

Bonaventura, C., Arumugam, M., Cashon, R., Bonaventura, J., and Moopenn, W. F. (1994). Chloride masks the opposing positive charges in HbA and Hb Hinsdale (β139 Asn-Lys) that can modulate cooperativity as well as oxygen affinity. *Journal of Molecular Biology*, **239**, 561–8.

Bonaventura, C. and Bonaventura, J. (1980). Competition in oxygen-linked anion binding to normal and variant human hemoglobins. *Hemoglobin*, **4**, 275–89.

Bonaventura, C., Sullivan, B., and Bonaventura, J. (1974). Effect of pH and anions on functional properties of hemoglobin from lemur, *Fulvus fulvus*. *Journal of Biological Chemistry*, **249**, 3768–75.

Brantley, R. E., Smerdon, S. J., Wilkinson, A. J., Singleton, E. W., and Olson, J. S. (1993). The mechanism of autooxidation of myoglobin. *Journal of Biological Chemistry*, **268**, 6995–7010.

Bringas, M., Petruk, A. A., Estrin, D. A., Capece, L., and Marti, M. A. (2017). Tertiary and quaternary structural basis of oxygen affinity in human hemoglobin as revealed by multiscale simulations. *Scientific Reports*, **7**, 10926.

Brunori, M., Falcioni, G., Fortuna, G., and Giardina, B. (1975). Effect of anions on oxygen binding properties of hemoglobin components from trout (*Salmo irideus*). *Archives of Biochemistry and Biophysics*, **168**, 512–19.

Bunn, H. F. (1971). Differences in interaction of 2,3-diphosphoglycerate with certain mammalian hemoglobins. *Science*, **172**, 1049–50.

Bunn, H. F. (1980). Regulation of hemoglobin-function in mammals. *American Zoologist*, **20**, 199–211.

Bunn, H. F. and Forget, B. G. (1986). *Hemoglobin: Molecular, Genetic and Clinical Aspects*, Philadelphia, PA, W. B. Saunders Company.

Busch, M. R. and Ho, C. (1990). Effects of anions on the molecular basis of the Bohr effect of hemoglobin. *Biophysical Chemistry*, **37**, 313–22.

Busch, M. R., Mace, J. E., Ho, N. T., and Ho, C. (1991). Roles of the β146 histidyl residue in the molecular basis of the Bohr effect of hemoglobin. A proton nuclear-magnetic resonance study. *Biochemistry*, **30**, 1865–77.

Camardella, L., Caruso, C., Davino, R., et al. (1992). Hemoglobin of the Antarctic fish *Pagothenia bernacchii*. Amino acid sequence, oxygen equilibria and crystal structure of its carbonmonoxy derivative. *Journal of Molecular Biology*, **224**, 449–60.

Campbell, K. L., Storz, J. F., Signore, A. V., et al. (2010). Molecular basis of a novel adaptation to hypoxic-hypercapnia in a strictly fossorial mole. *BMC Evolutionary Biology*, **10**, 214.

Cashon, R., Bonaventura, C., Bonaventura, J., and Focesi, A. (1986). The nicotinamide adenine dinucleotides as allosteric effectors of human hemoglobin. *Journal of Biological Chemistry*, **261**, 2700–5.

Chang, C. K., Simplaceanu, V., and Ho, C. (2002). Effects of amino acid substitutions at $\beta131$ on the structure and properties of hemoglobin: evidence for communication between $\alpha_1\beta_1$- and $\alpha_1\beta_2$-subunit interfaces. *Biochemistry*, **41**, 5644–55.

Chiancone, E., Norne, J. E., Forsen, S., Antonini, E., and Wyman, J. (1972). Nuclear magnetic resonance quadruple relaxation studies of chloride binding to human oxyhemoglobin and deoxyhemoglobin. *Journal of Molecular Biology*, **70**, 675–88.

Chothia, C., Wodak, S., and Janin, J. (1976). Role of subunit interfaces in allosteric mechanism of hemoglobin. *Proceedings of the National Academy of Sciences of the United States of America*, **73**, 3793–7.

Collman, J. P., Brauman, J. I., Halbert, T. R., and Suslick, K. S. (1976). Nature of O_2 and CO binding to metalloporphyrins and heme proteins. *Proceedings of the National Academy of Sciences of the United States of America*, **73**, 3333–7.

Dickerson, R. E. and Geis, I. (1983). *Hemoglobin: Structure, Function, Evolution, and Pathology*, Menlo Park, CA, Benjamin/Cummings.

Duhm, J. (1971). Effects of 2,3-diphosphoglycerate and other organic phosphate compounds on oxygen affinity and intracellular pH of human erythrocytes. *Pflügers Archiv-European Journal of Physiology*, **326**, 341–56.

Duhm, J. (1976). Dual effect of 2,3-diphosphoglycerate on Bohr effects of human blood. *Pflügers Archiv-European Journal of Physiology*, **363**, 55–60.

Eaton, W. A., Henry, E. R., Hofrichter, J., Bettati, S., Viappiani, C., and Mozzarelli, A. (2007). Evolution of allosteric models for hemoglobin. *IUBMB Life*, **59**, 586–99.

Eaton, W. A., Henry, E. R., Hofrichter, J., and Mozzarelli, A. (1999). Is cooperative oxygen binding by hemoglobin really understood? *Nature Structural Biology*, **6**, 351–8.

Fago, A., Carratore, V., Diprisco, G., et al. (1995). The cathodic hemoglobin of *Anguilla anguilla*—amino acid sequence and oxygen equilibria of a reverse Bohr effect hemoglobin with high oxygen-affinity and high phosphate sensitivity. *Journal of Biological Chemistry*, **270**, 18897–902.

Fan, J. S., Zheng, Y., Choy, W. Y., et al. (2013). Solution structure and dynamics of human hemoglobin in the carbonmonoxy form. *Biochemistry*, **52**, 5809–20.

Fang, T. Y., Zou, M., Simplaceanu, V., Ho, N. T., and Ho, C. (1999). Assessment of roles of surface histidyl residues in the molecular basis of the Bohr effect and of beta 143 histidine in the binding of 2,3-bisphosphoglycerate in human normal adult hemoglobin. *Biochemistry*, **38**, 13423–32.

Fermi, G. (1975). Three-dimensional Fourier synthesis of human deoxyhemoglobin at 2.5 A resolution. Refinement of atomic model. *Journal of Molecular Biology*, **97**, 237–56.

Fermi, G., Perutz, M. F., Shaanan, B., and Fourme, R. (1984). The crystal-structure of human deoxyhemoglobin at 1.74 A resolution. *Journal of Molecular Biology*, **175**, 159–74.

Ferry, G. 2007. *Max Perutz and the Secret of Life*, Cold Spring Harbor, NY, Cold Spring Harbor Laboratory Press.

Gelin, B. R. and Karplus, M. (1977). Mechanism of tertiary structural change in hemoglobin. *Proceedings of the National Academy of Sciences of the United States of America*, **74**, 801–5.

Gelin, B. R., Lee, A. W. M., and Karplus, M. (1983). Hemoglobin tertiary structural change on ligand-binding. Its role in the co-operative mechanism. *Journal of Molecular Biology*, **171**, 489–559.

Gong, Q. G., Simplaceanu, V., Lukin, J. A., et al. (2006). Quaternary structure of carbonmonoxyhemoglobins in solution: structural changes induced by the allosteric effector inositol hexaphosphate. *Biochemistry*, **45**, 5140–8.

Grispo, M. T., Natarajan, C., Projecto-Garcia, J., et al. (2012). Gene duplication and the evolution of hemoglobin isoform differentiation in birds. *Journal of Biological Chemistry*, **287**, 37647–58.

Gronenborn, A. M., Clore, G. M., Brunori, M., et al. (1984). Stereochemistry of ATP and GTP bound to fish haemoglobins. A transferred nuclear overhauser enhancement, ^{31}P-Nuclear Magnetic Resonance, oxygen equilibrium and molecular modelling study. *Journal of Molecular Biology*, **178**, 731–42.

Guesnon, P., Poyart, C., Bursaux, E., and Bohn, B. (1979). Binding of lactate and chloride ions to human adult hemoglobin. *Respiration Physiology*, **38**, 115–29.

Hamilton, M. N. and Edelstein, S. J. (1974). Cat hemoglobin—pH-dependence of cooperativity and ligand-binding. *Journal of Biological Chemistry*, **249**, 1323–9.

Hazard, E. S. and Hutchison, V. H. (1982). Distribution of acid-soluble phosphates in the erythrocytes of selected speices of amphibians. *Comparative Biochemistry and Physiology A-Physiology*, **73**, 111–24.

Heidner, E. J., Ladner, R. C., and Perutz, M. F. (1976). Structure of horse carbonmonoxyhemoglobin. *Journal of Molecular Biology*, **104**, 707–22.

Ho, C. and Russu, I. M. (1987). How much do we know about the Bohr effect of hemoglobin? *Biochemistry*, **26**, 6299–305.

Horvath, S. M., Malenfant, A., Rossi, F., and Rossi-Bernardi, L. (1977). Oxygen-affinity of concentrated human hemoglobin solutions and human blood. *American Journal of Hematology*, **2**, 343–54.

Hume, D. (1779). *Dialogues Concerning Natural Religion*. London.

Imai, K. (1982). *Allosteric Effects in Haemoglobin*, Cambridge, UK, Cambridge University Press.

Imai, K., Fushitani, K., Miyazaki, G., et al. (1991). Site-directed mutagenesis in hemoglobin—functional role of tyrosine 42(C7)α at the $\alpha_1\beta_2$ interface. *Journal of Molecular Biology*, **218**, 769–78.

Imai, K., Shih, D. T., Tame, J., Nagai, K., and Miyazaki, G. (1989). Structural and functional consequences of amino acid substitutions in hemoglobin as manifested in natural and artificial mutants. *Protein Sequences and Data Analysis*, **2**, 81–6.

Imaizumi, K., Imai, K., and Tyuma, I. (1979). Linkage between the four-step binding of oxygen and the binding of heterotropic anionic ligands in hemoglobin. *Journal of Biochemistry*, **86**, 1829–40.

Isaacks, R. E. and Harkness, D. R. (1980). Erythrocyte organic phosphates and hemoglobin function in birds, reptiles, and fishes. *American Zoologist*, **20**, 115–29.

Ito, N., Komiyama, N. H., and Fermi, G. (1995). Structure of deoxyhemoglobin of the Antarctic fish *Pagothenia bernacchii* with an analysis of the structural basis of the Root effect by comparison of the liganded and unliganded hemoglobin structures. *Journal of Molecular Biology*, **250**, 648–58.

Ivaldi, G., David, O., Paradossi, V., et al. (1999). Hb Bologna-St. Orsola beta 146(HC3)His→Tyr: a new high oxygen affinity variant with halved Bohr effect and highly reduced reactivity towards 2,3-diphosphoglycerate. *Hemoglobin*, **23**, 353–9.

Janecka, J. E., Nielsen, S. S. E., Andersen, S. D., et al. (2015). Genetically based low oxygen affinities of felid hemoglobins: lack of biochemical adaptation to high-altitude hypoxia in the snow leopard. *Journal of Experimental Biology*, **218**, 2402–9.

Janin, J. and Wodak, S. J. (1985). Reaction pathway for the quaternary structure change in hemoglobin. *Biopolymers*, **24**, 509–26.

Jensen, F. B. (1989). Hydrogen ion equilibria in fish hemoglobins. *Journal of Experimental Biology*, **143**, 225–34.

Jensen, F. B. (2001). Hydrogen ion binding properties of tuna haemoglobins. *Comparative Biochemistry and Physiology A-Molecular and Integrative Physiology*, **129**, 511–7.

Jensen, F. B. (2004). Red blood cell pH, the Bohr effect, and other oxygenation-linked phenomena in blood O_2 and CO_2 transport. *Acta Physiologica Scandinavica*, **182**, 215–27.

Jensen, F. B. and Weber, R. E. (1985). Proton and oxygen equilibria, their anion sensitivities and interrelationships in tench hemoglobin. *Molecular Physiology*, **7**, 41–50.

Jones, E. M., Monza, E., Balakrishnan, G., et al. (2014). Differential control of heme reactivity in α and β subunits of hemoglobin: a combined Raman spectroscopic and computational study. *Journal of the American Chemical Society*, **136**, 10325–39.

Kendrew, J. C., Bodo, G., Dintzis, H. M., et al. (1958). Three-dimensional model of the myoglobin molecule obtained by X-ray analysis. *Nature*, **181**, 662–6.

Kendrew, J. C., Dickerson, R. E., Strandberg, B. E., et al. (1960). Structure of myoglobin—three-dimensional Fourier synthesis at 2 A resolution. *Nature*, **185**, 422–7.

Kilmartin, J. V. (1977). Bohr effect of human hemoglobin. *Trends in Biochemical Sciences*, 2, 247–9.

Kilmartin, J. V., Fogg, J. H., and Perutz, M. F. (1980). Role of C-terminal histidine in the alkaline Bohr effect of human hemoglobin. *Biochemistry*, **19**, 3189–93.

Kilmartin, J. V., Imai, K., Jones, R. T., et al. (1978). Role of Bohr group salt bridges in cooperativity in hemoglobin. *Biochimica Et Biophysica Acta*, **534**, 15–25.

Kilmartin, J. V. and Rossi-Bernardi, L. (1969). Inhibition of CO_2 combination and reduction of Bohr effect in haemoglobin chemically modified at its alpha-amino groups. *Nature*, **222**, 1243–6.

Kilmartin, J. V. and Rossi-Bernardi, L. (1973). Interaction of hemoglobin with hydrogen ions, carbon dioxide, and organic phosphates. *Physiological Reviews*, **53**, 836–90.

Kilmartin, J. V. and Wootton, J. F. (1970). Inhibition of Bohr effect after removal of C-terminal histidines from haemoglobin beta-chains. *Nature*, **228**, 766–7.

Koshland, D. E., Nemethy, G., and Filmer, D. (1966). Comparison of experimental binding data and theoretical models in proteins containing subunits. *Biochemistry*, **5**, 365–85.

Laberge, M., Kovesi, I., Yonetani, T., and Fidy, J. (2005). R-state hemoglobin bound to heterotropic effectors: models of the DPG, IHP and RSR13 binding sites. *FEBS Letters*, **579**, 627–32.

Ladner, R. C., Heidner, E. J., and Perutz, M. F. (1977). Structure of horse methemoglobin at 2.0 Å resolution. *Journal of Molecular Biology*, **114**, 385–414.

Lecomte, J. T. J., Vuletich, D. A., and Lesk, A. M. (2005). Structural divergence and distant relationships in proteins: evolution of the globins. *Current Opinion in Structural Biology*, **15**, 290–301.

Lee, A. W. M. and Karplus, M. (1983). Structure-specific model of hemoglobin cooperativity. *Proceedings of the National Academy of Sciences of the United States of America-Biological Sciences*, **80**, 7055–9.

Lee, A. W. M. and Karplus, M. (1988). Analysis of proton release in oxygen binding by hemoglobin. Implications for the cooperative mechanism. *Biochemistry*, **27**, 1285–301.

Lesk, A. M. and Chothia, C. (1980). How different amino acid sequences determine similar protein structures—structure and evolutionary dynamics of the globins. *Journal of Molecular Biology*, **136**, 225–70.

Lesk, A. M., Janin, J., Wodak, S., and Chothia, C. (1985). Hemoglobin—the surface buried between the $\alpha_1\beta_1$ and

$\alpha_2\beta_2$ dimers in the deoxy and oxy structures. *Journal of Molecular Biology*, **183**, 267–70.

Liddington, R., Derewenda, Z., Dodson, G., and Harris, D. (1988). Structure of the liganded T-state of hemoglobin identifies the origin of cooperative oxygen binding. *Nature*, **331**, 725–8.

Liong, E. C., Dou, Y., Scott, E. E., Olson, J. S., and Phillips, G. N. (2001). Waterproofing the heme pocket—role of proximal amino acid side chains in preventing hemin loss from myoglobin. *Journal of Biological Chemistry*, **276**, 9093–100.

Lukin, J. A. and Ho, C. (2004). The structure-function relationship of hemoglobin in solution at atomic resolution. *Chemical Reviews*, **104**, 1219–30.

Lukin, J. A., Kontaxis, G., Simplaceanu, V., et al. (2003). Quaternary structure of hemoglobin in solution. *Proceedings of the National Academy of Sciences of the United States of America*, **100**, 517–20.

Lukin, J. A., Simplaceanu, V., Zou, M., Ho, N. T., and Ho, C. (2000). NMR reveals hydrogen bonds between oxygen and distal histidines in oxyhemoglobin. *Proceedings of the National Academy of Sciences of the United States of America*, **97**, 10354–8.

Mairbaurl, H. and Weber, R. E. (2012). Oxygen transport by hemoglobin. *Comprehensive Physiology*, **2**, 1463–89.

Margaria, R. and Green, A. A. (1933). The first dissociation constant, pK '(1), of carbonic acid in hemoglobin solutions and its relation to the existence of a combination of hemoglobin with carbon dioxide. *Journal of Biological Chemistry*, **102**, 611–34.

Maruyama, T., Watt, K. W. K., and Riggs, A. (1980). Hemoglobins of the tadpole of the bullfrog, *Rana catesbeiana*. Amino acid sequence of the alpha chain of a major component. *Journal of Biological Chemistry*, **255**, 3285–93.

Mathews, A. J. and Olson, J. S. (1994). Assignment of rate constants for O_2 and CO binding to α-subunit and β-subunit within R-state and T-state human hemoglobin. *Methods in Enzymology*, **232**, 363–86.

Mathews, A. J., Olson, J. S., Renaud, J. P., Tame, J., and Nagai, K. (1991). The assignment of carbon monoxide association rate constants to the α subunit and β subunit in native and mutant human deoxyhemoglobin tetramers. *Journal of Biological Chemistry*, **266**, 21631–9.

Maxwell, J. C. and Caughey, W. S. (1976). Infrared study of NO bonding to heme B and hemoglobin A. Evidence for inositol hexaphosphate induced cleavage of proximal histidine-to-iron bonds. *Biochemistry*, **15**, 388–96.

Mihailescu, M. R. and Russu, I. M. (2001). A signature of the T->R transition in human hemoglobin. *Proceedings of the National Academy of Sciences of the United States of America*, **98**, 3773–7.

Miyazaki, G., Morimoto, H., Yun, K. M., et al. (1999). Magnesium(II) and zinc(II)-protoporphyrin IX's stabilize the lowest oxygen affinity state of human hemoglobin even more strongly than deoxyheme. *Journal of Molecular Biology*, **292**, 1121–36.

Monod, J., Wyman, J., and Changeux, J. P. (1965). On the nature of allosteric transitions—a plausible model. *Journal of Molecular Biology*, **12**, 88–118.

Nagai, K., Luisi, B., Shih, D., et al. (1987). Distal residues in the oxygen binding site of hemoglobin studied by protein engineering. *Nature*, **329**, 858–60.

Nielsen, M. S. and Weber, R. E. (2007). Antagonistic interaction between oxygenation-linked lactate and CO_2 binding to human hemoglobin. *Comparative Biochemistry and Physiology A-Molecular and Integrative Physiology*, **146**, 429–34.

Nigen, A. M., Manning, J. M., and Alben, J. O. (1980). Oxygen-linked binding sites for inorganic anions to hemoglobin. *Journal of Biological Chemistry*, **255**, 5525–9.

O'Donnell, S., Mandaro, R., Schuster, T. M., and Arnone, A. (1979). X-ray diffraction and solutions studies of specifically carbamylated human hemoglobin A—evidence for the location of a proton-linked and oxygen-linked chloride binding site at valine 1. *Journal of Biological Chemistry*, **254**, 2204–8.

Olianas, A., Messana, I., Sanna, M. T., et al. (2005). Two sites for GTP binding in cathodic haemoglobins from Anguilliformes. *Comparative Biochemistry and Physiology B-Biochemistry & Molecular Biology*, **141**, 400–7.

Olson, J. S., Mathews, A. J., Rohlfs, R. J., et al. (1988). The role of the distal histidine in myoglobin and hemoglobin. *Nature*, **336**, 265–6.

Olson, J. S. and Phillips, G. N. (1997). Myoglobin discriminates between O_2, NO, and CO by electrostatic interactions with the bound ligand. *Journal of Biological Inorganic Chemistry*, **2**, 544–52.

Paoli, M., Dodson, G., Liddington, R. C., and Wilkinson, A. J. (1997). Tension in haemoglobin revealed by Fe-His(F8) bond rupture in the fully liganded T-state. *Journal of Molecular Biology*, **271**, 161–7.

Paoli, M., Liddington, R., Tame, J., Wilkinson, A., and Dodson, G. (1996). Crystal structure of T state haemoglobin with oxygen bound at all four haems. *Journal of Molecular Biology*, **256**, 775–92.

Park, C. M. (1970). Dimerization of deoxyhemoglobin and of oxyhemoglobin. Evidence for cleavage along the same plane. *Journal of Biological Chemistry*, **245**, 5390–4.

Park, S. Y., Yokoyama, T., Shibayama, N., Shiro, Y., and Tame, J. R. H. (2006). 1.25 angstrom resolution crystal structures of human haemoglobin in the oxy, deoxy and carbonmonoxy forms. *Journal of Molecular Biology*, **360**, 690–701.

Parkhurst, L. J., Goss, D. J., and Perutz, M. F. (1983). Kinetic and equilibrium studies on the role of the beta 147 histidine in the root effect and cooperativity in carp hemoglobin. *Biochemistry*, **22**, 5401–9.

Perrella, M., Kilmartin, J. V., Fogg, J., and Rossi-Bernardi, L. (1975). Identification of high and low affinity CO_2-binding sites of human hemoglobin. *Nature*, **256**, 759–61.

Perutz, M. F. (1965). Structure and function of haemoglobin. 1. A tentative atomic model of horse oxyhaemoglobin. *Journal of Molecular Biology*, **13**, 646–68.

Perutz, M. F. (1970). Stereochemistry of cooperative effects in haemoglobin. *Nature*, **228**, 726–39.

Perutz, M. F. (1972). Nature of heme-heme interaction. *Nature*, **237**, 495–9.

Perutz, M. F. (1978). Hemoglobin structure and respiratory transport. *Scientific American*, **239**, 92–125.

Perutz, M. F. (1979). Regulation of oxygen-affinity of hemoglobin—influence of structure of the globin on the heme iron. *Annual Review of Biochemistry*, **48**, 327–86.

Perutz, M. F. (1983). Species adaptation in a protein molecule. *Molecular Biology and Evolution*, **1**, 1–28.

Perutz, M. F. (1989a). Mechanisms of cooperativity and allosteric regulation in proteins. *Quarterly Reviews of Biophysics*, **22**, 139–236.

Perutz, M. F. (1989b). Myoglobin and hemoglobin—role of distal residues in reactions with heme ligands. *Trends in Biochemical Sciences*, **14**, 42–4.

Perutz, M. F. (1990). Mechanisms regulating the reactions of human hemoglobin with oxygen and carbon monoxide. *Annual Review of Physiology*, **52**, 1–25.

Perutz, M. F., Fermi, G., Luisi, B., Shaanan, B., and Liddington, R. C. (1987). Stereochemistry of cooperative mechanisms in hemoglobin. *Cold Spring Harbor Symposia on Quantitative Biology*, **52**, 555–65.

Perutz, M. F., Fermi, G., Poyart, C., Pagnier, J., and Kister, J. (1993). A novel allosteric mechanism in hemoglobin. Structure of bovine deoxyhemoglobin, absence of specific chloride-binding sites and origin of the chloride-linked Bohr effect in bovine and human hemoglobin. *Journal of Molecular Biology*, **233**, 536–45.

Perutz, M. F., Fermi, G., and Shih, T. B. (1984). Structure of deoxyhemoglobin Cowtown His HC3(146)β Leu. Origin of the alkaline Bohr effect and electrostatic interactions in hemoglobin. *Proceedings of the National Academy of Sciences of the United States of America*, **81**, 4781–4.

Perutz, M. F. and Imai, K. (1980). Regulation of oxygen-affinity of mammalian hemoglobins. *Journal of Molecular Biology*, **136**, 183–91.

Perutz, M. F., Kendrew, J. C., and Watson, H. C. (1965). Structure and function of haemoglobin. 2. Some relations between polypeptide chain configuration and amino acid sequence. *Journal of Molecular Biology*, **13**, 669–78.

Perutz, M. F., Kilmartin, J. V., Nagai, K., Szabo, A., and Simon, S. R. (1976). Influence of globin structures on state of heme ferrous low-spin derivatives. *Biochemistry*, **15**, 378–87.

Perutz, M. F., Kilmartin, J. V., Nishikura, K., et al. (1980). Identification of residues contributing to the Bohr effect of human hemoglobin. *Journal of Molecular Biology*, **138**, 649–70.

Perutz, M. F., Muirhead, H., Cox, J. M., and Goaman, L. C. G. (1968). Three-dimensional Fourier synthesis of horse oxyhaemoglobin at 2.8 A resolution—atomic model. *Nature*, **219**, 131–9.

Perutz, M. F., Muirhead, H., Mazzarella, L., et al. (1969). Identification of residues responsible for alkaline Bohr effect in haemoglobin. *Nature*, **222**, 1240–3.

Perutz, M. F., Rossmann, M. G., Cullis, A. F., et al. (1960). Structure of haemoglobin—three-dimensional Fourier synthesis at 5.5 A resolution, obtained by X-ray analysis. *Nature*, **185**, 416–22.

Perutz, M. F., Shih, D. T. B., and Williamson, D. (1994). The chloride effect in human hemoglobin. A new kind of allosteric mechanism. *Journal of Molecular Biology*, **239**, 555–60.

Perutz, M. F. and Ten Eyck, L. F. (1971). Stereochemistry of cooperative effects in hemoglobin. *Cold Spring Harbor Symposia on Quantitative Biology*, **36**, 295–310.

Perutz, M. F., Wilkinson, A. J., Paoli, M., and Dodson, G. G. (1998). The stereochemical mechanism of the cooperative effects in hemoglobin revisited. *Annual Review of Biophysics and Biomolecular Structure*, **27**, 1–34.

Pettigrew, D. W., Romeo, P. H., Tsapis, A., et al. (1982). Probing the energetics of proteins through structural perturbation. Sites of regulatory energy in human hemoglobin. *Proceedings of the National Academy of Sciences of the United States of America*, **79**, 1849–53.

Phillips, S. E. V. (1980). Structure and refinement of oxymyoglobin at 1.6 A resolution. *Journal of Molecular Biology*, **142**, 531–54.

Phillips, S. E. V. and Schoenborn, B. P. (1981). Neutron diffraction reveals oxygen-histidine hydrogen bond in oxymyoglobin. *Nature*, **292**, 81–2.

Rapoport, S. and Guest, G. M. (1941). Distribution of acid-soluble phosphorus in the blood cells of various vertebrates. *Journal of Biological Chemistry*, **138**, 269–82.

Riccio, A., Tamburrini, M., Giardina, B., and Di Prisco, G. (2001). Molecular dynamics analysis of a second phosphate site in the hemoglobins of the seabird, south polar skua. Is there a site-site migratory mechanism along the central cavity? *Biophysical Journal*, **81**, 1938–46.

Richard, V., Dodson, G. G., and Mauguen, Y. (1993). Human deoxyhemoglobin-2,3-diphosphoglycerate complex low-salt structure at 2.5 Angstrom resolution. *Journal of Molecular Biology*, **233**, 270–4.

Riggs, A. F. (1988). The Bohr effect. *Annual Review of Physiology*, **50**, 181–204.

Rollema, H. S., Debruin, S. H., Janssen, L. H. M., and Vanos, G. A. J. (1975). Effect of potassium-chloride on

Bohr effect of human hemoglobin. *Journal of Biological Chemistry*, **250**, 1333–9.

Rosemeyer, M. A. and Huehns, E. R. (1967). On the mechanism of dissociation of haemoglobin. *Journal of Molecular Biology*, **25**, 253–73.

Rossi-Bernardi, L. and Roughton, F. J. (1967). Specific influence of carbon dioxide and carbamate compounds on buffer power and Bohr effects in human haemoglobin solutions. *Journal of Physiology-London*, **189**, 1–29.

Sahu, S. C., Simplaceanu, V., Gong, Q., et al. (2007). Insights into the solution structure of human deoxy-hemoglobin in the absence and presence of an allosteric effector. *Biochemistry*, **46**, 9973–80.

Sahu, S. C., Simplaceanu, V., Gong, Q. G., et al. (2006). Orientation of deoxyhemoglobin at high magnetic fields: structural insights from RDCs in solution. *Journal of the American Chemical Society*, **128**, 6290–1.

Scott, A. F., Bunn, H. F., and Brush, A. H. (1977). Phylogenetic distribution of red cell 2,3-diphosphoglyc-erate and its interaction with mammalian hemoglobins. *Journal of Experimental Zoology*, **201**, 269–88.

Shaanan, B. (1983). Structure of human oxyhaemoglobin at 2.1 A resolution. *Journal of Molecular Biology*, **171**, 31–59.

Shibayama, N., Ohki, M., Tame, J. R. H., and Park, S. Y. (2017). Direct observation of conformational population shifts in crystalline human hemoglobin. *Journal of Biological Chemistry*, **292**, 18258–69.

Shibayama, N., Sugiyama, K., Tame, J. R. H., and Park, S. Y. (2014). Capturing the hemoglobin allosteric transition in a single crystal form. *Journal of the American Chemical Society*, **136**, 5097–105.

Shih, D. T. B., Luisi, B. F., Miyazaki, G., Perutz, M. F., and Nagai, K. (1993). A mutagenic study of the allosteric linkage of His(HC3)146β in hemoglobin. *Journal of Molecular Biology*, **230**, 1291–6.

Shih, T. B., Jones, R. T., Bonaventura, J., Bonaventura, C., and Schneider, R. G. (1984). Involvement of His HC3(146)β in the Bohr effect of human hemoglobin. Studies of native and N-ethylmaleimide-treated hemoglobin A and hemoglobin Cowtown (β146His→Leu). *Journal of Biological Chemistry*, **259**, 967–74.

Silva, M. M., Rogers, P. H., and Arnone, A. (1992). A third quaternary structure of human hemoglobin A at 1.7 A resolution. *Journal of Biological Chemistry*, **267**, 17248–56.

Springer, B. A., Egeberg, K. D., Sligar, S. G., et al. (1989). Discrimination between oxygen and carbon monoxide and inhibition of autoxidation by myoglobin—site-directed mutagenesis of the distal histidine. *Journal of Biological Chemistry*, **264**, 3057–60.

Springer, B. A., Sligar, S. G., Olson, J. S., and Phillips, G. N. (1994). Mechanisms of ligand recognition in myoglobin. *Chemical Reviews*, **94**, 699–714.

Steinberg, M. H. and Nagel, R. L. (2009). Unstable hemo-globins, hemoglobins with altered oxygen affinity, Hemoglobin M, and other variants of clinical and biological interest. *In:* Steinberg, M. H., Forget, B. G., Higgs, D. R., and Weatherhall, D. J. (eds.) *Disorders of Hemoglobin: Genetics, Pathophysiology, and Clinical Management*, 2nd edition. Cambridge, UK, Cambridge University Press.

Storz, J. F. and Moriyama, H. (2008). Mechanisms of hemoglobin adaptation to high-altitude hypoxia. *High Altitude Medicine & Biology*, **9**, 148–57.

Sun, D. Z. P., Zou, M., Ho, N. T., and Ho, C. (1997). Contribution of surface histidyl residues in the α-chain to the Bohr effect of human normal adult hemoglobin: roles of global electrostatic effects. *Biochemistry*, **36**, 6663–73.

Szabo, A. and Karplus, M. (1972). Mathematical model for structure-function relations in hemoglobin. *Journal of Molecular Biology*, **72**, 163–97.

Takano, T. (1977). Structure of myoglobin refined at 2.0 A resolution. 2. Structure of deoxymyoglobin from sperm whale. *Journal of Molecular Biology*, **110**, 569–84.

Taketa, F., Mauk, A. G., and Lessard, J. L. (1971). β-chain amino termini of cat hemoglobins and response to 2,3-diphosphoglycerate and adenosine triphosphate. *Journal of Biological Chemistry*, **246**, 4471–6.

Tamburrini, M., Riccio, A., Romano, M., Giardina, B., and Di Prisco, G. (2000). Structural and functional analysis of the two haemoglobins of the Antarctic seabird Catharacta maccormicki—characterization of an additional phos-phate binding site by molecular modelling. *European Journal of Biochemistry*, **267**, 6089–98.

Tame, J., Shih, D. T. B., Pagnier, J., Fermi, G., and Nagai, K. (1991). Functional role of the distal valine (E11) residue of α subunits in human hemoglobin. *Journal of Molecular Biology*, **218**, 761–7.

Tame, J. R. H. (1999). What is the true structure of liganded haemoglobin? *Trends in Biochemical Sciences*, **24**, 372–7.

Tame, J. R. H., Wilson, J. C., and Weber, R. E. (1996). The crystal structures of trout Hb I in the deoxy and car-bonmonoxy forms. *Journal of Molecular Biology*, **259**, 749–60.

Unzai, S., Eich, R., Shibayama, N., Olson, J. S., and Morimoto, H. (1998). Rate constants for O_2 and CO binding to the α and β subunits within the R and T states of human hemoglobin. *Journal of Biological Chemistry*, **273**, 23150–9.

Van Beek, G. G. M. and De Bruin, S. H. (1979). The pH-dependence of the binding of D-glycerate 2,3-bisphos-phate to deoxyhemoglobin and oxyhemoglobin. *European Journal of Biochemistry*, **100**, 497–502.

Van Beek, G. G. M. and De Bruin, S. H. (1980). Identification of the residues involved in the oxygen-linked chloride ion binding sites in human deoxyhemoglobin and oxyhemoglobin. *European Journal of Biochemistry*, **105**, 353–60.

Van Beek, G. G. M., Zuiderweg, E. R. P., and De Bruin, S. H. (1979). The binding of chloride ions to ligated and unligated human hemoglobin and its influence on the Bohr effect. *European Journal of Biochemistry*, **99**, 379–83.

Wajcman, H. and Galacteros, F. (2005). Hemoglobins with high oxygen affinity leading to erythrocytosis. New variants and new concepts. *Hemoglobin*, **29**, 91–106.

Watt, K. W. K., Maruyama, T., and Riggs, A. (1980). Hemoglobins of the tadpole of the bullfrog, *Rana catesbeiana*. Amino acid sequence of the beta chain of a major component. *Journal of Biological Chemistry*, **255**, 3294–301.

Watt, K. W. K. and Riggs, A. (1975). Hemoglobins of the tadpole of the bullfrog, *Rana catesbeiana*. Structure and function of isolated components. *Journal of Biological Chemistry*, **250**, 5934–44.

Weber, R. E., Campbell, K. L., Fago, A., Malte, H., and Jensen, F. B. (2010). ATP-induced temperature independence of hemoglobin-O_2 affinity in heterothermic billfish. *Journal of Experimental Biology*, **213**, 1579–85.

Weber, R. E., Fago, A., Malte, H., Storz, J. F., and Gorr, T. A. (2013). Lack of conventional oxygen-linked proton and anion binding sites does not impair allosteric regulation of oxygen binding in dwarf caiman hemoglobin. *American Journal of Physiology-Regulatory Integrative and Comparative Physiology*, **305**, R300–12.

Weber, R. E. and Jensen, F. B. (1988). Functional adaptations in hemoglobins from ectothermic vertebrates. *Annual Review of Physiology*, **50**, 161–79.

Weber, R. E. and Lykkeboe, G. (1978). Respiratory adaptations in carp blood. Influences of hypoxia, red cell organic phosphates, divalent cations and CO_2 on hemoglobin-oxygen affinity. *Journal of Comparative Physiology B*, **128**, 127–37.

Weber, R. E., Lykkeboe, G., and Johansen, K. (1976). Physiological properties of eel hemoglobin—hypoxic acclimation, phosphate effects and multiplicity. *Journal of Experimental Biology*, **64**, 75–88.

Weber, R. E., Ostojic, H., Fago, A., et al. (2002). Novel mechanism for high-altitude adaptation in hemoglobin of the Andean frog Telmatobius peruvianus. *American Journal of Physiology-Regulatory Integrative and Comparative Physiology*, **283**, R1052–60.

Weber, R. E., Wells, R. M. G., and Rossetti, J. E. (1983). Allosteric interactions governing oxygen equilibria in the hemoglobin system of the spiny dogfish, *Squalus acanthias*. *Journal of Experimental Biology*, **103**, 109–20.

Wyman, J. and Allen, D. W. (1951). The problem of the heme interactions in hemoglobin and the basis of the Bohr effect. *Journal of Polymer Science*, **7**, 499–518.

Yokoyama, T., Chong, K. T., Miyazaki, G., et al. (2004). Novel mechanisms of pH sensitivity in tuna hemoglobin. A structural explanation of the root effect. *Journal of Biological Chemistry*, **279**, 28632–40.

Yuan, Y., Simplaceanu, V., Ho, N. T., and Ho, C. (2010). An investigation of the distal histidyl hydrogen bonds in oxyhemoglobin: effects of temperature, pH, and inositol hexaphosphate. *Biochemistry*, **49**, 10606–15.

Yuan, Y., Tam, M. F., Simplaceanu, V., and Ho, C. (2015). New look at hemoglobin allostery. *Chemical Reviews*, **115**, 1702–24.

Zheng, G., Schaefer, M., and Karplus, M. (2013). Hemoglobin Bohr effects: atomic origin of the histidine residue contributions. *Biochemistry*, **52**, 8539–55.

CHAPTER 5

Evolution of the vertebrate globin gene family

5.1 Gene duplication and the evolution of novel protein functions

Gene duplication is known to play an extremely important role in the evolution of new protein functions. As stated by Li (1997): "Gene duplication is the most important mechanism for generating new genes and new biochemical processes that have facilitated the evolution of complex organisms from primitive ones." Following the complete duplication of a protein-coding gene, functional redundancy between the two daughter copies will often entail a relaxation of selective constraints that permits the accumulation of degenerative mutations in one or both copies (Zhang, 2003, Lynch and Katju, 2004). In the majority of cases, one of the two gene duplicates will be rendered functionless by inactivating mutations. However, in a small minority of cases, the fixation of previously forbidden mutations may lead to the acquisition of a novel function and/or expression pattern in one copy or the other. In such cases, both duplicate copies may be selectively retained in the genome, and they can then evolve new functions or divide up ancestral functions.

The diversification of the vertebrate globin gene family provides an especially vivid illustration of the role of gene duplication in promoting evolutionary innovation. In this chapter we will explore several important case studies. To provide context, we will briefly explore the structural and functional diversity of globins in the three domains of life (Bacteria, Archaea, and Eukaryota). We will then

meet the less familiar members of the vertebrate globin gene family—enigmatic cousins of Hb and Mb—and we will unravel their duplicative origins and evolutionary history. We will then see how the proto hemoglobin (*Hb*) and myoglobin (*Mb*) genes originated via whole-genome duplication in the common ancestor of vertebrates. This duplication event facilitated a physiological division of labor between O_2-binding proteins with distinct roles in respiratory gas transport. Finally, we will see how comparative genomic analyses of vertebrate globins have shed light on mechanism and process in gene family evolution.

5.2 Phylogenetic insights into gene family evolution

Phylogenetic reconstructions permit inferences about the branching relationships among homologous members of a multigene family that have diversified via successive rounds of duplication and divergence. In comparisons among different species, phylogenetic reconstructions provide a means of distinguishing different types of homology. Specifically, the congruence or lack of congruence between a species tree and the gene tree contained within it enables us to distinguish "paralogous" genes, which trace their common ancestry to duplication events, and "orthologous" genes, which trace their common ancestry to speciation events (that is, they descend from a common ancestral gene by phylogenetic splitting at the organismal level) (Fig. 5.1).

Hemoglobin: Insights into Protein Structure, Function, and Evolution. Jay F. Storz, Oxford University Press (2019).
© Jay F. Storz 2019. DOI: 10.1093/oso/9780198810681.001.0001

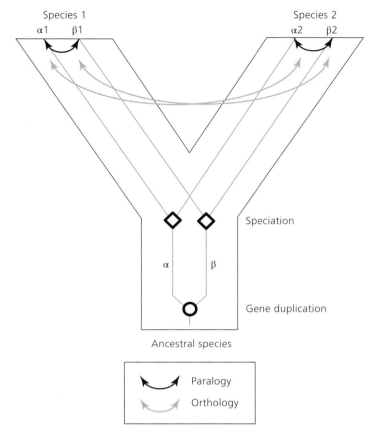

Fig. 5.1. Phylogenetic reconstructions reveal the branching relationships among members of a multigene family that have diversified via successive rounds of duplication and divergence. By drawing a distinction between the topologies of species trees and the gene tress contained within them, we can distinguish two types of homologous relationships between members of multigene families: paralogy and orthology. Whereas paralogous genes trace their common ancestry to duplication events, orthologous genes trace their common ancestry to speciation events.

Modified from Storz (2016).

The reconstruction of phylogenetic relationships among homologous members of a protein family is essential for understanding the pathways by which particular structural or functional properties evolved. For example, it is possible to reconstruct the history of evolutionary change in particular features of proteins by mapping observed character states onto the branch tips of a phylogenetic tree that depicts the genealogical history of the underlying genes. This provides a basis for inferring character states of internal nodes of the tree (ancestral states), and can therefore suggest when a particular property evolved by identifying the phylogenetic interval in which the ancestral character state transitioned to the derived state.

5.3 The ancient history of the globin superfamily

Globins are found in bacteria, archaea, and eukaryotes, indicating that the protein superfamily traces its roots to the last universal common ancestor. The structural unity of these diverse proteins is reflected in the globin fold, a diagnostic three-dimensional packing of α-helices that encapsulate the heme group. This folding pattern is highly conserved among homologous globins from all domains of life in spite of extensive divergence in amino acid sequence (Lesk and Chothia, 1980, Lecomte et al., 2005). Two structural subclasses of the globin fold are distinguished: the "3-on-3" fold (of which verte-brate Mbs and Hbs are canonical examples) and the "2-on-2" fold (Pesce et al., 2000, Wittenberg et al., 2002, Vinogradov et al., 2005, Vinogradov et al., 2013, Gell, 2018) (Plate 6). The 3-on-3 designation refers to the fold formed by the stacking of antiparallel A-G-H and B-E-F α-helices. The 2-on-2 designation refers to the arrangement of the B-E and G-E helical pairs. The 2-on-2 globins are also referred to as the truncated Hb (trHb) class, since the A, C, D, and F helices are reduced or absent.

Representatives of the 3-on-3 and 2-on-2 globin subclasses are found in all domains of life. Among the 3-on-3 globins, the family of "Mb-like" globins are found in bacteria and eukaryotes, but not archaea. Two subfamilies of 3-on-3 Mb-like globins, the flavo-hemoglobins and single-domain globins, are found exclusively in bacteria and some eukaryotes (algae and fungi). These globins serve as nitric oxide dioxy-genase (NOD) enzymes that protect against nitrosa-tive stress by converting NO to nitrate (Gardner, 2005, Gardner and Gardner, 2008). The flavohemoglobins are chimeric proteins that have an N-terminal glo-bin domain and a C-terminal FAD/NAD binding domain. The second main group of 3-on-3 globins, the sensor globins, are not found in eukaryotes. They typically function as sensors of O_2, NO, or CO (Martinkova et al., 2013). Similar to the flavohemo-globins, many of the 2-on-2 globins are chimeric proteins that combine a globin domain with a dis-tinct protein domain with a redox or signaling func-tion (Bonamore et al., 2007; Hade et al., 2017).

5.4 The diversity of animal globins and the repurposing of ancestral functions

In the animal kingdom (Metazoa), globin proteins display an underlying unity in tertiary structure that belies an extraordinary diversity in biochemical properties and physiological functions. Animal glob-ins are expressed in a diverse range of tissues and cell types including muscle and nerve cells as well as red blood cells, and in many invertebrates, extracel-lular globins are present in vascular, coelomic, and/ or perienteric fluids (Weber and Vinogradov, 2001). This diversity in anatomical and cellular sites of expression is associated with a correspondingly wide variety of quaternary structures. Animal globins function as single-domain monomers, single-domain multisubunit proteins, or multidomain proteins with 2–18 covalently linked globin domains per chain (Vinogradov, 1985, Riggs, 1991, Vinogradov et al., 1993, Terwilliger, 1998, Royer et al., 2001, Weber and Vinogradov, 2001, Royer et al., 2005). It is always the individual subunits of these oligomeric proteins that adopt the characteristic globin fold. Some of the most extraordinary elaborations of qua-ternary structure are found in the extracellular globins of invertebrates. The monomeric Mb and tetrameric Hb of jawed vertebrates (~17 kDa and ~64 kDa, respectively) are dwarfed by the multi-subunit "hexagonal bilayer" globins of annelid and vestimentiferan worms (~3,600 kDa), which are composed of 144 globin chains and 36 linker pep-tides (Lamy et al., 1996, Royer et al., 2006), and the multidomain globins of bivalve molluscs (~800–12,000 kDa) (Terwilliger and Terwilliger, 1978).

The diversity of primary and quaternary struc-tures displayed by animal globins underlies an equally rich diversity of function, ranging from reversible O_2 binding that is central to the familiar O_2-storage/transport functions of Mb and Hb, as well as functions related to intracellular O_2 transfer, O_2 sensing, scavenging of reactive O_2 and nitrogen species, redox signaling, and enzymatic functions involving nitric oxide (NO) oxygenase and reduc-tase activities (Weber and Vinogradov, 2001, Pesce et al., 2002, Fago et al., 2004, Hankeln et al., 2005, Kakar et al., 2010, Helbo et al., 2013, Burmester and Hankeln, 2014, Gell, 2018).

It is a safe bet that globins in the common ances-tor of eukaryotes performed functions that were not directly related to O_2 storage or transport (Hardison, 1998, Vinogradov et al., 2007, Vinogradov and Moens, 2008, Wajcman et al., 2009, Blank and Burmester, 2012, Koch and Burmester 2016). There are many unresolved questions about the ancient origins of globins, and their possible physiological functions during a time in Earth's history when the overwhelming majority of living organisms would have regarded O_2 as a lethal toxin. Hardison (1998) suggested a hypothesis about the co-option of hemoproteins for new functions in cellular respiration once the advent of photosynthesis produced a rising tide of O_2 in Earth's atmosphere:

"Given the capacity of oxygen to damage various cellular components, oxygen-binding hemoproteins may have functioned initially to protect cells from this reactive species. Once the utility of oxygen as an electron acceptor was realized in the evolution of respiratory chains, hemoproteins could serve as electron-transfer agents (leading to contemporary cytochromes) and oxygen-bound hemoproteins could serve as terminal electron acceptors . . . The intracellular oxygen-transport properties may have arisen from a need to scavenge scarce oxygen to provide it for the respiratory chain . . . "

The diverse functions of bacterial globins, such as enzymatic scavenging of toxic O_2 and NO (Forrester and Foster, 2012) and sensing and signaling in response to O_2, NO, or CO (Vinogradov et al., 2013), may provide the best clues as to the possible ancestral functions of animal globins. Given the deep roots of the globin superfamily, the physiological functions of Hb and other globins in contemporary animals likely represent repurposed modifications of distinct ancestral functions.

5.5 The full repertoire of vertebrate globins

During the last half century, Hb and Mb played leading roles in research efforts to understand relationships between protein structure and function. Since the dawn of the new millennium, investigations into the structure, function, and evolution of globin proteins have been reinvigorated by the discovery of new members of the globin protein superfamily in humans and other vertebrates (Hankeln et al., 2005, Hankeln and Burmester, 2008, Burmester and Hankeln, 2009, Burmester and Hankeln, 2014).

The canonical globin gene repertoire of jawed vertebrates comprises eight main members, each of which may be represented by multiple, structurally distinct paralogs with distinct expression domains that are specific to different tissues or cell types: neuroglobin (*Ngb*), cytoglobin (*Cygb*), androglobin (*Adgb*), globin X (*GbX*), globin Y (*GbY*), globin E (*GbE*), *Mb*, and the α- and β-type subunits of *Hb*. Whereas some cartilaginous fishes have lost *Ngb*, almost all bony vertebrates that have been examined to date possess copies of *Ngb*, *Cygb*, and *Adgb* (Burmester et al., 2000, Awenius et al., 2001, Burmester

et al., 2002, Trent and Hargrove, 2002, Burmester et al., 2004, Fuchs et al., 2004, Kugelstadt et al., 2004, Wystub et al., 2004, Fuchs et al., 2005, Hankeln et al., 2005, Roesner et al., 2005, Hankeln and Burmester, 2008, Burmester and Hankeln, 2009, Hoffmann et al., 2010a, Hoogewijs et al., 2012, Schwarze and Burmester, 2013, Burmester and Hankeln, 2014, Fabrizius et al., 2016). Within lobe-finned fishes, the coelacanth (*Latimera chalumnae*) possesses the full complement of globins that are found in other bony vertebrates (Schwarze and Burmester, 2013) but available evidence suggests that both *Ngb* and *Cygb* may have been secondarily lost in the West African lungfish, *Protopterus annectens* (Koch et al., 2016).

In globins such as Mb and Hb that play a role in O_2 storage and/or transport, the heme iron is coordinated to five ligands: the four coplanar pyrrole nitrogens of the porphyrin ring and the side chain of His F8 on the proximal side of the heme plane. The axial site on the distal side of the heme plane is therefore available for reversible ligand binding (Fig. 5.2A,B). A distinguishing feature of Ngb, Cygb, and GbX (and many other 3-on-3 eukaryotic globins) is that the heme iron is reversibly coordinated to an additional sixth ligand on the distal side of the heme plane: the imidazole side chain of the distal His E7 (Fig. 5.2C). Since the reversible binding of exogenous ligands such as O_2, NO, or CO have to compete with the His E7 side chain for the same axial site of the heme iron, "hexacoordinate" globins like Ngb, Cygb, and GbX typically have much higher ligand affinities than "pentacoordinate" globins like Mb and Hb (Kakar et al., 2010). The nature of heme coordination is a primary determinant of the ligand reactivities of globin proteins and can therefore provide hints about possible physiological functions (Kakar et al., 2010, Kiger et al., 2011).

The monomeric Ngb protein is mainly expressed in the neurons of the central and peripheral nervous system, and in some endocrine tissues, whereas the homodimeric Cygb protein is expressed in fibroblasts and related cell types and in distinct nerve cells in the central and peripheral nervous systems (Hankeln and Burmester, 2008, Burmester and Hankeln, 2009, Burmester and Hankeln, 2014, Fabrizius et al., 2016, Reuss et al., 2016a, Reuss et al., 2016b). Experiments with transgenic mice indicate that Ngb plays a

(A)   (B)   (C)

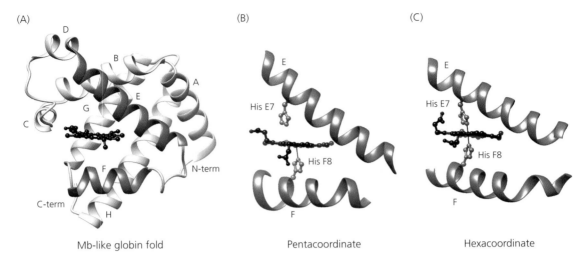

Mb-like globin fold   Pentacoordinate   Hexacoordinate

Fig. 5.2. Structures of representative pentacoordinate and hexacoordinate globins. (A) A representative Mb-like globin fold showing the position of the heme group between the E and F helices. (B) The structure of ferric (Fe^{3+}) sperm whale Mb (2MBW.pdb) shows a pentacoordinate heme iron with a ligand-accessible distal binding site. (C) The structure of hexacoordinate neuroglobin (1QIF.pdb) shows that the distal binding site of the heme iron is coordinated by the side chain of the distal histidine. The eight α-helices are labeled A through H, along with the N- and C-termini. In panels B and C, His E7 and His F8 are the distal and proximal histidines, respectively (see Chapter 4).

neuroprotective role during hypoxic or ischemic episodes, which suggests possible roles in signal transduction or the scavenging of oxygen/nitrogen radicals (Greenberg et al., 2008). Recent evidence suggests a possible NOD-like enzymatic function for Cygb, as it has been shown to regulate NO signaling and bioactivity in the vascular system (Halligan et al., 2009a, Halligan et al., 2009b, Liu et al., 2017, Jourd'heuil et al., 2017). The more recently discovered *Adgb* gene is especially enigmatic, as it is a chimeric fusion gene; the encoded protein has an N-terminal calpain-like domain, an internal globin domain that has undergone an internal shuffling of α-helical subdomains, and an IQ calmodulin-binding motif (Hoogewijs et al., 2012). In mammals, the *Adgb* gene is preferentially expressed in testis. The heme coordination chemistries and other structural features of Ngb, Cygb, and Adgb suggest that these globins may perform redox-regulated signaling functions or O$_2$-sensing functions that mediate oxygen-dependent protein activities, but the primary physiological functions of these globin proteins are still mostly shrouded in mystery (Greenberg et al., 2008, Mimura et al., 2010, Nishi et al., 2011, Oleksiewicz et al., 2011, Ascenzi et al., 2014, Burmester and Hankeln, 2014, Singh et al., 2014, Reeder, 2017).

In addition to the *Ngb*, *Cygb*, *Adgb*, *Hb*, and *Mb* genes that have been retained in all or most of the major lineages of bony vertebrates, several paralogous globins have been discovered that have far more restricted phyletic distributions. For example, one or more copies of *GbX* have been documented in lampreys and most examined bony vertebrates other than mammals and archosaurs (represented by birds and crocodilians) (Roesner et al., 2005, Fuchs et al., 2006, Blank et al., 2011b, Dröge and Makalowski, 2011, Blank and Burmester, 2012, Schwarze and Burmester, 2013, Schwarze et al., 2014, Opazo et al. 2015b). GbX of vertebrates is a monomeric, membrane-bound globin that possesses N-terminal acylation sites, which suggests possible roles in cellular signaling and/or protection against the oxidation of membrane lipids (Blank et al., 2011b, Blank and Burmester, 2012, Koch and Burmester, 2016). Intriguingly, GbX has also been shown to have a nitrite reductase activity in fish red blood cells (Corti et al., 2016). The *GbY* gene has been documented in the genomes of cartilaginous fishes, early-branching lineages of ray-finned fishes (represented by the spotted gar, *Lepisosteus oculatus*), lobe-finned fishes (represented by the coelacanth), and several tetrapod taxa such as lizards, turtles, crocodilians, and monotreme mammals (represented by the platypus, *Ornithorhynchus anatinus*), but the gene does not appear to be present in lampreys, teleost fishes, birds, or therian mammals (marsupials and eutherians) (Hoffmann et al., 2010a, Hoffmann et al., 2010b, Hoffmann et al., 2011, Schwarze and Burmester, 2013, Schwarze et al., 2014, Schwarze et al., 2015). In adult *Xenopus*, *GbY* is expressed in a broad range of tissue types (Fuchs et al., 2006), but its physiological function has yet to be elucidated. The *GbE* gene was initially found only in birds (Kugelstadt et al., 2004, Blank et al., 2011a, Hoffmann et al., 2010a, Hoffmann et al., 2011), but has since been found in the genomes of lobe-finned fishes and some reptiles (Schwarze and Burmester, 2013, Schwarze et al., 2015). The monomeric GbE protein appears to perform a Mb-like function in regulating oxygen supply to photoreceptor cells in the avascular retina (Blank et al., 2011a), although a role in regulating cellular redox homeostasis is also possible.

The physiological functions of Adgb, Ngb, GbX, and some of the vertebrate-specific globins such as Cygb, GbY, and GbE are not yet fully understood, but ongoing investigations into their structures, ligand reactivities, biochemical activities, and expression patterns are gradually yielding clues (Burmester and Hankeln, 2014).

5.6 Origins of vertebrate globins

Our understanding of the structural and functional diversity of vertebrate globins is enhanced by considering their ancient origins and genealogical relationships. Let us then trace the evolutionary history of globins back to their ancient roots. As with any historical analysis, when we delve deep enough into the past we eventually reach a point where the trail runs cold (e.g., when the sequences of interest have diverged to the point where they are no longer alignable). In the case of some vertebrate globins, we can confidently trace their duplicative origins back to the common ancestor of deuterostomes and protostomes (~700–900 million years ago) and, based on similarities of tertiary structure (Herman et al., 2014), we can readily discern ancient affinities between eukaryotic globins and their homologs in bacteria and archaea. However, inferences regarding the specific branch-

ing relationships among such anciently diverged proteins are glimpsed "through a glass, darkly."

In our historical survey of vertebrate globins, we will choose a branch point in the family tree (an internal node in the phylogeny) from which we can trace descendent gene lineages that gave rise to the set of globins possessed by contemporary vertebrates. The common ancestor of deuterostomes provides a good place to anchor our survey. From that ancestral starting point, we will then travel forward in time, zooming in on the set of chordate-specific globins (which are nested within the more inclusive set of deuterostome globins) and then the set of vertebrate-specific globins (nested within the chordate globins).

Comparative genomic analyses have revealed tremendous variability in the size and membership composition of the globin gene repertoires of deuterostomes, reflecting a complex history of lineage-specific gene duplications and deletions. The most recent common ancestor of deuterostomes appears to have possessed a repertoire of at least four distinct globin paralogs (a minimum estimate), and different subsets of this ancestral complement of genes have been retained in each of the descendent deuterostome phyla that survive to the present day. The phyletic distribution of globin genes among deuterostome taxa is clearly attributable to a winnowing of ancestral diversity such that different sets of structurally distinct paralogs have been retained in different lineages, and each of these paralogs seeded different lineage-specific expansions of functional diversity through repeated rounds of gene duplication and divergence (Hoffmann et al., 2012a). Globin gene repertoires diversified independently in each of the five main deuterostome lineages: (1) echinoderms (a phylum represented by starfish, sea urchins, and sea cucumbers); (2) hemichordates (a phylum represented by acorn worms and other worm-like marine organisms); (3) cephalochordates (a chordate subphylum represented by amphioxus, or lancelets); (4) tunicates (a chordate subphylum represented by sea squirts and other marine, filter-feeding invertebrates); and (5) vertebrates, the third chordate subphylum. An estimated phylogeny of deuterostome globins is shown in Fig. 5.3, and a diagrammatic summary is shown in Fig. 5.4.

The clade containing vertebrate *Ngb* shows an especially intriguing pattern. Orthologs of *Ngb* have

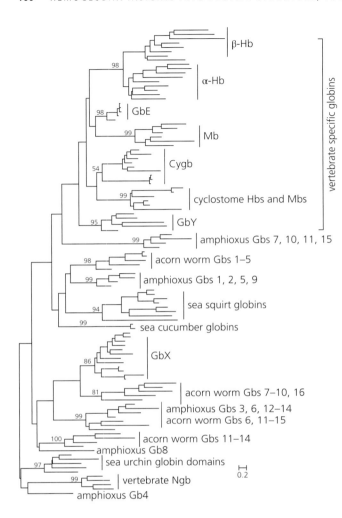

Fig. 5.3. Estimated phylogeny of deuterostome globins. Using an alignment of nucleotide sequences from a diverse set of deuterostome taxa, the phylogenetic tree was estimated in MEGA version 7 (Kumar et al., 2016) using the LG model of amino acid substitution. Among-site variation in substitution rate was modeled using a Γ distribution with five categories of sites. Support for the nodes was evaluated with 1,000 bootstrap pseudoreplicates.

been secondarily lost in tunicates (sea squirts) and hemichordates (acorn worms) and have been retained as single-copy genes in cephalochordates and in the majority of vertebrates (*Ngb* has been secondarily lost in cyclostomes and in at least some lineages of cartilaginous fishes) (Schwarze et al., 2014, Opazo et al., 2015b). Remarkably, the *Ngb* ortholog in sea urchin has undergone multiple rounds of internal domain duplication so that it encodes a polypeptide with sixteen covalently linked globin domains (Bailly and Vinogradov, 2008, Hoffmann et al., 2012a).

The same pattern of attrition and lineage-specific diversification is apparent if we zoom in on the set of globins specific to chordates (Fig. 5.5). In the absence of gene losses, each clade of orthologous

sequences should independently recapitulate the expected organismal phylogeny. Accordingly, a phylogeny of orthologous globin genes from representatives of the three chordate subphyla (Vertebrata, Tunicata, and Cephalochordata) would be expected to place the sea squirt globins sister to the vertebrate-specific globins, since tunicates and vertebrates share a more recent common ancestor with one another than with cephalochordates (Delsuc et al., 2006, Putnam et al., 2008; Cannon et al., 2016; Rouse et al., 2016). Contrary to this expectation, phylogenetic analyses consistently place a single clade of amphioxus globins ("amphioxus Gbs 7, 10, 11, and 15") sister to the vertebrate-specific globins, whereas the complete set of tunicate globins ("sea squirt Gbs 1–4") form a clade that is

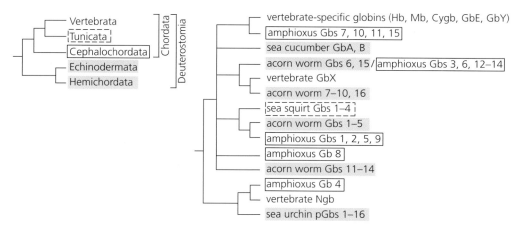

Fig. 5.4. Summary of phylogenetic relationships among globin genes from representative deuterostome taxa. The inset on top shows the organismal phylogeny.

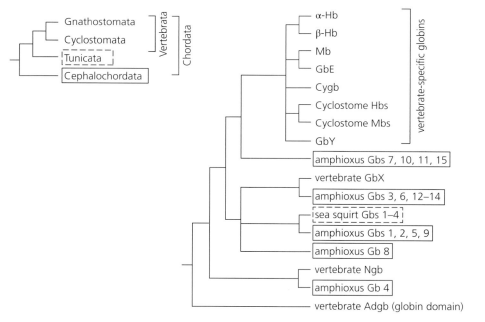

Fig. 5.5. Phylogeny of chordate globins. To understand the origins of vertebrate globins in evolutionary context, it is useful to examine their phylogenetic relationships with globins from representatives of the other two chordate subphyla: the sea squirt (*Ciona intestinalis*, a tunicate) and amphioxus (*Branchiostoma floridae*, a cephalochordate). Two of the vertebrate globin paralogs, GbX and Ngb, derive from independent duplication events that occurred prior to divergence between deuterostomes and protostomes. The remaining members of the vertebrate globin gene repertoire are all products of vertebrate-specific duplication events (Hoffmann et al. 2012b).

Based on data from Ebner et al. (2010), Storz et al. (2011b), Hoffmann et al. (2012a,b), and Opazo et al. (2015b).

nested within a diverse set of globins from amphioxus, echinoderms, and hemichordates (Figs. 5.3, 5.4). Thus, the topology of the globin tree is not congruent with the known organismal phylogeny due to lineage-specific gene duplications and deletions. The inferred phylogenetic relationships indicate that orthologs of *GbX*, *Ngb*, and the progenitor of all vertebrate-specific globins were secondarily lost from the sea squirt genome (it remains to be seen if this is true for all tunicates) (Storz et al., 2011; Hoffmann et al., 2012a; Opazo et al. 2015b) (Figs. 5.3–5.5). Likewise, orthologs of the sea squirt globins were lost in the stem lineage of vertebrates. The cephalochordate amphioxus possesses an especially diverse repertoire of globin genes, as sequences from this taxon are represented in each of the different globin gene lineages that trace back to the chordate common ancestor (Fig. 5.5).

Let us now zoom in on the repertoire of vertebrate globins. The *Ngb*, *Adgb*, and *GbX* genes originated prior to the split between protostomes and deuterostomes, and were inherited by both descendent lineages (Burmester et al., 2000, Blank and Burmester, 2012, Hoffmann et al., 2012a, Hoogewijs et al., 2012, Burmester and Hankeln, 2014). Phylogenetic analyses have revealed that vertebrate *Ngb* shows phylogenetic affinities with a diversity of nerve globins in annelid worms, echinoderms, and cephalochordates (amphioxus), and vertebrate *GbX* shows affinities with an equally diverse set of putatively N-acylated globins in protostomes and nonvertebrate deuterostomes (Ebner et al., 2010, Blank and Burmester, 2012, Hoffmann et al., 2012a) (Figs. 5.3–5.5). The remaining members of the vertebrate globin gene repertoire are products of vertebrate-specific duplication events (Hoffmann et al., 2010a, Hoffmann et al., 2011, Storz et al., 2011, Hoffmann et al., 2012a, Hoffmann et al., 2012b, Schwarze and Burmester, 2013, Storz et al., 2013, Schwarze et al., 2015) (Fig. 5.6).

The last common ancestor of vertebrates appears to have possessed a globin gene repertoire that included copies of *Adgb*, *Ngb*, as many as four paralogous copies of *GbX*, and the progenitor of the entire set of vertebrate-specific globins (jawed vertebrate *α/β-Hbs*, *Cygb*, *GbE*, *GbY*, *Mb*, and cyclostome *Hbs* and *Mbs*) (Hoffmann et al., 2010a, Blank and Burmester, 2012, Hoffmann et al., 2012a,

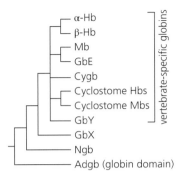

Fig. 5.6. Phylogeny of vertebrate globins. Phylogenetic reconstructions arrange vertebrate globins into those that derive from vertebrate-specific duplications (Cygb, GbE, GbY, and the Mb and Hb genes of jawed vertebrates and cyclostomes) and those that derive from duplications that predate the divergence between deuterostomes and protostomes (Adgb, GbX, and Ngb). Cyclostomes possess a surprisingly diverse repertoire of globin genes, but *Cygb* and *GbX* are the only orthologous globins shared between cyclostomes and jawed vertebrates (Hoffmann et al., 2010a, Schwarze et al., 2014, Fago et al., 2018).

Based on data from Hoffmann et al. (2010a), Blank and Burmester (2012), Hoffmann et al. (2012a, 2012b), Schwarze et al. (2014), and Opazo et al. (2015b).

Hoffmann et al., 2012b, Schwarze and Burmester, 2013, Schwarze et al., 2014, Opazo et al., 2015b, Schwarze et al., 2015). During the course of vertebrate evolution *Mb*, *GbE*, *GbY*, *GbX*, and *Ngb* have each been lost in at least two different lineages (Fig. 5.7). Given that *GbX* is present in the genomes of turtles and squamate reptiles, the principle of parsimony suggests that this gene must have been deleted independently in mammals and archosaurs. *Cygb* and *Adgb* are the only globins that have been retained in all vertebrate taxa examined to date, and *Cygb* and *GbX* are the only orthologous globins shared between jawed vertebrates and jawless fishes (cyclostomes) (Schwarze et al., 2014, Opazo et al., 2015b, Fago et al., 2018).

In jawed vertebrates, Hb and Mb appear to be essentially indispensable (Antarctic ice fishes notwithstanding) and—as discussed in section 5.7—analogous globin proteins have been co-opted to perform similar O_2-transport functions in cyclostomes. Surprisingly, however, the *Mb* gene has also been deleted in amphibians and some teleost lineages (Fuchs et al., 2006, Xi et al., 2007, Hoffmann et al., 2011, MacQueen et al., 2014).

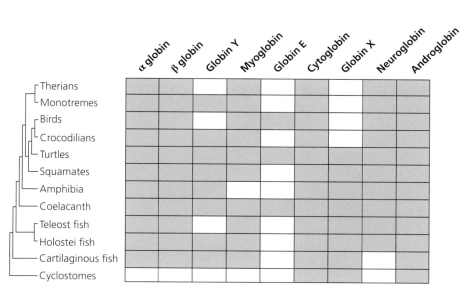

Fig. 5.7. Phyletic distribution of paralogous globin genes in the genomes of vertebrates, revealing evidence for multiple lineage-specific gene losses. Although the *Mb* gene is indicated as being present in teleost fishes, it should be noted that the gene has been secondarily lost in several lineages, as mentioned in the text. Similarly, the α- and/or β-globin genes have been secondarily lost in Antarctic ice fishes in the family Channichthyidae (section 2.6).

Globins that have restricted phyletic distributions within the vertebrates, like *GbE*, often represent the products of surprisingly ancient duplication events. Although initial genomic surveys suggested that *GbE* represented a bird-specific globin, phylogenetic analyses combined with assessments of conserved synteny (the preserved gene order in homologous chromosomal regions of different species) revealed that *GbE* and *Mb* represent the paralogous products of a tandem gene duplication that occurred in the stem lineage of jawed vertebrates (Hoffmann et al., 2011, Hoffmann et al., 2012b) (Fig. 5.8). The antiquity of the gene suggested the possibility that orthologs of *GbE* might be found in non-avian vertebrate taxa that had yet to be surveyed. Sure enough, it was not long before comparative genomic analyses identified orthologs of the hitherto "bird-specific" globin in the coelacanth (Schwarze and Burmester, 2013) and turtle (Schwarze et al., 2015).

In summary, phylogenetic reconstructions arrange vertebrate globins into those that derive from vertebrate-specific duplications (*Cygb*, *GbE*, *GbY*, and the independently derived *Mb*s and *Hb*s of jawed vertebrates and cyclostomes) and those that derive from far more ancient duplication events that predate the divergence between deuterostomes and protostomes (*Adgb*, *GbX*, and *Ngb*). Tracing the evolutionary history of deuterostome globins reveals evidence for the repeated culling of ancestral diversity, followed by lineage-specific diversification of surviving gene lineages via repeated rounds of duplication and divergence. Having now considered the evolution of vertebrate globins in a broad, historical context, we next consider the genetic and evolutionary mechanisms responsible for producing the observed functional diversity.

5.7 Gene duplication, genome duplication, and the origin of Hb as an O₂ carrier

Two rounds of whole-genome duplication in the stem lineage of vertebrates played an important role in promoting the diversification of the globin gene family (Storz et al., 2011, Hoffmann et al., 2012b, Storz et al., 2013, Opazo et al., 2015b). The progenitors of the *Hb* and *Mb* gene lineages originated as products of one such genome duplication event (Hoffmann et al., 2012b). Retention of the proto-*Hb*

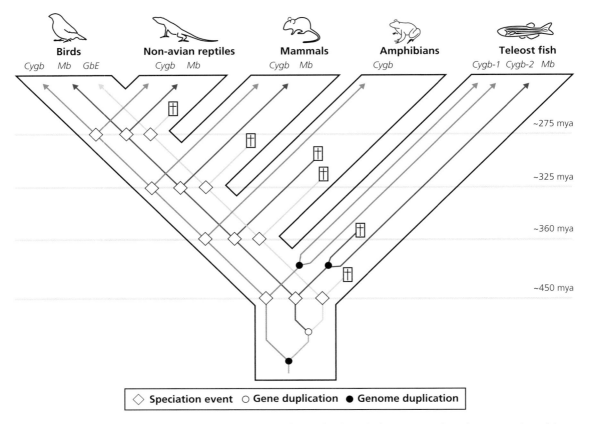

Fig. 5.8. A model for the duplicative origins and evolutionary history of the *Cygb*, *GbE*, and *Mb* genes in jawed vertebrates. Progenitors of the *Cygb* and *Mb* gene lineages were produced by a whole-genome duplication event in the stem lineage of vertebrates. Subsequently, the *GbE* and *Mb* genes originated originated via tandem gene duplication in the common ancestor of jawed vertebrates (prior to the split between cartilaginous fishes and the common ancestor of ray-finned fishes and tetrapods). One or more copies of the *Cygb* gene have been retained in all jawed vertebrate lineages, and a single copy of *Mb* has been retained in all major lineages other than the amphibians, whereas *GbE* has only been retained in the lineages leading to modern birds and turtles. The loss of the *GbE* gene in teleost fishes probably occurred prior to the teleost-specific genome duplication, which gave rise to the *Cygb-1* and *Cygb-2* paralogs.

Reproduced from Hoffmann et al. (2011).

and *Mb* genes in the ancestor of jawed vertebrates set the stage for a physiological division of labor between O₂-carrier and O₂-storage functions, and subsequent duplication of the proto-*Hb* gene gave rise to the progenitors of the α- and β-type globins (Fig. 5.9). The duplication event that gave rise to the proto α- and β-globin genes occurred ~450 million years ago, prior to the divergence between cartilaginous fishes and the common ancestor of bony vertebrates (Euteleostomi, which includes ray-finned fishes, lobe-finned fishes, and tetrapods) (Goodman et al., 1987, Storz et al., 2011, Hoffmann et al., 2012b, Storz et al., 2013, Storz, 2016). Functional

divergence of the proto α- and β-globin genes permitted the formation of multimeric Hbs composed of unlike subunits (α₂β₂). The evolution of this heteromeric quaternary structure was central to the emergence of Hb as a specialized O₂-transport protein because it provided a mechanism for cooperative O₂ binding and allosteric regulatory control. As explained in Chapters 2–4, both of these features require a coupling between the effects of ligand binding at individual subunits and the interactions between subunits in the quaternary assembly.

The ancestral linkage arrangement of the proto α- and β-globin genes is still retained in the genomes

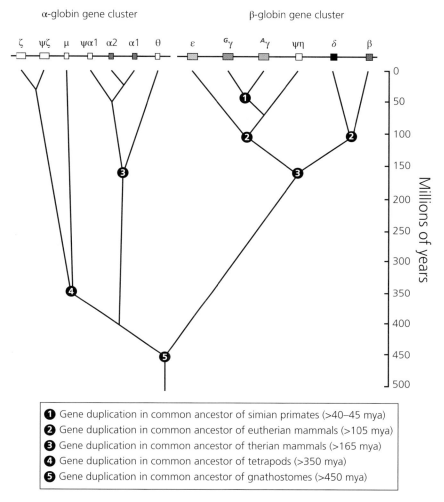

Fig. 5.9. Phylogenetic diversification of the α- and β-globin gene subfamilies. The human α- and β-globin gene clusters are shown at the top. Pseudogenes are denoted by the ψ symbol. In the human α-globin gene cluster, for example, "$\psi\zeta$" denotes an inactivated copy of the embryonic ζ-globin gene. The tree depicts phylogenetic relationships among the paralogous gene duplicates. The inferred timing of duplication events is indicated on the vertical axis. Note that the human "μ-globin" gene is orthologous to the α^D-globin gene of other tetrapods, as discussed in the text.

Modified from Storz (2016).

of cartilaginous fishes, ray-finned fishes, and amphibians (Fuchs et al., 2006, Opazo et al., 2013, Opazo et al. 2015b). In amniote vertebrates, by contrast, the α- and β-type globin genes are located on different chromosomes (Hoffmann et al., 2010b, Hardison, 2012, Hoffmann et al., 2012b). In the human genome, the α-globin gene cluster is located on chromosome 16 and the β-globin gene cluster is located on chromosome 11 (Fig. 5.10). This reflects the fact that the ancestral β-globin gene was

transposed to a new chromosomal location in the lineage leading to modern amniotes (Hardison, 2012). Intriguingly, an "orphaned" β-type globin gene (ω-globin) is still found in association with the tandemly linked α-type globin genes in the genomes of monotremes and marsupials (Wheeler et al., 2004, Hoffmann et al., 2008b, Opazo et al., 2008b).

Phylogenetic evidence indicates that erythroid-specific, O_2-transport Hbs evolved independently

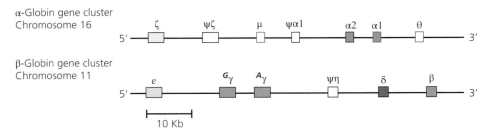

Fig. 5.10. Structure of the human α- and β-globin gene clusters.

from different ancestral precursor proteins in the two deepest branches of the vertebrate family tree: jawed vertebrates and jawless fishes (cyclostomes, represented by lampreys and hagfish) (Hoffmann et al., 2010a, Schwarze et al., 2014). The independent evolution of O_2-transport Hbs in these two anciently diverged vertebrate lineages involved the convergent co-option of distinct globin precursors to perform similar respiratory functions in circulating red blood cells. In the Hbs of both jawed vertebrates and cyclostomes, multisubunit quaternary structures provide the basis for cooperative O_2-binding and allosteric regulation, but differences in numerous structural details belie their independent origins. In the tetrameric Hbs of jawed vertebrates, cooperativity stems from an oxygenation-linked transition in quaternary structure between high- and low-affinity conformations. In the Hbs of cyclostomes, by contrast, cooperativity stems from an oxygenation-linked dissociation of low-affinity homo- and/or heterodimers into high-affinity monomers (Brittain and Wells, 1986, Brittain et al., 1989, Fago and Weber, 1995, Qiu et al., 2000, Fago et al., 2001, Heaslet and Royer, 2001). Thus, the O_2-transport Hbs of jawed vertebrates and cyclostomes represent superficially similar but structurally distinct design solutions to the challenge of maintaining cellular O_2 supply in support of aerobic metabolism (Hoffmann et al., 2010a).

5.8 Whole-genome duplication and the functional diversification of vertebrate globins

The role of gene duplication in promoting phenotypic novelty has been the subject of especially intense speculation in the context of vertebrate origins and evolution (Ohno, 1970, Holland, 2003). It has been hypothesized that two consecutive rounds of whole-genome duplication (WGD) in the stem lineage of vertebrates provided genetic raw materials for the innovation of numerous vertebrate-specific features (Ohno, 1970, Holland et al., 1994, Shimeld and Holland, 2000, Wada and Makabe, 2006, Zhang and Cohn, 2008, Braasch et al., 2009b, Larhammar et al., 2009, Van de Peer et al., 2009). In addition to the two rounds of WGD in the stem lineage of vertebrates, an additional WGD occurred in the stem lineage of teleost fishes (Meyer and Schartl, 1999; Taylor et al., 2001, 2003). Genomic evidence suggests that this teleost-specific WGD may have done much to fuel the phenotypic diversification of this morphologically diverse and speciose vertebrate group (Braasch et al., 2006, Braasch et al., 2007, Braasch et al., 2009b, Braasch et al., 2009a, Sato et al., 2009). Although it has proven difficult to document causal links between vertebrate-specific or teleost-specific innovations and specific WGD events (Van de Peer et al. 2009), Hoffmann et al. (2012b) demonstrated that precursors of key globin proteins that evolved specialized functions in different aspects of oxidative metabolism and oxygen signaling pathways (Hb, Mb, and Cygb) represent paralogous products of two successive rounds of WGD in the vertebrate common ancestor. These findings support the hypothesis that WGDs helped fuel key innovations in vertebrate evolution.

The first clue that WGDs may have spurred the diversification of vertebrate globins was provided by phylogenetic evidence that four distinct clades of vertebrate-specific globins were sister to a single clade of globins in the amphioxus genome (Storz et al., 2011, Hoffmann et al., 2012a, Hoffmann et al., 2012b). The four main clades of vertebrate-specific globins included: (1) *Cygb* and cyclostome *Hbs*; (2) *Mb* + *GbE*; (3) the α- and β-chain *Hb* subunits of jawed vertebrates; and (4) *GbY*. At face value, the fact that four clades of vertebrate-specific globins are sister to a single clade of amphioxus globins is consistent with the hypothesis that those four clades represent the paralogous products of two rounds of WGD in the stem lineage of vertebrates, as postulated by the so-called "2R" hypothesis (for "two rounds" of WGD) (Meyer and Schartl, 1999, Dehal and Boore, 2005, Putnam et al., 2008). However, following two rounds of WGD, only a small minority of gene families would be expected to retain all four of the newly created paralogs, and the 4:1 phylogenetic pattern would be gradually obscured by small-scale gene duplications and deletions that occur

after each round of WGD. For this reason, conclusive inferences about the role of WGDs in fueling the expansion of multigene families typically require the integration of molecular phylogenetic analyses with comparative genomic analyses of conserved synteny (Pébusque et al., 1998, Abi-Rached et al., 2002, Horton et al., 2003, Dehal and Boore, 2005). Fig. 5.11 illustrates how the effects of two successive WGDs should be reflected in the physical linkage arrangement of paralogous genes—specifically, the fourfold pattern of intragenomic macrosynteny among paralogous chromosomal segments.

To assess whether the four above-mentioned clades of vertebrate-specific globins represent the

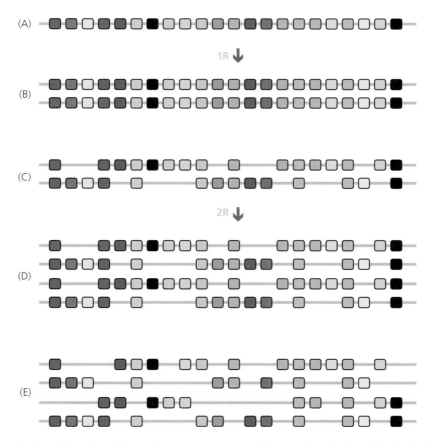

Fig. 5.11. Diagram depicting the hypothetical effects of two consecutive genome duplications (1R and 2R), as reflected in the physical linkage arrangement of paralogous genes—specifically, the fourfold pattern of intragenomic synteny. (A) Hypothetical chromosome in the vertebrate common ancestor; (B) the first genome duplication produces a complete set of paralogs in identical order; (C) many paralogous gene copies are subsequently deleted from the genome; (D) the second genome duplication produces yet another set of paralogs in identical order, with multigene families that retained two copies now present in four; (E) with the passage of time, additional gene losses ensue, thereby obscuring the fourfold pattern of synteny.

Reproduced from Storz et al. (2013).

products of two successive WGDs (as predicted by the 2R hypothesis), Hoffmann et al. (2012b) used an integrated genomic/phylogenetic approach to test the following predictions:

(1) Representatives of the four clades of vertebrate-specific globins should be embedded in unlinked chromosomal segments that share similar, inter-digitated arrangements of paralogous genes ("paralogons");

(2) The globin-defined paralogons should be united by "4:1" gene families (quartets of paralogs that coduplicated with the globin genes) and—in various combinations—by 3:1 and 2:1 gene families that trace their duplicative origins to the stem lineage of vertebrates.

(3) The globin-defined paralogons identified in vertebrate genomes should exhibit a fourfold pattern of conserved macrosynteny relative to the genomes of nonvertebrate chordates like amphioxus. This fourfold pattern would reflect the fact that the globin-defined paralogons represent the quadruplicated products of the same proto-chromosome in the chordate common ancestor.

With regard to predictions 1 and 2, the four globin-defined paralogons may not always contain the same flanking tracts of paralogous gene dupli-cates (due to stochastic gene losses, as illustrated in Fig. 5.11), but they should be united by multiple gene families that trace their duplicative origins to the stem lineage of vertebrates. For example, we would expect that a number of globin-linked genes are members of 4:1 gene families, where each of the four duplicate copies are located on a different globin-defined paralogon. As just mentioned, the problem with identifying 4:1 gene families is that only a small subset of gene families would be expected to retain all four duplicate gene copies pro-duced by two rounds of WGD, and subsequent gene turnover would further obscure the signal. However, the same globin-defined paralogons that are united by 4:1 gene families would also be united by 3:1 and 2:1 gene families in each of the possible configur-ations. Members of such gene families are located on either three or two of the four globin-defined par-alogons, respectively. The implication is that the missing members of the expected gene quartet were deleted after the first or second rounds of WGD.

The patterns described by predictions 1 and 2 could potentially be produced by large-scale seg-mental duplications as well as WGDs, but the pat-tern described by prediction 3 would be difficult to reconcile with any alternative to the 2R hypothesis. If the four clades of vertebrate-specific globins are products of the 2R WGD, then the globin-defined paralogons should all be derived from a single link-age group in the pre-WGD karyotype of the verte-brate ancestor.

To test these predictions, Hoffmann et al. (2012b) examined the genomic locations of the vertebrate-specific globin genes and characterized large-scale patterns of conserved macrosynteny in complete genome sequences from several representative vertebrate taxa. These analyses revealed that the *Cygb* gene, the *Mb/GbE* gene pair, and the α-*Hb/GbY* gene pair are each embedded in clearly demarcated paralogons (Fig. 5.12). The "*Mb*" paralogon is defined by the tandemly linked *Mb* and *GbE* genes. Analysis of phylogenetic relationships and conserved syn-teny revealed that these two genes represent prod-ucts of a tandem gene duplication that occurred in the ancestor of jawed vertebrates. The *GbE* gene was secondarily lost from most vertebrate lineages, but the ancestral linkage arrangement of the *Mb* and *GbE* genes is still retained in the genomes of birds, turtles, and coelacanth (Hoffmann et al., 2011; Schwarze and Burmester, 2013, Schwarze et al., 2015). The "*Hb*" paralogon is defined by the α-globin gene cluster of amniotes and by the tandemly linked α- and β-globin gene sets in teleost fishes and amphibians. Similar to the case with the *Mb/GbE* gene pair, combined analysis of phylogenetic rela-tionships and conserved synteny revealed that the proto-*Hb* gene (the single-copy progenitor of the α- and β-globin subfamilies) and the *GbY* gene represent the products of a tandem gene duplica-tion that occurred in the ancestor of jawed verte-brates. In fact, the ancestral linkage arrangement of these genes is still retained in the genomes of ele-phant shark, spotted gar, *Xenopus*, anole lizard, and platypus, as *GbY* is located downstream from the 3' end of the α-globin gene cluster in each of these taxa (Fuchs et al., 2006, Hoffmann et al., 2010b; Opazo et al., 2013; Opazo et al. 2015b).

The observed threefold pattern of conserved macrosynteny involving the *Cygb*, *Mb*, and *Hb*

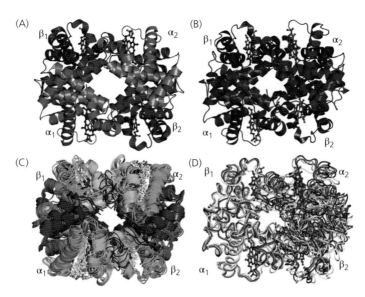

Plate 1. Structures of human Hb. (A) Crystal structure of deoxy Hb (PDB code 2DN2); (B) Crystal structure of HbCO (2DN3); (C) Ten lowest energy solution structures of HbCO obtained by NMR spectroscopy (2M6Z); (D) Superimposition of different R-type crystal structures of HbCO (2DN3, red; 1BBB, magenta; 1MKO, green; and 1YZI, cyan) with the average solution structure obtained by means of NMR (light gray). Structures were aligned according to the $\alpha_1\beta_1$ dimer.

Reproduced from Yuan et al. (2015), with permission from the American Chemical Society. See Chapter 2, section 2.1.

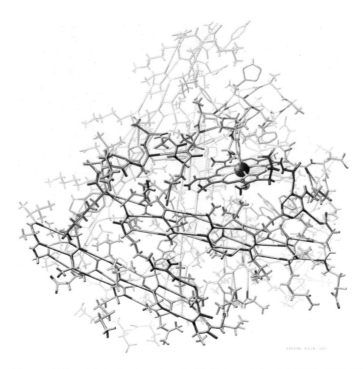

Plate 2. Skeletal drawing of the myoglobin protein by the pioneering scientific illustrator, Irving Geis (1908–1997). This illustration, which appeared in a 1961 issue of *Scientific American*, was the first full side-chain drawing of any protein molecule. Today, such renderings are created by computer.

Copyright 1961 by Scientific American. See Chapter 2, section 2.1.

Plate 3. Icefish in the family Channichthyidae inhabit the freezing, ice-laden waters surrounding the continental shelf of Antarctica. These physiologically enigmatic fish do not express Hb and therefore have colorless blood. (A) Icefish larva (Uwe Kils, Wikimedia Commons) and (B) adult crocodile icefish, *Chionodraco hamatus* (Marrabbio2, Wikimedia Commons). See Chapter 2, section 2.6.

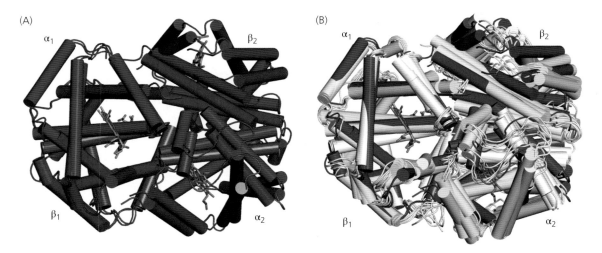

Plate 4. Hb allostery was originally viewed in terms of a rigid two-structure model (A), but is now interpreted in terms of a dynamic ensemble of structures (B). Panel A shows superimposed structures of human Hb in the T-state (unliganded conformation, PDB code 4HHB, blue) and in the R-state (liganded, conformation, PDB code 2DN3, red). Panel B shows the ten lowest energy solution structures of HbCO obtained via NMR (PDB code 2M6Z, gray) superimposed with the canonical T and R structures determined by X-ray crystallography. Structures were aligned according to the $\alpha_1\beta_1$ dimer.

Reproduced from Yuan et al. (2015), with permission from the American Chemical Society. See Chapter 4, section 4.3.

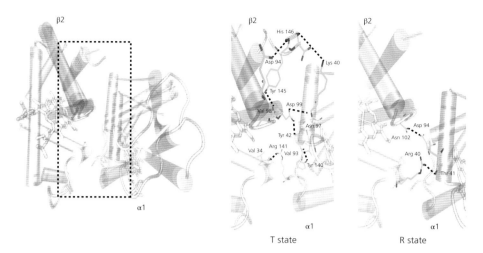

Plate 5. Network of atomic contacts at interdimer ($\alpha_1\beta_2/\alpha_2\beta_1$) interfaces involving the C and G helices and FG corner. These interdimer "sliding" contacts involve α_1C-β_2FG and α_1FG-β_2C, and the minor contacts involve α_1C-β_2C and α_1FG-β_2FG. The complete set of subunit contacts are detailed in Table 4.1. During the oxygenation-linked transition from the T-state to the R-state, some interactions at the $\alpha_1\beta_2/\alpha_2\beta_1$ interface are broken and new interactions are formed. As stated by Perutz (1978), this intersubunit interface "acts as a snap-action switch, with two alternative stable positions, each braced by a different set of hydrogen bonds." Intermediate positions of the interface are sterically prevented, consistent with the all-or-nothing conformational transition envisioned by the MWC model. See Chapter 4, section 4.3.

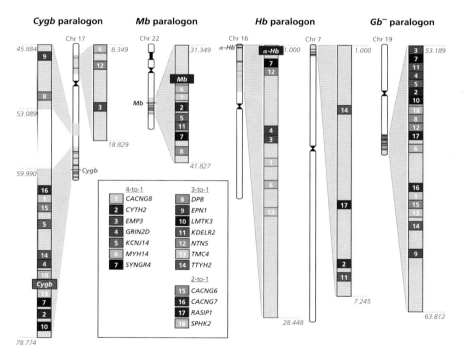

Plate 7. Fourfold pattern of conserved macrosynteny between the four globin-defined paralogons in the human genome (including the *Gb⁻* paralogon) and "linkage group 15" of the reconstructed proto-karyotype of the chordate common ancestor (shaded regions). This pattern of conserved macrosynteny demonstrates that the *Cygb*, *Mb*, *Hb*, and *Gb⁻* paralogons trace their duplicative origins to the same proto-chromosome of the chordate common ancestor and provides conclusive evidence that each of the four paralogons are products of a genome quadruplication in the stem lineage of vertebrates. Shared gene duplicates that map to secondarily translocated segments of the *Mb* paralogon (on chromosome 12) and the *Hb* paralogon (on chromosomes 7 and 17) are not pictured.

Reproduced from Hoffmann et al. (2012b). See Chapter 5, section 5.8.

Mb-like family 3-on-3 Hb fold

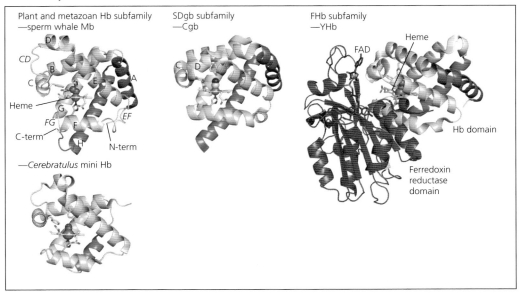

Plant and metazoan Hb subfamily
—sperm whale Mb

SDgb subfamily
—Cgb

FHb subfamily
—YHb

—*Cerebratulus* mini Hb

Globin-coupled sensor family 3-on-3 Hb fold

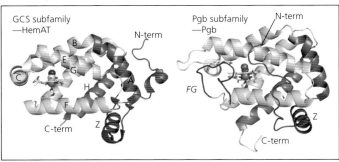

GCS subfamily
—HemAT

Pgb subfamily
—Pgb

Truncated Hbs 2-on-2 Hb fold

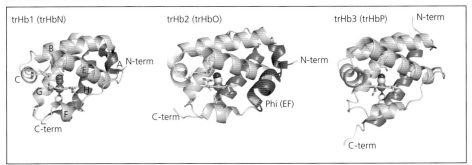

trHb1 (trHbN)

trHb2 (trHbO)

trHb3 (trHbP)

Plate 6. Conserved and variable features of globin tertiary structure across different domains of life. The depicted tertiary structures are representative of major lineages comprising the globin protein superfamily. Most globins from the plant and metazoan "Hb subfamily" are highly similar to sperm whale Mb (pdb 2mgm), the exemplar of the Mb fold. Variations on the theme include the *Cerebratulus lacteus* mini Hb (1kr7). Other families and subfamilies are represented by Cgb from *Campylobacter jejuni* (2wy4), YHb from *Saccharomyces cerevisiae* (4g1v), HemAT from *Bacilus subtilis* (1or4), Pgb from *Methanosarcina acetivorans* (2veb), trHb1 (trHbN) from *Tetrahymena pyriformis* (3aq5), trHb2 (trHbO) from *Mycobacterium tuberculosis* (1ngk), and trHb3 (HbP) from *Campylobacter jejuni* (2ig3). Conserved α-helices that comprise the canonical 3-on-3 tertiary structure are color-coded as follows: A (red-brown), B (pink), E (yellow/tan), F (green), G (cyan), H (blue). Functionally important loops (CD, EF, FG) are also labeled. Conserved elements of the 2-on-2 globin fold are color-coded in the same fashion as the 3-on-3 globins. Additional elements of secondary structure that are present in Mb but not in some other globins are shown in gray. Additional structural elements that are unique to specific globin subfamilies are shown in magenta.

Modified from Gell (2017). See Chapter 5, section 5.3.

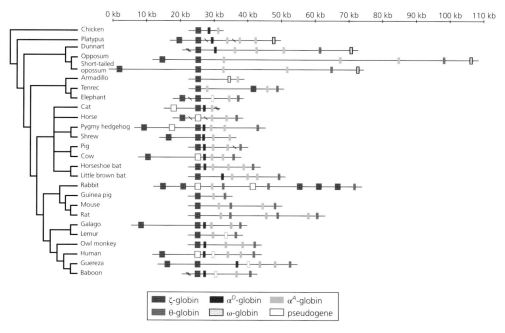

Plate 8. Variation in the size and membership composition of the α-globin gene family in mammals.

Reproduced from Hoffmann et al. (2008b).

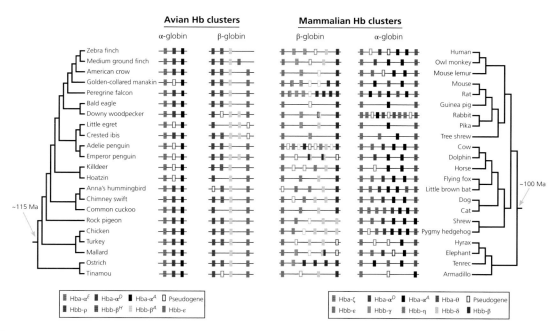

Plate 9. Dramatic differences in stability of the globin gene clusters in birds and mammals. Genes in the mammalian clusters have undergone a much higher rate of turnover. Shown are diagrammatic depictions of the chromosomal organization of the α- and β-type globin genes. In the case of the α-type globin genes, birds and mammals share orthologous copies of the αD- and αA-globin genes (αD-globin is annotated as μ-globin in the human genome assembly). Likewise, the avian π-globin and the mammalian ζ-globin genes are 1:1 orthologs. In contrast, the genes in the avian and mammalian β-globin gene clusters are derived from independent duplications of one or more β-type globin genes that were inherited from the common ancestor of tetrapod vertebrates (Hoffmann et al., 2010).

Reproduced from Zhang et al. (2014). See Chapter 5, section 5.9.1.

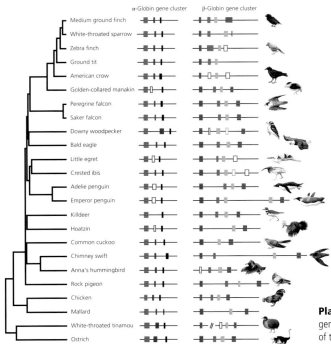

Plate 10. Genomic structure of the avian α- and β-type globin gene clusters. The phylogeny depicts a time-calibrated supertree of twenty-four species representing all the major avian lineages.

Reproduced from Opazo et al. (2015). See Chapter 6, section 6.6.1.

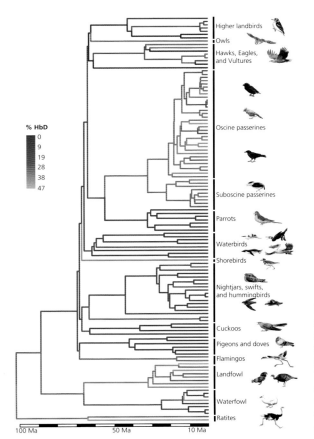

Plate 11. Inferred evolutionary changes in relative expression level of the HbD isoform (percentage of total Hb) during the diversification of birds. Expression data are based on experimental measures of protein abundance in the definitive red blood cells of 122 bird species (n = 1–30 individuals per species, 267 specimens in total). Terminal branches are color-coded according to the measured HbD expression level of each species, and internal branches are color-coded according to maximum-likelihood estimates of ancestral character states. Branch lengths are proportional to time.

Reproduced from Opazo et al. (2015). See Chapter 6, section 6.6.1.

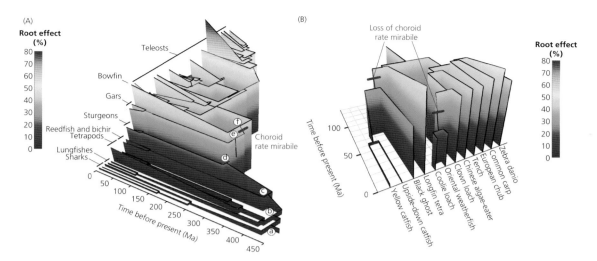

Plate 12. Evolution of the Root effect in jawed vertebrates. Ancestral values at internal nodes of the tree were estimated using linear parsimony (for methodological details, see Berenbrink et al. [2005]). (A) Rotation of the three-dimensional projection shows the gradual increase of the Root effect in early ray-finned fishes (nodes c–f) after divergence from the line leading to modern-day lobe-finned fishes (including tetrapods). The red hash mark denotes the origin of the choroid *rete mirabile* in the branch leading to the common ancestor of bowfin and teleosts. Note that the choroid *rete* originated only after the Root effect had increased to more than 40 percent. (B) Zoomed-in view of a clade of teleost fishes showing secondary reductions of the Root effect in Ostariophysi. Secondary reductions of the Root effect only occur in lineages in which the choroid *rete* has been lost (denoted by red hash marks).

Ma = million years.

Modified from Berenbrink (2007). See Chapter 7, section 7.2.1.

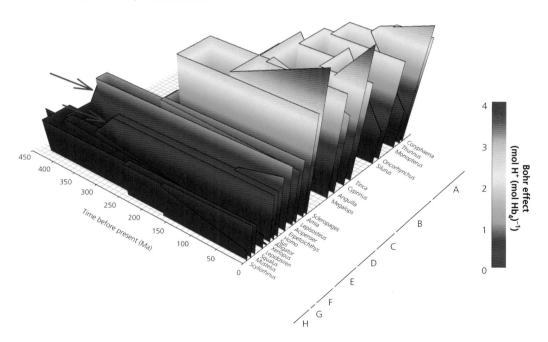

Plate 13. Evolution of the Bohr effect in jawed vertebrates. Values refer to the maximal alkaline Bohr effect of organic phosphate-free hemolysates in 0.1 M KCl at physiological temperature, determined by titration of hemolysates equilibrated under N_2 and O_2 atmospheres and calculated as the number of additional protons bound upon deoxygenation of Hb at constant pH. Ancestral values at internal nodes of the tree were estimated using linear parsimony (for methodological details, see Berenbrink et al. [2005]). Red arrows denote two independent increases of the Bohr effect in the stem lineage of tetrapods (F) and early ray-finned fishes (E).

Modified from Berenbrink et al. (2005). See Chapter 7, section 7.2.1.

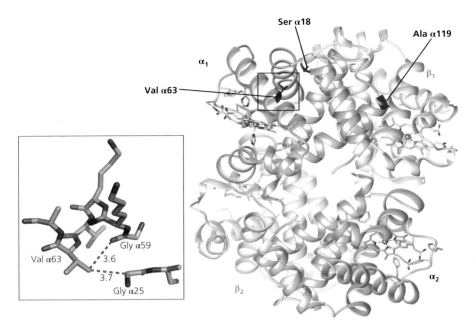

Plate 14. Structural model showing bar-headed goose Hb in the deoxy state (PDB1hv4), along with locations of each of the three amino substitutions that occurred in the bar-headed goose lineage after divergence from the common ancestor of other *Anser* species. The inset graphic shows the environment of the Val α63 residue. When valine replaces the ancestral alanine at this position, the larger volume of the side chain causes minor steric clashes with two neighboring glycine residues, Gly α25 and Gly α59. The distances between non-hydrogen atoms (depicted by dotted lines) are given in Å.

Reproduced from Natarajan et al. (2018). See Chapter 9, section 9.5.3.

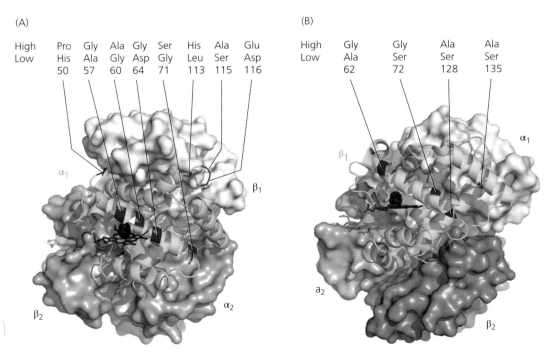

Plate 15. Amino acid polymorphisms in the Hbs of deer mice (*Peromyscus maniculatus*) contribute to genetic differences in Hb-O₂ affinity between high-altitude populations in the Rocky Mountains and low-altitude populations in prairie grassland. Eight sites in the α-chain (A) and four sites in the β-chain (B) exhibit striking altitudinal differences in allele frequency (Storz et al. 2010).

Reproduced from Storz et al. (2009). See Chapter 9, section 9.5.4.

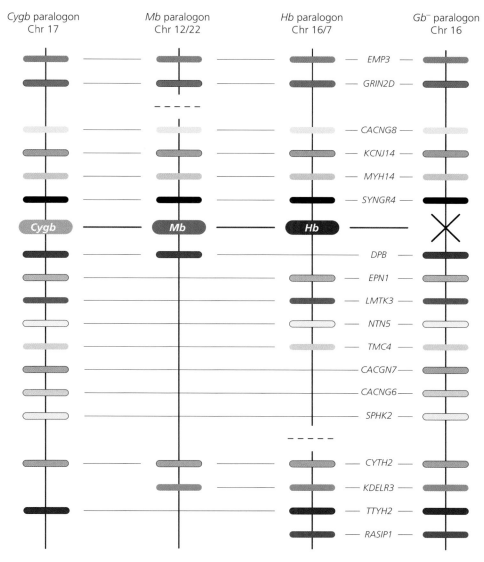

Fig. 5.12. Graphical depiction of gene duplicates that are shared between the *Gb⁻* paralogon and the remaining three globin-defined paralogons (*Cygb*, *Mb*, and *Hb*) in the human genome. There are seven 4:1 gene families that unite the *Gb⁻* paralogon with the *Cygb*, *Mb*, and *Hb* paralogons, there are seven 3:1 gene families that unite the *Gb⁻* paralogon with two of the three globin-defined paralogons, and there are four 2:1 gene families that unite the *Gb⁻* paralogon with a single globin-defined paralogon. On each chromosome, annotated genes are depicted as bars with different shading. The "missing" globin gene on the *Gb⁻* paralogon is denoted by an "X." The shared paralogs are depicted in collinear arrays for display purposes only, as there is substantial variation in gene order among the four paralogons. For clarity of presentation, genes that are not shared between the *Gb⁻* paralogon and any of the three globin-defined paralogons are not shown. In the human genome, the *Gb⁻* paralogon on chromosome 19 shares multiple gene duplicates with fragments of the *Hb* paralogon on chromosomes 16 and 7, and fragments of the *Mb* paralogon on chromosomes 12 and 22. Members of the *EPN1*, *LMTK3*, and *KCNJ14* gene families that map to the *Hb* paralogon have been secondarily translocated from chromosome 16.

Modified from Hoffmann et al. (2012b).

paralogons can be reconciled with the expected fourfold "tetraparalogon" pattern predicted by the 2R hypothesis by invoking the secondary loss of one of the four paralogous globin genes that would have been produced by two successive WGDs. Consistent with this secondary-loss scenario, a detailed bioinformatic analysis of the human genome identified a clearly demarcated segment of human chromosome 19 that shares multiple gene duplicates with

one or more of the other three globin-defined paralogons (Hoffmann et al., 2012b) (Fig. 5.12). Thus, the sets of linked genes comprising these 4:1, 3:1, and 2:1 gene families appear to have coduplicated with the *Cygb*, *Mb*, and *Hb* genes and, as predicted by the 2R hypothesis, phylogenetic analysis revealed that the duplications occurred prior to the divergence of tetrapods and teleost fishes (Fig. 5.13). These results implicate human chromosome 19 as the genomic

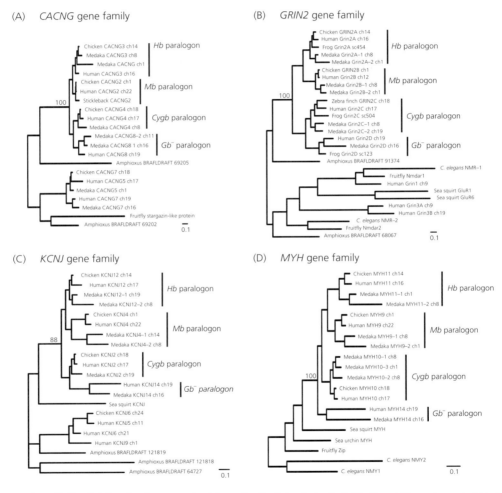

Fig. 5.13. Maximum likelihood phylogenies of representative 4:1 gene families that unite the *Cygb*, *Mb*, *Hb*, and *Gb⁻* paralogons. Individual members of the *CACNG*, *Grin2*, *KCNJ*, and *MYH* gene families (panels A–D, respectively) are located on each of the four globin-defined paralogons. As the tree topologies indicate, paralogous members of the same gene family always form a monophyletic group relative to the putative ortholog in nonvertebrate chordates (amphioxus or sea squirt). In each of the four trees, bootstrap support values are shown for the key node that unites all vertebrate-specific genes as a monophyletic group. These phylogenies (and those for many other globin-linked gene duplicates) are consistent with the genome-duplication hypothesis, and indicate that each of the gene families diversified prior to the divergence between tetrapods and teleost fishes.

Reproduced from Storz et al. (2011b).

location of the missing fourth paralogon, dubbed the "globin-minus" (*Gb⁻*) paralogon since the associated globin gene must have been secondarily lost. The identification of the *Gb⁻* paralogon reveals the tell-tale pattern of fourfold conserved macrosynteny that is predicted by the 2R hypothesis (Hoffmann et al., 2012b).

The final line of evidence implicating the 2R WGD in the diversification of vertebrate globins was provided by a comparative analysis of conserved synteny between the genomes of human and amphioxus. This comparison revealed that the *Hb*, *Mb*, *Cygb*, and *Gb⁻* paralogons represent the quadruplicated products of the same linkage group in the reconstructed proto-karyotype of the chordate common ancestor (Fig. 5.14, Plate 7). In combination with the phylogenetic reconstructions and the observed linkage arrangements of paralogous genes, the fact that the globin-defined paralogons trace their duplicative origins to the same ancestral chordate "proto-chromosome" provides conclusive evidence that three of the four main

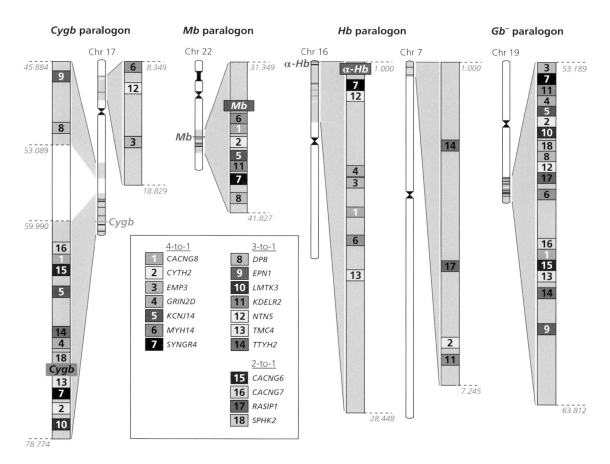

Fig. 5.14. Fourfold pattern of conserved macrosynteny between the four globin-defined paralogons in the human genome (including the *Gb⁻* paralogon) and "linkage group 15" of the reconstructed proto-karyotype of the chordate common ancestor (shaded regions). This pattern of conserved macrosynteny demonstrates that the *Cygb*, *Mb*, *Hb*, and *Gb⁻* paralogons trace their duplicative origins to the same proto-chromosome of the chordate common ancestor and provides conclusive evidence that each of the four paralogons are products of a genome quadruplication in the stem lineage of vertebrates. Shared gene duplicates that map to secondarily translocated segments of the *Mb* paralogon (on chromosome 12) and the *Hb* paralogon (on chromosomes 7 and 17) are not pictured.

Reproduced from Hoffmann et al. (2012b). See Plate 7.

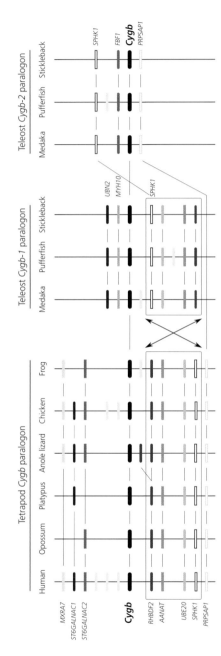

Fig. 5.15. Patterns of conserved synteny in the chromosomal region that harbors the *Cygb* gene in jawed vertebrates. Horizontal lines denote orthologous relationships.

Reproduced from Hoffmann et al. (2011).

lineages of vertebrate-specific globins originated via two successive WGD events in the stem lineage of vertebrates.

Whereas all tetrapod vertebrates examined to date possess a single copy of the *Cygb* gene, most teleost fishes possess two copies, and zebra fish possess three copies. Analysis of conserved synteny revealed that the two *Cygb* paralogs possessed by all teleosts, *Cygb-1* and *Cygb-2*, are embedded in clearly demarcated paralogons that derive from a teleost-specific WGD (Fig. 5.15). In fact, a genomic analysis of conserved synteny demonstrated that both of the *Cygb*-defined paralogons descend from the same ancestral chromosome in the reconstructed teleost proto-karyotype that was duplicated in the teleost-specific WGD (Hoffmann et al., 2011). The additional round of WGD in teleost fishes also contributed to the diversification of the α- and β-globin gene clusters (Opazo et al., 2013), but available data suggest that the *Ngb*, *GbX*, and *Mb* genes reverted to the single-copy state in all or most teleost lineages.

In summary, the three WGD-derived globin proteins (the monomeric Mb, the homodimeric Cygb, and the heterotetrameric Hb) each evolved highly distinct specializations of function involving different heme coordination chemistries and ligand affinities, and they also evolved distinct expression domains that are specific to completely different cell types. The physiological division of labor between Hb and Mb represents a key innovation in the O_2-transport system that was central to the evolution of aerobic energy metabolism in early vertebrates. This supports the hypothesis that WGDs can play a pivotal role in the evolution of phenotypic novelty. As we learn more about the functional specializations of the diverse globin genes in cyclostomes, additional WGD-derived innovations may come to light.

5.9 Mechanisms of gene family evolution

5.9.1 Concerted evolution

In contrast to the sequence divergence between globin genes that are expressed during different stages of development (discussed in Chapter 6), paralogous genes that are coexpressed during the same

stage of development are often identical or nearly identical in sequence. For example, most mammals possess two to three tandemly duplicated copies of adult α-globin genes that have identical coding sequences and therefore encode identical polypeptides (Hoffmann et al., 2008b) (Fig. 5.16, Plate 8). This pattern is typically attributable to a history of gene conversion—a form of nonreciprocal recombinational exchange between duplicated genes. Gene conversion is basically a "copy-and-paste" mechanism that results in one gene (the donor) overwriting the sequence of a duplicate gene copy (the recipient) (Fig. 5.17A). Recurrent gene conversion between duplicated genes results in the gradual homogenization of sequence variation among paralogous members of the same gene family, which gives rise to a pattern referred to as concerted evolution. Patterns of concerted evolution are well documented in the globin gene family, as phylogeny reconstructions often reveal that paralogous globin genes within the genome of a single species are more similar to one another than they are to their orthologous counterparts in closely related species (Hoffmann et al., 2008b, Hoffmann et al., 2008a, Opazo et al., 2008b, Opazo et al., 2008a, Opazo et al., 2009, Storz et al., 2008, Runck et al., 2009).

In the α- and β-globin gene clusters of vertebrates, there are several cases where the sequence similarity between tandemly duplicated globin genes is only partly attributable to concerted evolution between pre-existing paralogs; instead, it is often attributable to recent ancestry between the products of *de novo* gene duplications that occurred independently in different lineages (Hoffmann et al., 2008a, Hoffmann et al., 2008b, Opazo et al., 2008a, Opazo et al., 2008b). To distinguish between the effects of concerted evolution and gene turnover ("birth-and-death evolution") it is necessary to integrate phylogenetic information from multiple partitions of genomic sequence alignments. Because interparalog gene conversion is largely restricted to the coding regions of globin genes (Storz et al., 2007, Hoffmann et al., 2008a, Hoffmann et al., 2008b, Runck et al., 2009, Natarajan et al., 2015), orthologous and paralogous relationships can typically be determined by analyzing variation in flanking sequence and/or intronic sequence.

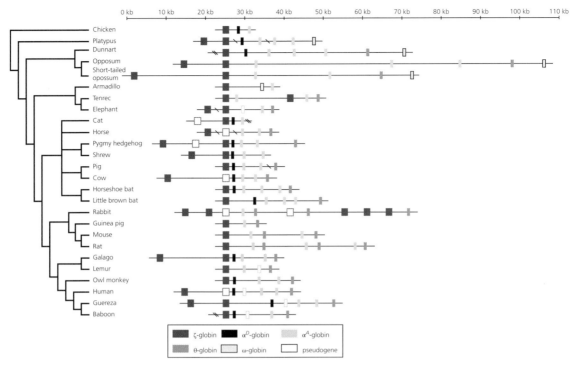

Fig. 5.16. Variation in the size and membership composition of the α-globin gene family in mammals.

Reproduced from Hoffmann et al. (2008b). See Plate 8.

5.9.2 The genomic revolving door: evolutionary changes in the size and membership composition of the globin gene family

Differences in the complement of genes between species are attributable to lineage-specific gene gains via duplication and lineage-specific losses via deletion or inactivation. These evolutionary changes in gene copy number are mainly caused by unequal crossing-over, whereby recombination between misaligned chromosomes results in gene deletion in one daughter strand and gene duplication in the other (Fig. 5.17B). Comparative genomic studies of globin gene family evolution have revealed surprisingly high rates of gain and loss among the α- and β-like globin genes in particular vertebrate groups (Hoffmann et al., 2008a, Hoffmann et al., 2008b, Opazo et al., 2008a, Opazo et al., 2008b, Opazo et al., 2009, Opazo et al., 2013, Opazo et al., 2015a, Hoffmann et al., 2018). This "genomic revolving door" (Demuth et al., 2006) of gene gain and loss has resulted in continual turnover in the

membership composition of the α- and β-globin gene clusters, and is reflected by extensive interspecific variation in the size and membership composition of the gene family. This gene turnover has been especially well documented in the α- and β-globin gene clusters of mammals (Fig. 5.18A, Plate 9A). In birds, by contrast, the size and membership composition of the α- and β-globin gene clusters have remained remarkably constant during ~100 million years of evolution, with most examined species retaining an identical complement of genes (Opazo et al., 2015a) (Fig. 5.18B, Plate 9B). Estimated rates of gene turnover in the α- and β-globin gene clusters of birds were roughly half the rate in the homologous gene clusters of mammals (Zhang et al., 2014).

As we will see in Chapter 6, evolutionary changes in the globin gene repertoire of mammals have given rise to variation in the functional diversity of Hb isoforms (isoHbs) that are expressed at different stages of development. In birds, the relative stasis of globin gene family evolution suggests that the

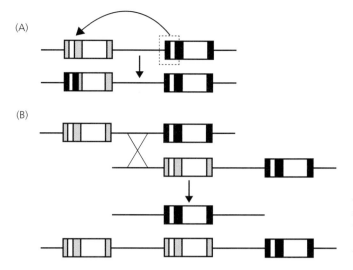

Fig. 5.17. Recombinational exchanges between homologous chromosomes. (A) Gene conversion is a form of non-reciprocal recombination that homogenizes sequence variation between tandemly duplicated genes. (B) Unequal crossing-over results in gene deletion in one recombinant chromosome and gene duplication in the other.

developmental regulation of Hb synthesis may be more highly conserved, with orthologous genes having similar stage-specific expression profiles and similar functional properties in disparate taxa. In comparison with the highly variable globin gene clusters of mammals and other amniote taxa, the general absence of redundant gene copies in the avian gene clusters suggests less opportunity for the acquisition of novel protein functions and/or expression patterns.

5.9.3 Formation of chimeric fusion genes

In addition to the role of unequal crossing-over in producing variation in gene copy number among species, this recombinational mechanism can also lead to the creation of chimeric genes via gene fusion. The β-globin gene cluster of eutherian mammals has proven to be an especially rich system for investigating the origins and evolutionary fates of chimeric fusion genes. In humans, pathological globin gene deletion mutants such as Hb Lepore, which causes $\delta\beta$-thalassemia (a type of hemolytic anemia caused by reduced synthesis of β-globin subunits), and Hb Kenya, which causes persistence of fetal Hb into adulthood, provide well-characterized examples of how gene duplications, gene deletions, and gene fusions can be generated as reciprocal exchange products of the same recombination event (Hoffmann et al., 2008a, Opazo et al., 2009, Gaudry

et al., 2014). For example, in humans, crossovers between misaligned copies of the closely linked δ- and β-globin genes result in a solitary δ/β fusion gene on one recombinant chromosome (the Hb Lepore deletion mutant) and the reciprocal β/δ fusion gene on the other recombinant chromosome (the "anti-Lepore" duplication mutant) (Fig. 5.19). In the former case, the δ/β fusion gene on the Lepore chromosome is solely responsible for the synthesis of the β-chain subunits of adult Hb. In the latter case, the β/δ fusion gene on the "anti-Lepore" chromosome is flanked by fully functional copies of the parental δ-globin gene on the 5' side and the parental β-globin gene on the 3' side. Heterozygotes for the Hb Lepore deletion produce adult red blood cells that contain normal $a_2\beta_2$ Hb tetramers (HbA), as well as lesser quantities of $a_2(\delta/\beta)_2$ tetramers (Hb Lepore) that incorporate products of the δ/β fusion gene. The HbA and Hb Lepore isoforms are not present in equal concentrations in circulating red blood cells because the δ/β fusion gene is under the control of a weak δ-type promoter (Antoniou and Grosveld, 1990) and is therefore transcribed at a much lower rate than the normal β-globin gene.

Whereas heterozygous carriers of the Lepore δ/β fusion gene typically suffer a relatively mild form of thalassemia, homozygotes for the δ/β fusion gene suffer more severe forms of erythrocytic dysfunction caused by an imbalance of a- and β-chain synthesis. The dosage imbalance results in insoluble

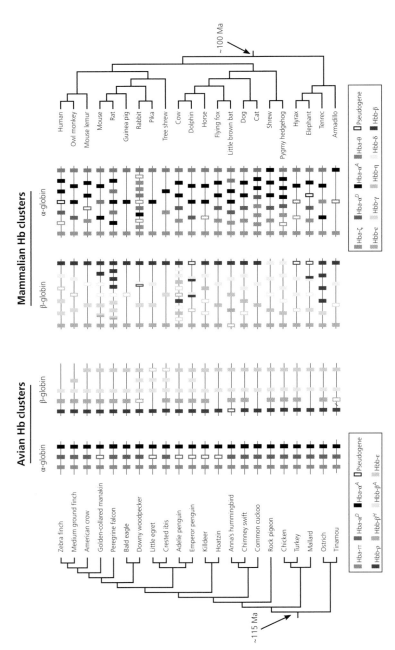

Fig. 5.18. Dramatic differences in stability of the globin gene clusters in birds and mammals. Genes in the mammalian clusters have undergone a much higher rate of turnover. Shown are diagrammatic depictions of the chromosomal organization of the α- and β-type globin genes that are expressed during different stages of development. In the case of the α-type globin genes, birds and mammals share orthologous copies of the α^D- and α^A-globin genes (α^D-globin is annotated as μ-globin in the human genome assembly). Likewise, the avian π-globin and the mammalian ζ-globin genes are 1:1 orthologs. In contrast, the genes in the avian and mammalian β-globin gene clusters are derived from independent duplications of one or more β-type globin genes that were inherited from the common ancestor of tetrapod vertebrates (Hoffmann et al., 2010).

Reproduced from Zhang et al. (2014) with permission from the American Association for the Advancement of Science. See Plate 9.

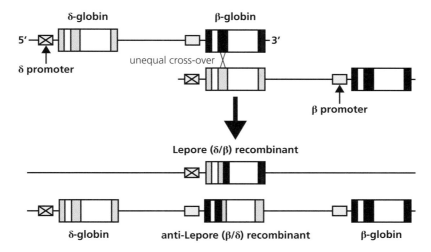

Fig. 5.19. Unequal crossing-over between a misaligned pair of δ- and β-globin gens can produce Lepore and anti-Lepore recombinant chromosomes.

Modified from Gaudry et al. (2014).

aggregations of oxidized α-chain monomers and their cytotoxic breakdown products (iron, heme, and hemichrome) in erythroid precursor cells and mature erythrocytes, which leads to premature hemolysis (Rachmilewitz and Schrier, 2001). In contrast to the complications associated with inheritance of the Hb Lepore deletion chromosome (and the associated δ/β fusion gene), the inheritance of the reciprocal exchange product of the same crossover event—the 5′ δ − β/δ − β 3′ anti-Lepore chromosome—is not associated with any hematological pathology because the recombinant chromosome retains a transcriptionally active β-globin gene in addition to the δ- and β/δ-globin genes (Fig. 5.19). The difference in fitness between the Hb Lepore deletion mutant and the anti-Lepore duplication mutant—which appears to stem entirely from differences in the transcriptional efficacy of the β- and δ-type promoters—provides a compelling explanation for why the two reciprocal β/δ and δ/β fusion genes are not equally common among contemporary mammals. Independently derived β/δ fusion genes (which occur on an anti-Lepore chromosome and have β-type promoters) are present in numerous mammalian groups (Opazo et al., 2009, Gaudry et al., 2014), whereas the reciprocal δ/β fusion gene (which occurs on the Lepore chromosome and has a δ-type promoter) is conspicuous by its absence. This illustrates how the evolutionary fates of chimeric fusion genes can be strongly influenced by their recombinational mode of origin.

5.10 Future directions in the survey of globin diversity

Historically, much of the uncertainty about phylogenetic relationships among vertebrate globins was attributable to inadequate taxon sampling. In recent years, the inclusion of a broader diversity of vertebrate globins in phylogenetic analyses has led to surprising findings regarding the antiquity of "taxon-specific" globins like *GbE*, which often turn out to have far more ancient origins than originally suspected (Hoffmann et al., 2011, Schwarze and Burmester, 2013), and the independent origins of functionally analogous globins in disparate taxa that were previously assumed to be orthologous. For example, it was previously assumed that *Mb* and the single-copy progenitor of the α/β-*Hb*s originated via duplication of an ancestral, single-copy globin gene prior to the divergence between cyclostomes and jawed vertebrates, such that the ancestors of these two groups inherited orthologous copies of the same "proto-*Hb*" gene (Goodman et al., 1975,

Goodman et al., 1987). According to this scenario, the *Hb* genes of cyclostomes would be sister to the clade of jawed vertebrate α- and β-globin genes: [*Mb* (cyclostome *Hb*, jawed vertebrate α/β-*Hb*)]. Only by including the full panoply of cyclostome-specific globins in phylogenetic analyses was it possible to infer that functionally analogous "*Hb*s" and "*Mb*s" evolved independently from different precursor globins that were present in the vertebrate ancestor (Hoffmann et al., 2010a, Schwarze et al., 2014).

The availability of complete genome sequences for an exponentially growing list of taxa not only enhances the density of taxon sampling for phylogenetic analyses, it also increases the sampling of paralogous gene lineages because some globin genes have very restricted phyletic distributions. As more complete genome sequences become available in coming years, it will be possible to trace the ancestry of globins back to more ancient branch points in the tree of life, and such insights should help to inform experimental studies of globin function.

The duplication and functional divergence of globin genes has promoted key physiological innovations in respiratory gas transport and other physiological functions during animal evolution. A combination of both tandem gene duplication and whole-genome duplication contributed to the diversification of the globin gene family in vertebrates. In Chapter 6 we will see how repeated rounds of gene duplication and divergence during the evolution of jawed vertebrates promoted the diversification of the subfamilies of genes that encode the different subunit chains of tetrameric Hb, leading to functional differentiation between Hb isoforms that are expressed during different stages of prenatal development and postnatal life.

References

Abi-Rached, L., Gilles, A., Shiina, T., Pontarotti, P., and Inoko, H. (2002). Evidence of en bloc duplication in vertebrate genomes. *Nature Genetics*, **31**, 100–5.

Antoniou, M. and Grosveld, F. (1990). β-globin dominant control region interacts differently with distal and proximal promoter elements. *Genes and Development*, **4**, 1007–13.

Ascenzi, P., Gustincich, S., and Marino, M. (2014). Mammalian nerve globins in search of functions. *IUBMB Life*, **66**, 268–76.

Awenius, C., Hankeln, T., and Burmester, T. (2001). Neuroglobins from the zebrafish *Danio rerio* and the pufferfish *Tetraodon nigroviridis*. *Biochemical and Biophysical Research Communications*, **287**, 418–21.

Bailly, X. and Vinogradov, S. (2008). The bilaterian sea urchin and the radial starlet sea anemone globins share strong homologies with vertebrate neuroglobins. *In:* Bolognesi, M., Di Prisco, G., and Verde, C. (eds.) *Dioxygen Binding and Sensing Proteins*, pp. 191–201. New York, NY, Springer Verlag.

Blank, M. and Burmester, T. (2012). Widespread occurrence of N-terminal acylation in animal globins and possible origin of respiratory globins from a membrane-bound ancestor. *Molecular Biology and Evolution*, **29**, 3553–61.

Blank, M., Kiger, L., Thielebein, A., et al. (2011a). Oxygen supply from the bird's eye perspective: globin E is a respiratory protein in the chicken retina. *Journal of Biological Chemistry*, **286**, 26507–15.

Blank, M., Wollberg, J., Gerlach, F., et al. (2011b). A membrane-bound vertebrate globin. *PloS One*, **6**, e25292.

Bonamore, A., Attill, A., Arenghi, F., et al. (2007). A novel chimera: the "truncated hemoglobin-antibiotic monooxygenase" from *Streptomyces avermitilis*. *Gene*, **398**, 52–61.

Braasch, I., Brunet, F., Volff, J.-N., and Schartl, M. (2009a). Pigmentation pathway evolution after whole-genome duplication in fish. *Genome Biology and Evolution*, **1**, 479–93.

Braasch, I., Salzburger, W., and Meyer, A. (2006). Asymmetric evolution in two fish-specifically duplicated receptor tyrosine kinase paralogons involved in teleost coloration. *Molecular Biology and Evolution*, **23**, 1192–202.

Braasch, I., Schartl, M., and Volff, J.-N. (2007). Evolution of pigment synthesis pathways by gene and genome duplication in fish. *BMC Evolutionary Biology*, **7**, 74.

Braasch, I., Volff, J.-N., and Schartl, M. (2009b). The endothelin system: evolution of vertebrate-specific ligand-receptor interactions by three rounds of genome duplication. *Molecular Biology and Evolution*, **26**, 783–99.

Brittain, T., O'Brien, A. J., Wells, R. M. G., and Baldwin, J. (1989). A study of the role of subunit aggregation in the expression of cooperative ligand-binding in the hemoglobin of the lamprey *Mordacia mordax*. *Comparative Biochemistry and Physiology B-Biochemistry & Molecular Biology*, **93**, 549–54.

Brittain, T. and Wells, R. M. G. (1986). Characterization of the changes in the state of aggregation induced by ligand-binding in the hemoglobin system of a primitive vertebrate, the hagfish *Eptatretus cirrhatus*. *Comparative Biochemistry and Physiology A-Physiology*, **85**, 785–90.

Burmester, T., Ebner, B., Weich, B., and Hankeln, T. (2002). Cytoglobin: a novel globin type ubiquitously expressed in vertebrate tissues. *Molecular Biology and Evolution*, **19**, 416–21.

Burmester, T., Haberkamp, M., Mitz, S., et al. (2004). Neuroglobin and cytoglobin: genes, proteins and evolution. *IUBMB Life*, **56**, 703–7.

Burmester, T. and Hankeln, T. (2009). What is the function of neuroglobin? *Journal of Experimental Biology*, **212**, 1423–8.

Burmester, T. and Hankeln, T. (2014). Function and evolution of vertebrate globins. *Acta Physiologica*, **211**, 501–14.

Burmester, T., Weich, B., Reinhardt, S., and Hankeln, T. (2000). A vertebrate globin expressed in the brain. *Nature*, **407**, 520–3.

Cannon, J. T., Cossermelli Vellutini, B., Smith, J., et al. (2016). Xenacoelomorpha is the sister group to Nephrozoa. *Nature*, **530**, 89–93.

Corti, P., Xue, J. M., Tejero, J., et al. (2016). Globin X is a six-coordinate globin that reduces nitrite to nitric oxide in fish red blood cells. *Proceedings of the National Academy of Sciences of the United States of America*, **113**, 8538–43.

Dehal, P. and Boore, J. L. (2005). Two rounds of whole genome duplication in the ancestral vertebrate. *PloS Biology*, **3**, 1700–8.

Delsuc, F., Brinkmann, H., Chourrout, D., and Philippe, H. (2006). Tunicates and not cephalochordates are the closest living relatives of vertebrates. *Nature*, **439**, 965–8.

Demuth, J. P., De Bie, T., Stajich, J. E., Cristianini, N., and Hahn, M. W. (2006). The evolution of mammalian gene families. *PloS One*, **1**, e85.

Dröge, J. and Makalowski, W. (2011). Phylogenetic analysis reveals wide distribution of globin X. *Biology Direct*, **6**, 54.

Ebner, B., Panopoulou, G., Vinogradov, S. N., et al. (2010). The globin gene family of the cephalochordate amphioxus: implications for chordate globin evolution. *BMC Evolutionary Biology*, **10**, 370.

Fabrizius, A., Andre, D., Laufs, T., et al. (2016). Critical re-evaluation of neuroglobin expression reveals conserved patterns among mammals. *Neuroscience*, **337**, 339–54.

Fago, A., Giangiacomo, L., d'Avino, R., et al. (2001). Hagfish hemoglobins. Structure, function, and oxygen-linked association. *Journal of Biological Chemistry*, **276**, 27415–23.

Fago, A., Hundahl, C., Malte, H., and Weber, R. E. (2004). Functional properties of neuroglobin and cytoglobin. Insights into the ancestral physiological roles of globins. *IUBMB Life*, **56**, 689–96.

Fago, A., Rohlfing, K., Petersen, E. E., Jendroszek, A., and Burmester, T. (2018). Functional diversification of sea lamprey globins in evolution and development. *Biochimica et Biophysica Acta*, **1866**, 283–91.

Fago, A. and Weber, R. E. (1995). The hemoglobin system of the hagfish *Myxine glutinosa*—Aggregation state and functional properties. *Biochimica Et Biophysica Acta-Protein Structure and Molecular Enzymology*, **1249**, 109–15.

Forrester, M. T. and Foster, M. W. (2012). Protection from nitrosative stress: a central role for microbial flavohemoglobin. *Free Radical Biology and Medicine*, **52**, 1620–33.

Fuchs, C., Burmester, T., and Hankeln, T. (2006). The amphibian globin gene repertoire as revealed by the *Xenopus* genome. *Cytogenetic and Genome Research*, **112**, 296–306.

Fuchs, C., Heib, V., Kiger, L., et al. (2004). Zebrafish reveals different and conserved features of vertebrate neuroglobin gene structure, expression pattern, and ligand binding. *Journal of Biological Chemistry*, **279**, 24116–22.

Fuchs, C., Luckhardt, A., Gerlach, F., Burmester, T., and Hankeln, T. (2005). Duplicated cytoglobin genes in teleost fishes. *Biochemical and Biophysical Research Communications*, **337**, 216–23.

Gardner, P. R. (2005). Nitric oxide dioxygenase function and mechanism of flavohemoglobin, hemoglobin, myoglobin and their associated reductases. *Journal of Inorganic Biochemistry*, **99**, 247–66.

Gardner, P. R. and Gardner, A. M. (2008). Nitric oxide dioxygenase: an ancient enzymic function of hemoglobin. *In*: Ghosh, A. (ed.) *The Smallest Biomolecules: Diatomics and Their Interactions with Heme Proteins*, pp. 290–326. New York, NY, Elsevier Science.

Gaudry, M. J., Storz, J. F., Butts, G. T., Campbell, K. L., and Hoffman, F. G. (2014). Repeated evolution of chimeric fusion genes in the β-globin gene family of laurasiatherian mammals. *Genome Biology and Evolution*, **6**, 1219–33.

Gell, D. A. (2018). Structure and function of haemoglobins. *Blood Cells, Molecules, and Disease*, **70**, 13–42.

Goodman, M., Czelusniak, J., Koop, B. F., Tagle, D. A., and Slightom, J. L. (1987). Globins—a case-study in molecular phylogeny. *Cold Spring Harbor Symposia on Quantitative Biology*, **52**, 875–90.

Goodman, M., Moore, G. W., and Matsuda, G. (1975). Darwinian evolution in the genealogy of hemoglobin. *Nature*, **253**, 603–8.

Greenberg, D. A., Jin, K., and Khan, A. A. (2008). Neuroglobin: an endogenous neuroprotectant. *Current Opinion in Pharmacology*, **8**, 20–4.

Hade, M. D., Kaur, J., Chakraborti, P. K., and Dikshit, K. L. (2017). Multidomain truncated hemoglobins: new members of the globin family exhibiting tandem repeats of globin units and domain fusion. *IUBMB Life*, **69**, 479–88.

Halligan, K. E., Jourd'heuil, F., and Jourd'heuil, D. (2009a). Cytoglobin regulates cell respiration and nitrosative stress through NO dioxygenation and co-localizes with inducible nitric oxide synthase during vascular injury. *Faseb Journal*, **23**(1 Suppl), 852.3–852.3.

Halligan, K. E., Jourd'heuil, F. L., and Jourd'heuil, D. (2009b). Cytoglobin is expressed in the vasculature and regulates cell respiration and proliferation via nitric oxide dioxygenation. *Journal of Biological Chemistry*, **284**, 8539–47.

Hankeln, T. and Burmester, T. (2008). Neuroglobin and cytoglobin. In: Ghosh, A. (ed.) *Smallest Biomolecules: Diatomics and Their Interactions with Heme Proteins*, pp. 203–18. New York, NY, Elsevier Science.

Hankeln, T., Ebner, B., Fuchs, C., et al. (2005). Neuroglobin and cytoglobin in search of their role in the vertebrate globin family. *Journal of Inorganic Biochemistry*, **99**, 110–19.

Hardison, R. (1998). Hemoglobins from bacteria to man: evolution of different patterns of gene expression. *Journal of Experimental Biology*, **201**, 1099–117.

Hardison, R. C. (2012). Evolution of hemoglobin and its genes. *Cold Spring Harbor Perspectives in Medicine*, **2**, a011627.

Heaslet, H. A. and Royer, W. E. (2001). Crystalline ligand transitions in Lamprey hemoglobin—structural evidence for the regulation of oxygen affinity. *Journal of Biological Chemistry*, **276**, 26230–6.

Helbo, S., Weber, R. E., and Fago, A. (2013). Expression patterns and adaptive functional diversity of vertebrate myoglobins. *Biochimica Et Biophysica Acta-Proteins and Proteomics*, **1834**, 1832–9.

Herman, J. L., Challis, C. J., Novák, A., Hein, J., and Schmidler, S. C. (2014). Simultaneous Bayesian estimation of alignment and phylogeny under a joint model of protein sequence and structure. *Molecular Biology and Evolution*, **31**, 2251–66.

Hoffmann, F. G., Opazo, J. C., Hoogewijs, D., et al. (2012a). Evolution of the globin gene family in deuterostomes: lineage-specific patterns of diversification and attrition. *Molecular Biology and Evolution*, **29**, 1735–45.

Hoffmann, F. G., Opazo, J. C., and Storz, J. F. (2008a). New genes originated via multiple recombinational pathways in the β-globin gene family of rodents. *Molecular Biology and Evolution*, **25**, 2589–600.

Hoffmann, F. G., Opazo, J. C., and Storz, J. F. (2008b). Rapid rates of lineage-specific gene duplication and deletion in the α-globin gene family. *Molecular Biology and Evolution*, **25**, 591–602.

Hoffmann, F. G., Opazo, J. C., and Storz, J. F. (2010a). Gene cooption and convergent evolution of oxygen transport hemoglobins in jawed and jawless vertebrates. *Proceedings of the National Academy of Sciences of the United States of America*, **107**, 14274–9.

Hoffmann, F. G., Opazo, J. C., and Storz, J. F. (2011). Differential loss and retention of cytoglobin, myoglobin, and globin-E during the radiation of vertebrates. *Genome Biology and Evolution*, **3**, 588–600.

Hoffmann, F. G., Opazo, J. C., and Storz, J. F. (2012b). Whole-genome duplications spurred the functional diversification of the globin gene superfamily in vertebrates. *Molecular Biology and Evolution*, **29**, 303–12.

Hoffmann, F. G., Storz, J. F., Gorr, T. A., and Opazo, J. C. (2010b). Lineage-specific patterns of functional diversification in the α- and β-globin gene families of tetrapod vertebrates. *Molecular Biology and Evolution*, **27**, 1126–38.

Hoffmann, F. G., Vandewege, M. W., Storz, J. F., and Opazo, J. C. (2018). Gene turnover and diversification of the α- and β-globin gene families in sauropsid vertebrates. *Genome Biology and Evolution*, **10**, 344–58.

Holland, P. W. H. (2003). More genes in vertebrates? *Journal of Structural and Functional Genomics*, **3**, 75–84.

Holland, P. W. H., Garciafernandez, J., Williams, N. A., and Sidow, A. (1994). Gene duplications and the origins of vertebrate development. *Development Supplement*, 125–33.

Hoogewijs, D., Ebner, B., Germani, F., et al. (2012). Androglobin: a chimeric globin in metazoans that is preferentially expressed in mammalian testes. *Molecular Biology and Evolution*, **29**, 1105–14.

Horton, A. C., Mahadevan, N. R., Ruvinsky, I., and Gibson-Brown, J. J. (2003). Phylogenetic analyses alone are insufficient to determine whether genome duplication(s) occurred during early vertebrate evolution. *Journal of Experimental Zoology Part B-Molecular and Developmental Evolution*, **299B**, 41–53.

Jourd'heuil, F. L., Xu, H. Y., Reilly, T., et al. (2017). The hemoglobin homolog cytoglobin in smooth muscle inhibits apoptosis and regulates vascular remodeling. *Arteriosclerosis Thrombosis and Vascular Biology*, **37**, 1944–55.

Kakar, S., Hoffmann, F. G., Storz, J. F., Fabian, M., and Hargrove, M. S. (2010). Structure and reactivity of hexacoordinate hemoglobins. *Biophysical Chemistry*, **152**, 1–14.

Kiger, L., Tilleman, L., Geuens, E., et al. (2011). Electron transfer function versus oxygen delivery: a comparative study for several hexacoordinated globins across the animal kingdom. *PloS One*, **6**, e20478.

Koch, J. and Burmester, T. (2016). Membrane-bound globin X protects the cell from reactive oxygen species. *Biochemical and Biophysical Research Communications*, **469**, 275–80.

Koch, J., Ludemann, J., Spies, R., et al. (2016). Unusual diversity of myoglobin genes in the lungfish. *Molecular Biology and Evolution*, **33**, 3033–41.

Kugelstadt, D., Haberkamp, M., Hankeln, T., and Burmester, T. (2004). Neuroglobin, cytoglobin, and a novel, eye-specific globin from chicken. *Biochemical and Biophysical Research Communications*, **325**, 719–25.

Kumar, S., Stecher, G., and Tamura, K. (2016). MEGA7: Molecular Evolutionary Genetics Analysis Version 7.0 for bigger datasets. *Molecular Biology and Evolution*, **33**, 1870–4.

Lamy, J. N., Green, B. N., Toulmond, A., et al. (1996). Giant hexagonal bilayer hemoglobins. *Chemical Reviews*, **96**, 3113–24.

Larhammar, D., Nordstrom, K., and Larsson, T. A. (2009). Evolution of vertebrate rod and cone phototransduction genes. *Philosophical Transactions of the Royal Society B-Biological Sciences*, **364**, 2867–80.

Lecomte, J. T. J., Vuletich, D. A., and Lesk, A. M. (2005). Structural divergence and distant relationships in proteins: evolution of the globins. *Current Opinion in Structural Biology*, **15**, 290–301.

Lesk, A. M. and Chothia, C. (1980). How different amino acid sequences determine similar protein structures—structure and evolutionary dynamics of the globins. *Journal of Molecular Biology*, **136**, 225–70.

Li, W.-H. (1997). *Molecular Evolution*, Sunderland, MA, Sinauer Associates, Inc.

Liu, X. P., El-Mahdy, M. A., Boslett, J., et al. (2017). Cytoglobin regulates blood pressure and vascular tone through nitric oxide metabolism in the vascular wall. *Nature Communications*, **8**, 14807.

Lynch, M. and Katju, V. (2004). The altered evolutionary trajectories of gene duplicates. *Trends in Genetics*, **20**, 544–9.

Macqueen, D. J., de la Serrana, D. G., and Johnston, I. A. (2014). Cardiac myoglobin deficit has evolved repeatedly in teleost fishes. *Biology Letters*, **10**, 20140225.

Martinkova, M., Kitanishi, K., and Shimizu, T. (2013). Heme-based globin-coupled oxygen sensors: linking oxygen binding to functional regulation of diguanylate cyclase, histidine kinase, and methyl-accepting chemotaxis. *Journal of Biological Chemistry*, **288**, 27702–11.

Meyer, A. and Schartl, M. (1999). Gene and genome duplications in vertebrates: the one-to-four (-to-eight in fish) rule and the evolution of novel gene functions. *Current Opinion in Cell Biology*, **11**, 699–704.

Mimura, I., Nangaku, M., Nishi, H., et al. (2010). Cytoglobin, a novel globin, plays an antifibrotic role in the kidney. *American Journal of Physiology-Renal Physiology*, **299**, F1120–33.

Natarajan, C., Hoffman, F. G., Lanier, H. C., et al. (2015). Intraspecific polymorphism, interspecific divergence, and the origins of function-altering mutations in deer mouse hemoglobin. *Molecular Biology and Evolution*, **32**, 978–97.

Nishi, H., Inagi, R., Kawada, N., et al. (2011). Cytoglobin, a novel member of the globin family, protects kidney fibroblasts against oxidative stress under ischemic conditions. *American Journal of Pathology*, **178**, 128–39.

Ohno, S. (1970). *Evolution by Gene Duplication*, New York, NY, Springer-Verlag.

Oleksiewicz, U., Liloglou, T., Field, J. K., and Xinarianos, G. (2011). Cytoglobin: biochemical, functional and clinical perspective of the newest member of the globin family. *Cellular and Molecular Life Sciences*, **68**, 3869–83.

Opazo, J. C., Butts, G. T., Nery, M. F., Storz, J. F., and Hoffmann, F. G. (2013). Whole-genome duplication and the functional diversification of teleost fish hemoglobins. *Molecular Biology and Evolution*, **30**, 140–53.

Opazo, J. C., Hoffman, F. G., Natarajan, C., et al. (2015a). Gene turnover in the avian globin gene family and evolutionary changes in hemoglobin isoform expression. *Molecular Biology and Evolution*, **32**, 871–87.

Opazo, J. C., Hoffmann, F. G., and Storz, J. F. (2008a). Differential loss of embryonic globin genes during the radiation of placental mammals. *Proceedings of the National Academy of Sciences of the United States of America*, **105**, 12950–5.

Opazo, J. C., Hoffmann, F. G., and Storz, J. F. (2008b). Genomic evidence for independent origins of β-like globin genes in monotremes and therian mammals. *Proceedings of the National Academy of Sciences of the United States of America*, **105**, 1590–5.

Opazo, J. C., Lee, A. P., Hoffmann, F. G., et al. (2015b). Ancient duplications and expression divergence in the globin gene superfamily of vertebrates: insights from the elephant shark genome and transcriptome. *Molecular Biology and Evolution*, **32**, 1684–94.

Opazo, J. C., Sloan, A. M., Campbell, K. L., and Storz, J. F. (2009). Origin and ascendancy of a chimeric fusion gene: the β/δ-globin gene of paenungulate mammals. *Molecular Biology and Evolution*, **26**, 1469–78.

Pébusque, M. J., Coulier, F., Birnbaum, D., and Pontarotti, P. (1998). Ancient large-scale genome duplications: phylogenetic and linkage analyses shed light on chordate genome evolution. *Molecular Biology and Evolution*, **15**, 1145–59.

Pesce, A., Bolognesi, M., Bocedi, A., et al. (2002). Neuroglobin and cytoglobin—Fresh blood for the vertebrate globin family. *Embo Reports*, **3**, 1146–51.

Pesce, A., Couture, M., DeWilde, S., et al. (2000). A novel two-over-two alpha-helical sandwich fold is characteristic of the truncated hemoglobin family. *EMBO Journal*, **19**, 2424–34.

Putnam, N. H., Butts, T., Ferrier, D. E. K., et al. (2008). The amphioxus genome and the evolution of the chordate karyotype. *Nature*, **453**, 1064–U3.

Qiu, Y., Maillett, D. H., Knapp, J., Olson, J. S., and Riggs, A. F. (2000). Lamprey hemoglobin. Structural basis of the Bohr effect. *Journal of Biological Chemistry*, **275**, 13517–28.

Rachmilewitz, E. A. and Schrier, S. L. (2001). Pathophysiology of β thalassemia. In: Steinberg, M. H., Forget, B. G., Higgs, D. R., and Nagel, R. L. (eds.) *Disorders of Hemoglobin: Genetics, Pathophysiology, and Clinical Management*, pp. 233–51. Cambridge, Cambridge University Press.

Reeder, B. J. (2017). Redox and peroxidase activities of the hemoglobin superfamily: relevance to health and disease. *Antioxidants and Redox Signaling*, **26**, 763–76.

Reuss, S., Banica, O., Elgurt, M., et al. (2016a). Neuroglobin expression in the mammalian auditory system. *Molecular Neurobiology*, **53**, 1461–77.

Reuss, S., Wystub, S., Disque-Kaiser, U., Hankeln, T., and Burmester, T. (2016b). Distribution of cytoglobin in the mouse brain. *Frontiers in Neuroanatomy*, **10**, 47.

Riggs, A. F. (1991). Aspects of the origin and evolution of non-vertebrate hemoglobins. *American Zoologist*, **31**, 535–45.

Roesner, A., Fuchs, C., Hankeln, T., and Burmester, T. (2005). A globin gene of ancient evolutionary origin in lower vertebrates: evidence for two distinct globin families in animals. *Molecular Biology and Evolution*, **22**, 12–20.

Rouse, G. W., Wilson, N., Carvajal, J. I., and Vrijenhoek, R. C. (2016). New deep-sea species of *Xenoturbella* and the position of Xenacoelomorpha. *Nature*, **530**, 94–7.

Royer, W. E., Knapp, J. E., Strand, K., and Heaslet, H. A. (2001). Cooperative hemoglobins: conserved fold, diverse quaternary assemblies and allosteric mechanisms. *Trends in Biochemical Sciences*, **26**, 297–304.

Royer, W. E., Sharma, H., Strand, K., Knapp, J. E., and Bhyravbhatla, B. (2006). Lumbricus erythrocruorin at 3.5 angstrom resolution: architecture of a megadalton respiratory complex. *Structure*, **14**, 1167–77.

Royer, W. E., Zhu, H., Gorr, T. A., Flores, J. F., and Knapp, J. E. (2005). Allosteric hemoglobin assembly: diversity and similarity. *Journal of Biological Chemistry*, **280**, 27477–80.

Runck, A. M., Moriyama, H., and Storz, J. F. (2009). Evolution of duplicated β-globin genes and the structural basis of hemoglobin isoform differentiation in *Mus*. *Molecular Biology and Evolution*, **26**, 2521–32.

Sato, Y., Hashiguchi, Y., and Nishida, M. (2009). Temporal pattern of loss/persistence of duplicate genes involved in signal transduction and metabolic pathways after teleost-specific genome duplication. *BMC Evolutionary Biology*, **9**, 127.

Schwarze, K. and Burmester, T. (2013). Conservation of globin genes in the "living fossil" *Latimeria chalumnae* and reconstruction of the evolution of the vertebrate globin family. *Biochimica Et Biophysica Acta-Proteins and Proteomics*, **1834**, 1801–12.

Schwarze, K., Campbell, K. L., Hankeln, T., et al. (2014). The globin gene repertoire of lampreys: convergent evolution of hemoglobin and myoglobin in jawed and jawless vertebrates. *Molecular Biology and Evolution*, **31**, 2708–21.

Schwarze, K., Singh, A., and Burmester, T. (2015). The full globin repertoire of turtles provides insights into vertebrate globin evolution and functions. *Genome Biology and Evolution*, **7**, 1896–913.

Shimeld, S. M. and Holland, P. W. H. (2000). Vertebrate innovations. *Proceedings of the National Academy of Sciences of the United States of America*, **97**, 4449–52.

Singh, S., Canseco, D. C., Manda, S. M., et al. (2014). Cytoglobin modulates myogenic progenitor cell viability and muscle regeneration. *Proceedings of the National Academy of Sciences of the United States of America*, **111**, E129–38.

Storz, J. F. (2016). Gene duplication and evolutionary innovations in hemoglobin-oxygen transport. *Physiology*, **31**, 223–32.

Storz, J. F., Baze, M., Waite, J. L., Hoffmann, F. G., Opazo, J. C., and Hayes, J. P. (2007). Complex signatures of selection and gene conversion in the duplicated globin genes of house mice. *Genetics*, **177**, 481–500.

Storz, J. F., Hoffmann, F. G., Opazo, J. C., and Moriyama, H. (2008). Adaptive functional divergence among triplicated α-globin genes in rodents. *Genetics*, **178**, 1623–38.

Storz, J. F., Opazo, J. C., and Hoffmann, F. G. (2011). Phylogenetic diversification of the globin gene superfamily in chordates. *IUBMB Life*, **63**, 313–22.

Storz, J. F., Opazo, J. C., and Hoffmann, F. G. (2013). Gene duplication, genome duplication, and the functional diversification of vertebrate globins. *Molecular Phylogenetics and Evolution*, **66**, 469–78.

Taylor, J. S., Braasch, I., Frickey, T., Meyer, A., and Van De Peer, Y. (2003). Genome duplication, a trait shared by 22,000 species of ray-finned fish. *Genome Research*, **13**, 382–90.

Taylor, J. S., Van De Peer, Y., Braasch, I., and Meyer, A. (2001). Comparative genomics provides evidence for an ancient genome duplication event in fish. *Philosophical Transactions of the Royal Society of London Series B-Biological Sciences*, **356**, 1661–79.

Terwilliger, N. B. (1998). Functional adaptations of oxygen-transport proteins. *Journal of Experimental Biology*, **201**, 1085–98.

Terwilliger, N. B. and Terwilliger, R. C. (1978). Oxygen binding domains of a clam (*Cardita borealis*) extracellular hemoglobin. *Biochimica et Biophysica Acta*, **537**, 77–85.

Trent, J. T. and Hargrove, M. S. (2002). A ubiquitously expressed human hexacoordinate hemoglobin. *Journal of Biological Chemistry*, **277**, 19538–45.

Van De Peer, Y., Maere, S., and Meyer, A. (2009). The evolutionary significance of ancient genome duplications. *Nature Reviews Genetics*, **10**, 725–32.

Vinogradov, S. N. (1985). The structure of invertebrate extracellular hemoglobins (erythrocruorins and chlorocruorins). *Comparative Biochemistry and Physiology B-Biochemistry and Molecular Biology*, **82**, 1–15.

Vinogradov, S. N., Hoogewijs, D., Bailly, X., et al. (2005). Three globin lineages belonging to two structural classes in genomes from the three kingdoms of life. *Proceedings of the National Academy of Sciences of the United States of America*, **102**, 11385–9.

Vinogradov, S. N., Hoogewijs, D., Bailly, X., et al. (2007). A model of globin evolution. *Gene*, **398**, 132–42.

Vinogradov, S. N. and Moens, L. (2008). Diversity of globin function: enzymatic, transport, storage, and sensing. *Journal of Biological Chemistry*, **283**, 8773–7.

Vinogradov, S. N., Tinajero-Trejo, M., Poole, R. K., and Hoogewijs, D. (2013). Bacterial and archaeal globins— a revised perspective. *Biochimica Et Biophysica Acta- Proteins and Proteomics*, **1834**(9), 1789–800.

Vinogradov, S. N., Walz, D. A., Pohajdak, B., et al. (1993). Adventitious variability—the amino acid sequences of nonvertebrate globins. *Comparative Biochemistry and Physiology B-Biochemistry and Molecular Biology*, **106**, 1–26.

Wada, H. and Makabe, K. (2006). Genome duplications of early vertebrates as a possible chronicle of the evolutionary history of the neural crest. *International Journal of Biological Sciences*, **2**, 133–41.

Wajcman, H., Kiger, L., and Marden, M. C. (2009). Structure and function evolution in the superfamily of globins. *Comptes Rendus Biologies*, **332**, 273–82.

Weber, R. E. and Vinogradov, S. N. (2001). Nonvertebrate hemoglobins: functions and molecular adaptations. *Physiological Reviews*, **81**, 569–628.

Wheeler, D., Hope, R. M., Cooper, S. J. B., Gooley, A. A., and Holland, R. A. B. (2004). Linkage of the β-like ω-globin gene to α-like globin genes in an Australian marsupial supports the chromosome duplication model for separation of globin gene clusters. *Journal of Molecular Evolution*, **58**, 642–52.

Wittenberg, J. B., Bolognesi, M., Wittenberg, B. A., and Guertin, M. (2002). Truncated hemoglobins: a new family of hemoglobins widely distributed in bacteria, unicellular eukaryotes, and plants. *Journal of Biological Chemistry*, **277**, 871–4.

Wystub, S., Ebner, B., Fuchs, C., Weich, B., Burmester, T., and Hankeln, T. (2004). Interspecies comparison of neuroglobin, cytoglobin and myoglobin: sequence evolution and candidate regulatory elements. *Cytogenetic and Genome Research*, **105**, 65–78.

Xi, Y., Obara, M., Ishida, Y., Ikeda, S., and Yoshizato, K. (2007). Gene expression and tissue distribution of cytoglobin and myoglobin in the Amphibia and Reptilia: possible compensation of myoglobin with cytoglobin in skeletal muscle cells of anurans that lack the myoglobin gene. *Gene*, **398**, 94–102.

Zhang, G. J. and Cohn, M. J. (2008). Genome duplication and the origin of the vertebrate skeleton. *Current Opinion in Genetics & Development*, **18**, 387–93.

Zhang, G., Li, C., Li, Q., et al. (2014). Comparative genomics reveals insights into avian genome evolution and adaptation. *Science*, **346**, 1311–20.

Zhang, J. Z. (2003). Evolution by gene duplication: an update. *Trends in Ecology & Evolution*, **18**, 292–8.

CHAPTER 6

Gene duplication and hemoglobin isoform differentiation

6.1 Developmental regulation of Hb synthesis

Having surveyed the entire superfamily of vertebrate globins in Chapter 5, here we will explore how repeated rounds of gene duplication and divergence promoted the diversification of the subfamilies of genes that encode the different subunit chains of tetrameric Hb. In all jawed vertebrates that have been investigated, the α- and β-type globin genes are ontogenetically regulated such that structurally and functionally distinct Hb isoforms (isoHbs) are expressed during different stages of prenatal development and postnatal life. In mammals, the arrangements of tandemly linked genes in the α- and β-globin gene clusters are colinear with the temporal order of expression during development (Forget and Hardison, 2009). For example, the human α-globin gene cluster is arranged: 5′-ζ (embryonic)-α_2 (fetal and adult)-α_1 (fetal and adult)-3′, and the human β-globin gene cluster is arranged: 5′-ϵ (embryonic)-$^G\gamma$ (fetal)-$^A\gamma$ (fetal)-δ (minor adult)-β (major adult)-3′ (Fig. 6.1A). This same general arrangement is also seen in the α- and β-globin gene clusters of other amniotes, although the individual identities of early- and late-expressed genes vary among taxa due to lineage-specific gene duplications and deletions (Hoffmann et al., 2008b, Opazo et al., 2008a, Opazo et al., 2008b, Hoffmann et al., 2010, Storz et al., 2011, Opazo et al., 2015, Hoffmann et al., 2018). Developmentally regulated members of the α- and β-globin gene clusters direct the synthesis of functionally distinct isoHbs in primitive (embryonic) erythrocytes derived from the yolk sac and in definitive (adult) erythroid cells derived from the bone marrow (Fig. 6.1B). The pre- and postnatal pattern of isoHb and globin subunit expression during human development is shown in Fig. 6.2.

Evolutionary changes in the developmental timing of isoHb expression are typically associated with changes in isoHb-specific oxygenation properties, with each different isoHb adapted to perform distinct O_2-scavenging/O_2-transport tasks during different stages of development (Weber, 1994, Weber, 1995, Brittain, 2002). Evolved changes in functional properties of the various pre- and postnatally expressed isoHbs are caused by amino acid substitutions in paralogous genes that encode the different α- and/or β-type subunits. As a representative example, amino acid sequence alignments of the full complement of α- and β-type human Hb chains are shown in Fig. 6.3.

During human embryogenesis, O_2 diffusion is sufficient to meet the metabolic demands of the developing embryo until day 15 postconception (Fantoni et al., 1981, Brittain, 2002). At this early stage of development, the embryonic α- and β-type globin genes (ζ- and ϵ-globin, respectively) are transcriptionally activated to produce Hb Gower I ($\zeta_2\epsilon_2$), which serves as the primary O_2-carrier (Huehns et al., 1964, Wood, 1976). After four weeks of gestation, the heart of the developing embryo becomes septated, the venous and arterial circulations are established, and the placenta begins to develop. During this phase, two additional embryonic isoHbs are synthesized, Hb Gower II ($\alpha_2\epsilon_2$) and Hb Portland ($\zeta_2\gamma_2$), along with fetal Hb (HbF,

Hemoglobin: Insights into Protein Structure, Function, and Evolution. Jay F. Storz, Oxford University Press (2019).
© Jay F. Storz 2019. DOI: 10.1093/oso/9780198810681.001.0001

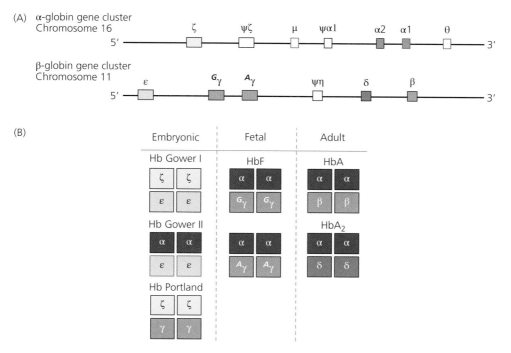

Fig. 6.1. The expression of α- and β-type globin genes is developmentally regulated, resulting in the synthesis of functionally distinct isoHbs. (A) Structure of the human α- and β-globin gene clusters. (B) The set of structurally distinct embryonic, fetal, and adult Hb isoHbs, with subunits encoded by each of the pre- and postnatally expressed α- and β-type genes.

Modified from Storz (2016).

$\alpha_2\gamma_2$). During the next six weeks, the placental circulation is established, the yolk sac gradually disappears, and the liver becomes the major site for hematopoiesis, producing definitive, enucleated erythrocytes containing a mix of HbF and adult Hb (HbA, $\alpha_2\beta_2$). After ~20 weeks of gestation, the bone marrow becomes established as a secondary site for hematopoiesis, producing mainly HbA (Wood, 1976). At birth, the neonatal circulation consists of erythrocytes containing ~70 percent HbF and ~30 percent HbA. In five-month old infants, the fraction of HbF in the blood falls to ~3 percent, and by two years of age, circulating erythrocytes derived from the bone marrow contain ~97 percent HbA and ~3 percent HbA$_2$ ($\alpha_2\delta_2$) (Fig. 6.2). This developmental pattern of Hb switching has been profitably exploited as a model system for understanding the transcriptional regulation of gene expression (Forget and Hardison, 2009, Sankaran et al., 2010, Hardison, 2012, Philipsen and Hardison, 2018).

6.2 Evolution of developmentally regulated Hb synthesis

The pattern of gene switching during human development described in section 6.1 is also observed in all other tetrapod vertebrates that have been examined to date (Wells, 1979, Weber et al., 1987, Alev et al., 2009, Storz et al., 2011). In the α-globin gene cluster, the physiological division of labor between early- and late-expressed genes was established in the common ancestor of tetrapod vertebrates and it appears to have been retained in nearly all descendent lineages. The ancestral arrangement of the tetrapod α-globin gene cluster is 5'-α^E-α^D-α^A-3' (Hoffmann and Storz, 2007, Hoffmann et al., 2010), where α^E is orthologous to the embryonic ζ-globin gene in humans and other mammals and α^A is orthologous to the adult α-globin gene. The duplicative history of these three α-type paralogs has proven difficult to unravel. The original view was that the embryonic α^E-globin gene and the adult-expressed progenitor of

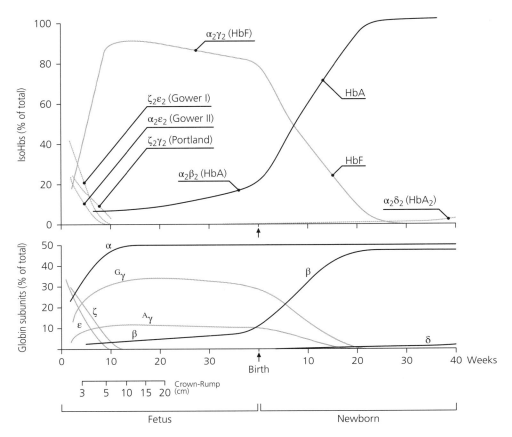

Fig. 6.2. Timeline for changes in the expression levels of tetrameric isoHbs and individual α- and β-type subunits during human development. Modified from Bunn and Forget (1986) and Storz (2016).

the α^A/α^D genes originated via tandem duplication of an ancestral proto α-globin gene in the stem lineage of tetrapods. According to this view, the α^D- and α^A-globin genes originated via a subsequent tandem duplication event in the stem lineage of amniotes, yielding the tree topology ($\alpha^E(\alpha^D,\alpha^A)$) (Cooper et al., 2006) (Fig. 6.4A). An alternative hypothesis is that the α^A-globin gene and the (presumably embryonic) progenitor of the α^E/α^D genes originated via tandem duplication of an ancestral proto α-globin gene in the stem lineage of tetrapods; this was then followed by a subsequent duplication event that gave rise to the α^E- and α^D-globins, yielding the topology (($\alpha^E,\alpha^D)\alpha^A$) (Hoffmann and Storz, 2007; Hoffmann et al. 2010) (Fig. 6.4B). Under this latter scenario both duplication events occurred in the stem lineage of tetrapods after

divergence from the ancestor of lobe-finned fishes approximately 370–430 million years ago.

In modern tetrapods, the α^E-globin gene appears to be expressed exclusively in larval/embryonic erythroid cells and the α^A-globin gene is expressed in definitive erythroid cells during later stages of prenatal development and postnatal life. In mammals, products of the α^D-globin gene (annotated as "μ-globin" in the human genome assembly) do not appear to be incorporated into functional Hb tetramers. However, the α^D-globin gene is expressed in both primitive and definitive erythroid cells of birds, turtles, and squamate reptiles (Alev et al., 2009, Storz et al., 2011, Storz et al., 2015). If shared amino acid states between the α^E- and α^D-globins of amniotes are explained by common descent from the same single-copy progenitor (or a history of

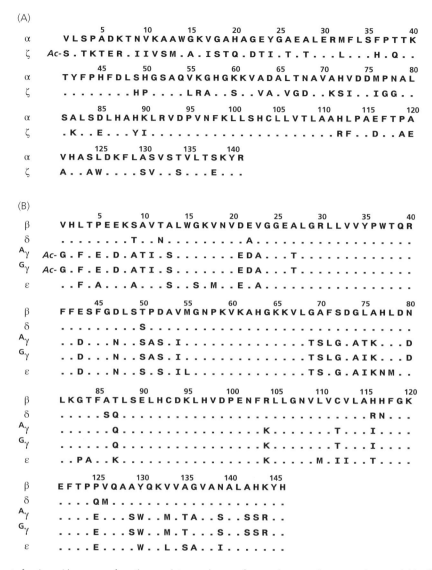

Fig. 6.3. Alignment of amino acid sequences from the complete complement of pre- and postnatally expressed α-type globins (A) and β-type globins (B) in humans. N-terminal residues are acetylated in the embryonic ζ chain and in the fetal γ chains.

interparalog $\alpha^E \rightarrow \alpha^D$ gene conversion), the intriguing implication is that distinct biochemical properties of isoHbs that incorporate α^D subunits may reflect a retained ancestral character state that harkens back to a primordial, embryonic function (Hoffmann and Storz, 2007, Grispo et al., 2012).

In contrast to the ancient functional diversification of α-type globin genes (Fig. 6.5A), the developmental regulation of gene expression in the β-globin gene cluster evolved independently in several different tetrapod lineages (Hoffmann et al., 2010, Hoffmann

et al., 2018). For example, in mammals and birds, the β-type globin genes that are solely expressed during the earliest stages of embryogenesis were independently derived from lineage-specific duplications of the same proto β-globin gene. In other words, the embryonic β-type globin genes of mammals and birds are not "1:1 orthologs" (Fig. 6.5B). Even within mammals, embryonic β-type globin genes appear to have originated independently as the products of lineage-specific duplication events in monotremes (egg-laying mammals) and in the common ancestor

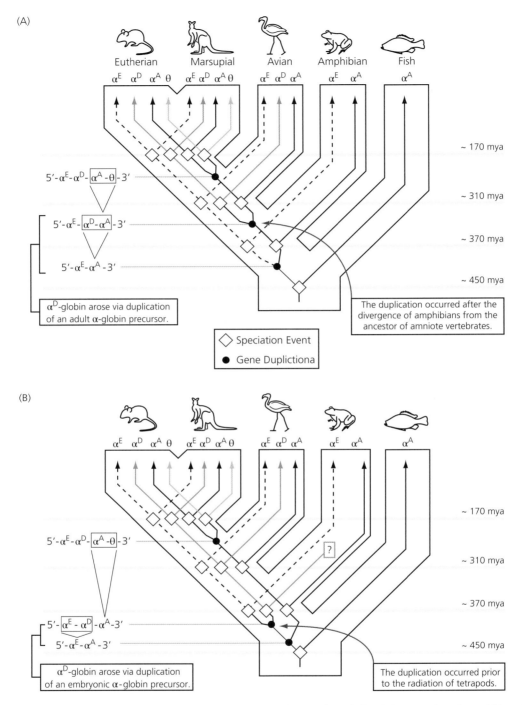

Fig. 6.4. Alternative hypotheses regarding the duplicative history of three α-type globin paralogs, α^E, α^D, and α^A in tetrapod vertebrates. (A) According to one hypothesis, the embryonic α^E-globin gene and the adult-expressed progenitor of the α^A/α^D genes originated via tandem duplication of an ancestral proto α-globin gene in the stem lineage of tetrapods. According to this view, the α^E- and α^A-globin genes originated via a subsequent tandem duplication event in the stem lineage of amniotes, yielding the tree topology ($\alpha^E(\alpha^D,\alpha^A)$). (B) An alternative hypothesis is that the α^A-globin gene and the (presumably embryonic) progenitor of the α^E/α^D genes originated via tandem duplication of an ancestral proto α-globin gene in the stem lineage of tetrapods; this was then followed by a subsequent duplication event that gave rise to the α^E- and α^D-globins, yielding the topology (($\alpha^E,\alpha^D)\alpha^A$). Under this latter scenario both duplication events occurred in the stem lineage of tetrapods after divergence from the ancestor of lobe-finned fishes approximately 370–430 million years ago. At present, the weight of phylogenetic evidence supports the latter scenario (Hoffmann and Storz 2007; Hoffmann et al. 2010).

Reproduced from Hoffmann and Storz (2007).

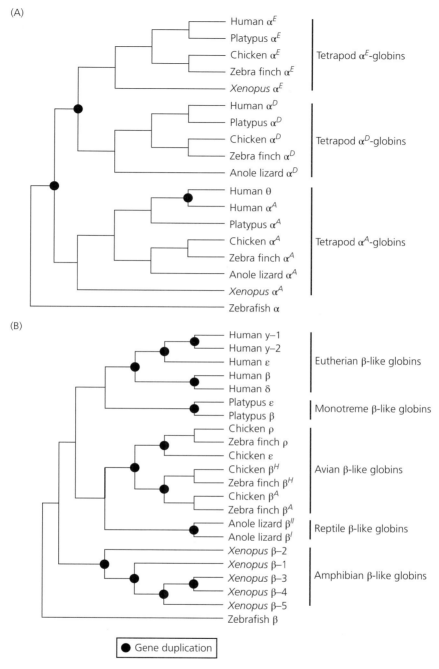

(A)

Human α^E
Platypus α^E
Chicken α^E
Zebra finch α^E
Xenopus α^E
⎤ Tetrapod α^E-globins

Human α^D
Platypus α^D
Chicken α^D
Zebra finch α^D
Anole lizard α^D
⎤ Tetrapod α^D-globins

Human θ
Human α^A
Platypus α^A
Chicken α^A
Zebra finch α^A
Anole lizard α^A
Xenopus α^A
⎤ Tetrapod α^A-globins

Zebrafish α

(B)

Human y–1
Human y–2
Human ε
Human β
Human δ
⎤ Eutherian β-like globins

Platypus ε
Platypus β
⎤ Monotreme β-like globins

Chicken ρ
Zebra finch ρ
Chicken ε
Chicken β^H
Zebra finch β^H
Chicken β^A
Zebra finch β^A
⎤ Avian β-like globins

Anole lizard β^{II}
Anole lizard β^I
⎤ Reptile β-like globins

Xenopus β–2
Xenopus β–1
Xenopus β–3
Xenopus β–4
Xenopus β–5
⎤ Amphibian β-like globins

Zebrafish β

● Gene duplication

Fig. 6.5. Diagrammatic phylogenies depicting the inferred relationships among members of the α- and β-globin gene subfamilies in tetrapods. In each tree, nodes depicted as filled symbols represent gene duplication events. The remaining nodes represent speciation events (phylogenetic splitting at the organismal level). (A) Phylogeny of α-type globin genes in representative tetrapod lineages. Note that the three paralogs (α^E-, α^D-, and α^A-globin) are reciprocally monophyletic relative to one another. As discussed in the text, the α^E- and α^D-globin genes are products of a duplication event that occurred in the stem lineage of tetrapods. Orthologs of the embryonic α^E-globin gene are known as α^L-globin in amphibians, π-globin in birds, and ζ-globin in mammals. The human ortholog of the α^D-globin gene is known as μ-globin. (B) Phylogeny of β-type globin genes in representative tetrapod lineages. Note that eutherian mammals, monotremes, birds, non-avian reptiles, and amphibians each inherited an ortholog of the same proto β-type gene, which then underwent one or more rounds of duplication and divergence to produce distinct repertoires of β-type globins in each descendent lineage. The depicted phylogenies are based on data reported in Hoffmann and Storz (2007), Opazo, Hoffmann et al. (2008), Hoffmann, Storz et al. (2010).

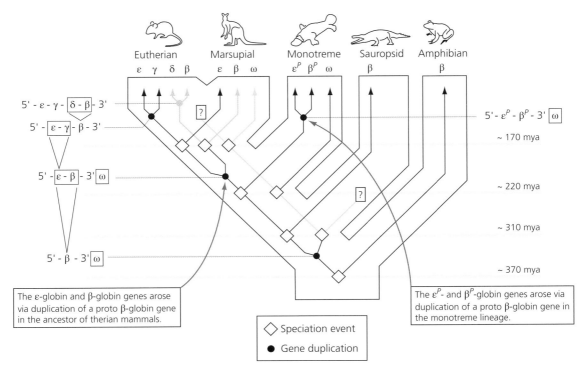

Fig. 6.6. An evolutionary hypothesis regarding the evolution of the β-globin gene family in mammals. According to this model, the ω-globin gene originated via duplication of an ancient β-globin gene that occurred before the divergence of birds and mammals but after the amniote/amphibian split. The ω-globin gene has been retained in contemporary monotremes and marsupials, but it has been lost independently in birds and placental mammals. In the common ancestor of marsupials and placental mammals, a pair of ε- and β-globin genes originated via duplication of a proto β-globin gene after the therian/monotreme split. In the placental mammal lineage, subsequent duplications of the ε- and β-globin genes gave rise to the prenatally expressed γ-globin and the adult-expressed δ-globin, respectively. In the monotreme lineage, a pair of β-like globin genes ($ε^P$- and $β^P$-globin) originated via duplication of a proto β-globin gene sometime before the divergence of the platypus and echidnas (the two monotreme lineages). The $β^P$-globin gene is expressed during adulthood, and, based on positional homology with other β-like globin genes, expression of the $ε^P$-globin gene is most likely restricted to embryonic erythroid cells.

Reproduced from Opazo et al. (2008b) with permission from the National Academy of Sciences.

of marsupials and eutherian mammals (Opazo et al., 2008b) (Fig. 6.6). Likewise, fetally expressed β-type globin genes originated independently in simian primates (New World monkeys, Old World monkeys, apes, and humans) and in ruminant artiodactyls (cattle, buffalo, antelope, sheep, goats, and allies). These are the only two groups of mammals that are known to express fetal isoHbs that are structurally and functionally distinct from all early-expressed embryonic isoHbs and late-expressed adult isoHbs. In most eutherian mammals, the γ-globin gene encodes the β-chain subunit of embryonic isoHbs, but in simian primates, duplicated copies of γ-globin, γ1 and γ2 (= $^Gγ$ and $^Aγ$ in humans) have been co-opted for fetal expression (Johnson et al., 1996, Johnson et al., 2000) (Fig. 6.7). In New

World monkeys, γ1 is expressed in nucleated erythroid cells derived from the embryonic yolk sac (the ancestral condition), but γ2 is expressed in enucleated erythroid cells derived from the fetal liver. In catarrhine primates (Old World monkeys, apes, and humans), both γ1 and γ2 are fetally expressed. This developmental switch was accompanied by a delay in the fetal expression of the β-globin gene, which is predominantly expressed during postnatal life in mammals. Goodman et al. (1987) suggested that the acquisition of a fetally expressed isoHb may have played an important role in the life history evolution of simian primates because it facilitated an extended duration of fetal development.

Whereas embryonic γ-globin genes were co-opted for fetal expression in simian primates, duplicate

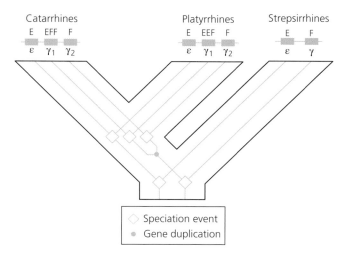

Fig. 6.7. Recruitment of duplicated γ-globin genes for fetal expression in simian primates: catarrhines (Old World monkeys and apes, including humans) and platyrrhines (New World monkeys). The ancestral single-copy γ-globin was tandemly duplicated in the common ancestor of catarrhines and platyrrhines after divergence from the ancestor of strepsirrhines (lemuriform primates such as lemurs, galagos, pottos, and lorises). In simian primates, both γ-globin paralogs evolved a switch from exclusive embryonic expression to a combined embryonic/fetal expression (in the case of γ1) and exclusive fetal expression (in the case of γ1). In catarrhines, insertion of a transposable element between the ε and γ1 genes increased the distance of both γ-globin genes from an upstream regulatory element (the locus control region), which resulted in a further shift in favor of fetal expression of γ1-globin relative to its orthologous counterpart in platyrrhines (Johnson et al., 2006). E, F, and A denote embryonic, fetal, and adult expression, respectively. For genes that are expressed in more than one stage, the predominant stage of expression is denoted by double letters. Note that the γ1 and γ2 genes of simian primates are referred to as ᴳγ and ᴬγ, respectively, in humans.

Based on data reported in Johnson et al. (2006).

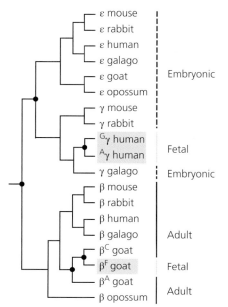

Fig. 6.8. Phylogenetic history of representative β-type globin genes in mammals. Internal nodes of the tree marked by filled symbols denote duplication events. The tree indicates that fetally expressed genes in simian primates (represented by the two γ-globin paralogs of human) and ruminant artiodactyls (represented by the βᶠ-globin gene of goat) derived from independent duplication events. Whereas embryonic γ-globin genes were co-opted for fetal expression in simian primates, duplicate copies of the adult β-globin gene were co-opted for fetal expression in ruminants.

copies of the adult β-globin gene were co-opted for fetal expression in ruminant artiodactyls (Townes et al., 1984, Schimenti and Duncan, 1985, Clementi et al., 1996). Thus, the stage-specific expression of fetal isoHbs evolved twice independently from different ancestral states (Fig. 6.8). In simian primates and ruminants, the co-option of γ- or β-globin genes for fetal expression was likely facilitated by the fact that redundant or semi-redundant copies of other early- or late-expressed β-type globin genes continued to perform their ancestral functions. The acquisition of fetally expressed isoHbs would not have been possible if the ancestor of simian primates had possessed only a single embryonic gene, or if the ancestor of ruminants had possessed only a single adult-expressed gene, as in contemporary monotremes and marsupials (Opazo et al., 2008a, Opazo et al., 2008b).

6.3 Fetal Hb and placental gas exchange

Compared with pulmonary O_2 uptake during postnatal life, O_2 uptake by the developing fetus is impaired by the low diffusive conductance of the placental vasculature. Since the rate of mass O_2 transfer is the product of the diffusive conductance and the mean difference in PO_2 across the diffusion barrier, the inherently low O_2 diffusive conductance of the placenta can be offset by increasing the mean unloading tension of maternal blood relative to the mean loading tension of fetal blood (Weber, 1994). In eutherian mammals, the requisite affinity differential is typically achieved by increasing blood-O_2 affinity in the fetal circulation (Huggett, 1927, Weber, 1994, Ingermann, 1997).

In humans, the fetally expressed isoHb, HbF ($\alpha_2\gamma_2$), and adult Hb, HbA ($\alpha_2\beta_2$), have similar intrinsic O_2 affinities at 37°C. However, in the presence of physiological concentrations of anions, HbF exhibits a higher O_2 affinity than HbA due to its reduced sensitivity to the organic phosphate 2,3-diphosphoglycerate, or DPG (Bauer et al., 1968, Tyuma and Shimizu, 1969, Bunn and Briehl, 1970, Tomita, 1981) (Fig. 6.9). During pregnancy, the resultant O_2 affinity difference between HbF in the fetal circulation and HbA in the maternal circulation facilitates O_2 transfer across the placental barrier. Since HbF and HbA have identical

α-type subunits, the different functional properties of the two isoHbs must be attributable to substitutions between the γ- and β-globin genes. The lower DPG sensitivity of HbF relative to HbA appears to be mainly attributable to the amino acid substitution γ143His→Ser, which eliminates two DPG binding sites per tetramer (Frier and Perutz, 1977), in combination with indirect effects of amino acid replacements that are structurally remote from the DPG-binding site (Chen et al., 2000). Additionally, acetylation of the N-terminal Gly 1γ(NA1) in fetal Hb, which replaces Val 1β(NA1) in adult Hb (Fig. 6.3), is also expected to reduce DPG binding. Apart from the isoHb differences in O_2 affinity in the presence of DPG, the oxygenation reaction of human fetal Hb is also less exothermic than that of adult Hb under physiological conditions: at pH 7.4, the enthalpy values (ΔH) are −6.5 and −9.0 kcal per mole of O_2, respectively. This indicates that placental O_2 exchange between the maternal and fetal circulations contributes to the dissipation of heat generated by fetal metabolism (Giardina et al., 1993).

In the case of ruminant artiodactyls, the fetal and adult isoHbs are equally unresponsive to DPG due to deletions of the N-terminal Val 1β(NA1) and a Val→Met substitution at the neighboring 2β(NA2) residue position (Breepoel et al., 1981a, Breepoel et al., 1981b). As explained in Chapter 4 (section 4.6.4), this modification of the β-chain N-terminus all but abolishes DPG binding. The fetally expressed isoHb (which incorporates the product of the "β^F-globin" gene duplicate) has a higher intrinsic O_2 affinity than the adult isoHb (which incorporates the product of β^A globin) at physiological temperature and pH, both in the presence and absence of anions (Blunt et al., 1971, Baumann et al., 1972, Weber et al., 1988, Clementi et al., 1996). Since the fetal and adult isoHbs share the same α-type subunits, the observed difference in Hb-O_2 affinity must be attributable to one or more amino acid substitutions that distinguish the β^F and β^A chains. As in the case with simian primates, the fetal Hb of ruminants has a lower overall heat of oxygenation than the adult Hb under physiological conditions (Clementi et al., 1996), a pattern of isoHb differentiation that facilitates heat transfer from fetal to maternal blood.

In other eutherian mammals examined to date, fetal and adult red cells express the same Hbs, and

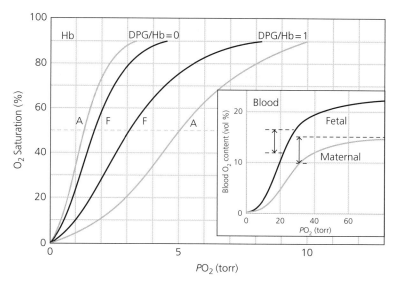

Fig. 6.9. O_2-equilibrium curves of human adult and fetal isoHbs (A and F, respectively). Data are shown for "stripped" Hbs (purified Hbs that are stripped of organic phosphates and other allosteric effectors) in the absence and presence of equimolar concentrations of 2,3-diphosphoglycerate (DPG:Hb = 0 and DPG:Hb = 1, respectively) at 20°C and pH 7.2 (the approximate intraerythrocytic pH value). Inset: O_2-equilibrium curves for maternal and fetal blood at 37°C and extracelluar pH 7.4 (corresponding to an intracellular pH of 7.2), illustrating the difference in arteriovenous O_2 content (double-headed arrows), as well as the higher O_2 affinity and higher O_2-carrying capacity of fetal blood.

Adapted with permission from Tomita (1981) and Weber (1994).

the requisite PO_2 difference between fetal and maternal blood is accomplished by reducing the concentration of DPG in fetal red cells (Dhindsa et al., 1972, Tweeddale, 1973, Jelkmann and Bauer, 1977, Petschow et al., 1978, Qvist et al., 1981). Differences in DPG concentration between fetal and adult red cells are attributable to regulatory changes in pyruvate kinase activity in some mammals (Jelkmann and Bauer, 1980, Franzke and Jelkmann, 1982). In the case of the Weddell seal (*Leptonychotes weddelli*), adult and fetal red cells express the same pair of major and minor isoHbs in the same relative concentrations. However, fetal blood has an appreciably higher O_2 affinity than that of pregnant females (Fig. 6.10) due to reduced red cell concentrations of DPG: 2.45 vs. 6.45 mM in fetal vs. maternal red cells, respectively (Qvist et al., 1981). Since there are no appreciable differences in mean cell Hb concentration in fetal vs. maternal red cells, the differences in intracellular DPG concentration translate into DPG/tetrameric Hb ratios of 1.07 vs. 0.41, respectively. Although the higher O_2 affinity of fetal blood relative to maternal blood appears to be almost universal among eutherian mammals, cats represent a notable

exception, as fetal and maternal red cells exhibit no appreciable differences in isoHb composition or anion concentration (Novy and Parer, 1969).

In addition to eutherian mammals, placental gas exchange is also common in other viviparous vertebrates. Among viviparous elasmobranchs and teleost fishes, the maintenance of a higher O_2 affinity in fetal blood relative to maternal blood is sometimes exclusively attributable to the stage-specific expression of fetal isoHbs that have higher intrinsic O_2 affinities than adult-expressed isoHbs, as in the big skate, *Beringraja binoculata* (Manwell, 1958), Pacific spiny dogfish, *Squalus suckleyi* (Manwell, 1963), swell shark, *Cephaloscyllium ventriosum* (King, 1994), and European eelpout, *Zoarces viviparous* (Weber and Hartvig, 1984). In some teleost species, the effects of fetal-adult isoHb differentiation may be further augmented by reduced concentrations of nucleotide triphosphates in fetal red cells, as in the striped seaperch, *Embiotoca lateralis* (Ingermann and Terwilliger, 1981a, Ingermann and Terwilliger, 1981b).

In viviparous squamates (lizards and snakes), the maintenance of a higher O_2 affinity in fetal blood relative to maternal blood is often attributable to

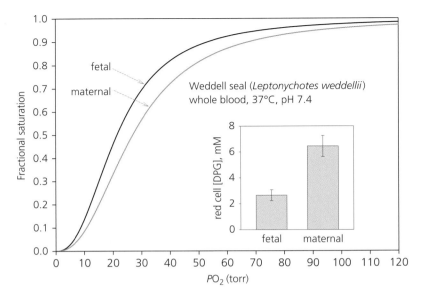

Fig. 6.10. O_2-equilibrium curves for fetal and adult whole blood from Weddell seal (*Leptonychotes weddellii*) at 37°C, pH 7.4. Inset shows intracellular concentrations of DPG (mean ± SD, $n = 10$ adults and 10 fetuses).

Based on data reported in Qvist et al. (1981).

differences in red cell concentrations of ATP (Grigg and Harlow, 1981, Birchard et al., 1984, Berner and Ingermann, 1988, Ragsdale and Ingermann, 1991). Studies of the Pacific rattlesnake (*Crotalus oreganus*) suggest that stage-specific expression of structurally distinct isoHbs with different phosphate sensitivities may also contribute to the O_2 affinity difference between fetal and maternal blood (Ragsdale and Ingermann, 1993, Ragsdale et al., 1995). One of the more exotic mechanisms for maintaining the maternal/fetal difference in blood-O_2 affinity has been described in garter snakes (*Thamnophis elegans*). In this species, the increased O_2 affinity of fetal blood appears to stem from an elevated fraction of metHb in fetal red cells (Pough, 1977). For any given ligation state, Hbs with one or more oxidized (Fe^{3+}) heme groups will have more R-state character than would otherwise be the case, so the remaining unliganded, ferrous (Fe^{2+}) hemes will have increased O_2 affinities. Thus, a modest increase in the fraction of metHb in fetal red cells can increase blood-O_2 affinity relative to maternal blood at the expense of a reduced O_2 carrying capacity.

In summary, viviparous vertebrates employ a diversity of mechanisms for maintaining the PO_2

differences between maternal and fetal circulations, only some of which involve genetically based differences in the oxygenation properties of isoHbs with stage-specific expression (Weber et al., 1988, Weber, 1994, Weber, 1995, Clementi et al., 1996, Ingermann, 1997). Genetically based differences between fetal and adult Hbs may be caused by differences in intrinsic affinity, as in ruminant artiodactyls, or differences in sensitivity to allosteric effectors, as in simian primates. Alternatively, in species that do not express functionally distinct isoHbs during prenatal development and postnatal life, the O_2 affinity difference between fetal and maternal circulations is achieved through changes in red cell concentrations of allosteric effectors that modulate Hb-O_2 binding.

6.4 Functional differentiation between larval and adult isoHbs in aquatic ectotherms

Differences in O_2-binding properties have also been documented between larval and adult isoHbs in fishes and amphibians (Weber and Jensen, 1988, Weber, 1994, Ingermann, 1997). In teleost fishes, larval isoHbs generally have higher O_2 affinities and smaller Bohr effects relative to those expressed in adulthood (Iuchi, 1973). Especially interesting examples of ontogenetic isoHb differentiation have been documented in amphibian taxa that undergo a metamorphic transition from aquatic larvae to a terrestrial adult stage. The aquatic, water-breathing larvae and terrestrial, air-breathing adults face very different respiratory challenges to O_2 uptake because they have different gas exchange organs (gills and lungs, respectively) and contend with respiratory media with different O_2 concentrations and diffusion rates. Among anurans, isoHbs expressed in tadpoles of the American bullfrog (*Rana catesbeiana*) have higher intrinsic O_2 affinities than adult isoHbs that are expressed after metamorphosis (McCutcheon, 1936, Riggs, 1951, Watt and Riggs, 1975). Similarly, in the Iberian ribbed newt (*Pleurodeles waltl*), isoHbs expressed in aquatic larvae have higher intrinsic O_2 affinities than those that are expressed in terrestrial adults (Flavin et al., 1978, Flavin et al., 1983). By contrast, in the California giant salamander (*Dicamptodon ensatus*), the blood of aquatic larvae exhibits a higher O_2 affinity than that of terrestrial adults due to a twofold lower ATP concentration in larval red blood cells (Wood, 1971).

6.5 Functional properties of embryonic Hbs

One consistent pattern among viviparous vertebrates is that—within a given species—isoHbs that are expressed during early embryogenesis have higher O_2 affinities and lower cooperativities than isoHbs expressed later in prenatal development or in postnatal life (Wells, 1979, Brittain and Wells, 1983, Weber et al., 1987, Weber, 1994, Brittain, 2002). In humans, for example, the embryonic isoHbs Gower I ($\zeta_2\epsilon_2$), Gower II ($\alpha_2\epsilon_2$), and Portland ($\zeta_2\gamma_2$) have uniformly higher O_2 affinities and lower cooperativities than the later expressed fetal and adult isoHbs (Hellegers and Schruefer, 1961, Tuchinda et al., 1975, Brittain et al., 1997, Brittain, 2002). Under physiologically relevant experimental conditions (37°C, pH 7.2, 100 mM Cl^-), the embryonic isoHbs exhibit relatively low cooperativity coefficients (n_{50} = 1.9–2.3) and P_{50} values of approximately 4, 12, and 6 torr for Gower I, Gower II, and Portland, respectively (Fig. 6.11). The low cooperativities of the embryonic isoHbs are indicated by the less sigmoidal shapes of their O_2-equilibrium curves relative to that of adult Hb (Fig. 6.11). An analysis of human embryonic Hbs in terms of the two-state MWC model indicates

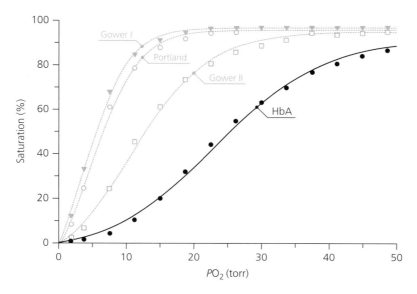

Fig. 6.11. O_2-equilibrium curves for human embryonic Hbs (Gower I, Gower II, and Portland) and adult Hb (HbA) at 37°C, pH 7.2, 100 mM Cl^-.
Modified from Brittain (2002).

Table 6.1 Estimated parameters of the two-state MWC model for embryonic and adult human Hbs

isoHb	K_R (torr)	K_T (torr)	L
HbA, $a_2\beta_2$ (adult)	0.3	21	5×10^5
Gower I, $\zeta_2\epsilon_2$ (embryonic)	0.5	11	2×10^3
Gower II, $a_2\epsilon_2$ (embryonic)	0.4	10	2×10^4
Portland, $\zeta_2\gamma_2$ (embryonic)	0.3	8	6×10^4

Data are from Brittain et al. (1997).

that the higher O_2 affinities of the embryonic isoHbs relative to adult Hb are attributable to reductions in the relative stability of the T-state (Brittain et al., 1997). This is indicated by lower values of the allosteric constant, L (= the ratio of the concentrations of ^{T}Hb and ^{R}Hb in the absence of O_2), in comparison with that of adult Hb (Table 6.1). Relative to adult Hb, the reduced T-state stabilities of the embryonic isoHbs appear to be mainly attributable to substitutions at $\zeta1$, $\zeta6$, and $\zeta38$ (affecting Gower I and Portland), $\epsilon9$ and $\epsilon67$ (affecting Gower I and Gower II), and $\gamma9$, $\gamma51$, and $\gamma67$ (affecting Portland) (Brittain, 2002) (Fig. 6.3). The substitution of Gln for Thr at $\zeta38$ likely accounts for much of the difference in T-state stability between adult Hb ($a_2\beta_2$) and the embryonic Gower I ($\zeta_2\epsilon_2$) and Portland ($\zeta_2\gamma_2$) isoHbs. The $a/\zeta38$(C3) residue is located in the $a_1\beta_2$ interdimer switch region (= "$\zeta_2\epsilon_2$" and "$\zeta_2\gamma_2$" switch regions in Hbs Gower I and Portland, respectively), and therefore plays a key role in mediating the allosteric T↔R transition in quaternary structure.

Relative to adult human Hb, the embryonic isoHbs also exhibit different patterns of responsiveness to allosteric effectors (Fig. 6.12). Over the physiologically relevant pH range, adult and fetal Hbs exhibit fairly similar Bohr effects (Doyle et al., 1989), whereas the O_2 affinities of some embryonic isoHbs are modulated by pH in a qualitatively distinct fashion (Hofmann et al., 1995b) (Fig. 6.12A). The embryonic Hb Gower II ($a_2\epsilon_2$) exhibits a pattern of pH sensitivity that is very similar to that of adult Hb ($a_2\beta_2$), although overall O_2 affinity is higher across the physiological pH range. The similar Bohr effects of these two isoHbs can be explained by the fact that they possess identical a-type chains, and

the distinct β-type chains (embryonic ϵ and adult β) share the same principal Bohr groups. Hb Gower I has a much smaller Bohr effect than Gower II and adult Hb. Since Hbs Gower I and Gower II share the same ϵ-chains, the lower pH sensitivity of Hb Gower I is mainly attributable to the elimination of a key Bohr group at the N-terminus of the a-type chain, as Val $a1$(NA1) is replaced with acetylated Ser $\zeta1$(NA1) (Fig. 6.3). Site-directed mutagenesis experiments have confirmed that converting Ac-Ser $\zeta1$(NA1) to Val increases the Bohr effect of the mutant Hb so that it is equal in magnitude to that of adult Hb (Scheepens et al., 1995). The embryonic Hb Portland ($\zeta_2\gamma_2$) exhibits a pattern of pH sensitivity that is qualitatively and quantitatively distinct from each of the others. Like Hb Gower I, Hb Portland has Ac-Ser $\zeta1$(NA1), and the responsiveness to changes in pH is also likely altered by the replacement of His with Ser at $\gamma143$(H21) (Brittain, 2002).

Adult Hb, Gower I, and Gower II all exhibit similar responses to DPG, in spite of variation in intrinsic O_2 affinities (Fig. 6.12B) (Hofmann et al., 1995a, Hofmann et al., 1995b). The similar DPG sensitivities of these three isoHbs likely reflect the fact that the canonical phosphate-binding sites are present in both the adult β chains and the embryonic ϵ chains, although the N-terminus of the ϵ chain is acetylated, which should slightly decrease DPG binding in a pH-dependent manner. By contrast, the embryonic Hb Portland—like fetal Hb—has a dramatically reduced DPG sensitivity which is primarily attributable to the γ-chain substitution at 143 mentioned earlier (Fig. 6.3).

The O_2 affinities of the embryonic and adult Hbs are also modulated in different ways by changes in Cl^- concentration (Hofmann et al., 1995a) (Fig. 6.12C). Hb Gower II is most similar to adult Hb with respect to Cl^- sensitivity, although the response is quantitatively much lower. Since these two isoHbs share the same a-chains, the difference in Cl^- sensitivity must be attributable to amino acid substitutions that distinguish the β and ϵ chains. This was confirmed by site-directed mutagenesis experiments which demonstrated that the substitution $\beta77$His→$\epsilon77$Asn is primarily responsible for the lower Cl^- sensitivity of Hb Gower II because it reduces the positive charge in the central cavity (Zheng et al., 1999). The low Cl^- sensitivity of Hb Portland is likely attributable to the

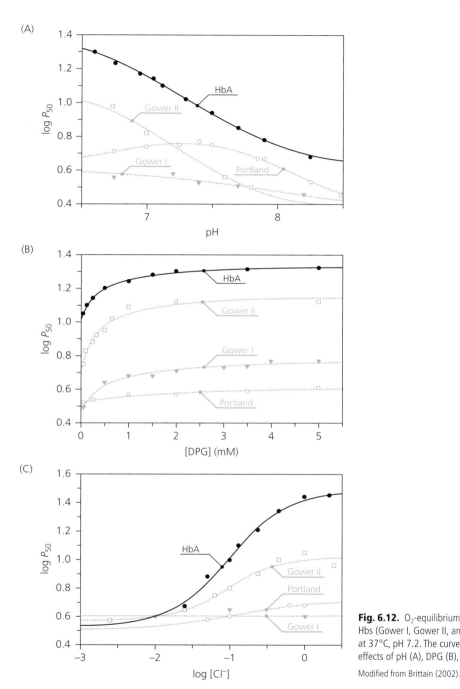

Fig. 6.12. O_2-equilibrium curves for human embryonic Hbs (Gower I, Gower II, and Portland) and adult Hb (HbA) at 37°C, pH 7.2. The curves show variation in the allosteric effects of pH (A), DPG (B), and Cl$^-$ ions (C).

Modified from Brittain (2002).

acetylation of γ1Ser as well as the substitutions α138Ser→ζ138Glu and β143His→γ143Ser, as each of these changes reduce positive charge in the central cavity. Hb Gower I is completely unresponsive to Cl$^-$ ions due to the effects of each of the above-mentioned

ϵ-chain substitutions shared with Hb Gower II and the above-mentioned ζ-chain substitutions shared with Hb Portland.

Similar to the pattern of isoHb differentiation documented in humans, embryonic isoHbs of other mammals also typically exhibit relatively high intrinsic O_2 affinities, low cooperativities, and low sensitivities to changes in pH and anion concentrations (Bauer et al., 1975, Wells, 1979, Wells and Brittain, 1981, Purdie et al., 1983, Wells and Brittain, 1983, Brittain et al., 1986, Weber et al., 1987). Pigs (*Sus scofra domesticus*) express a total of four embryonic isoHbs. Hbs Gower I ($\zeta_2\epsilon_2$) and Gower II ($\alpha_2\epsilon_2$) have the same subunit composition as their like-named human homologs, whereas Hbs Heide I and Heide II incorporate ζ- and α-subunits, respectively, in combination with products of an embryonic β-type globin gene that has no functional ortholog in humans. At physiological pH in both the presence and absence of DPG, the two isoHbs that are most highly expressed in the earliest stages of embryogenesis (Hbs Gower I and Heide I) have the highest O_2 affinities and the adult Hb has the lowest (Fig. 6.13). The two most early-expressed Hbs, Gower I and Heide I, have appreciably lower pH sensitivities than Gower II, Heide II, and adult Hb (Bohr coefficients, $\phi = -0.1$ vs. −0.3, respectively, in the absence of DPG) (Weber et al., 1987). Hbs Gower I and Heide I share the same ζ-chains, and—as with human Hb Portland—their low Bohr coefficients are explained by the acetylation of the N-terminal Ser $\zeta1(NA1)$, which eliminates a key proton-binding site.

6.6 Functional differentiation of coexpressed isoHbs

Most eutherian mammals possess multiple copies of α- and β-type globin genes that are coexpressed during postnatal life (Hoffmann et al., 2008a, Hoffmann et al., 2008b, Opazo et al., 2008a, Opazo et al., 2008b, Opazo et al., 2009, Runck et al., 2009, Gaudry et al., 2014, Natarajan et al., 2015a). Adult-expressed genes of the same subunit type typically have highly similar coding sequences and therefore encode identical or nearly identical polypeptides. Thus, in definitive red blood cells, isoHbs that incorporate the different α- and β-type subunits typically have very similar functional properties (Kleinschmidt et al., 1987, Runck et al., 2010, Storz et al., 2012, Janecka et al., 2015). One notable exception involves wild and domestic sheep (genus *Ovis*) which harbor allelic variation in the number of postnatally expressed β-globin gene duplicates, and which therefore express a variable set of β-chain isoHbs with different oxygenation properties (Dawson and Evans, 1966, Bauer and Jung, 1975).

In contrast to the typical pattern in mammals, most birds, reptiles, and amphibians coexpress multiple structurally and functionally distinct Hb isoforms during adult life (Storz et al., 2011, Grispo et al., 2012, Damsgaard et al., 2013, Opazo et al., 2015, Storz et al., 2015). Crocodilians are an exception, as

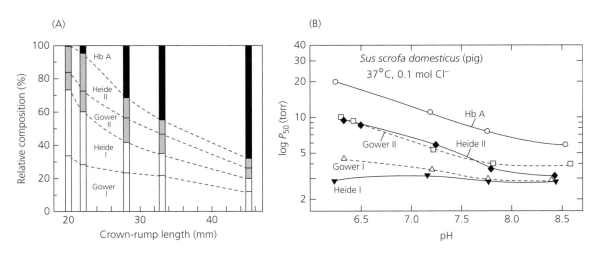

Fig. 6.13. Relative abundance and O_2-binding properties of embryonic Hbs (Gower I, Gower II, Heide I, and Heide II) and adult Hb (HbA) from pig, *Sus scofra domesticus*. (A) Temporal change in the relative abundance of the different pig isoHbs during early embryonic development. (B) Variation in the O_2 affinities and pH sensitivities of the embryonic and adult isoHbs of pig (37°C, pH 7.2).

Modified from Weber et al. (1987).

all species that have been examined to date express a single adult Hb (Weber and White, 1986, Weber and White, 1994, Weber et al., 2013). Birds typically express two main isoHbs in definitive red blood cells: HbA (the major isoHb, with α-chain subunits encoded by the α^A-globin gene) and HbD (the minor isoHb, with α-chain subunits encoded by the α^D-globin gene); both isoHbs incorporate the same

β-chain subunits (Fig. 6.14). In all bird species that have been examined to date, the minor HbD exhibits a substantially higher O_2 affinity than the major HbA in the presence of physiological concentrations of allosteric effectors (Grispo et al., 2012, Projecto-Garcia et al., 2013, Cheviron et al., 2014, Galen et al., 2015, Natarajan et al., 2015b, Kumar et al., 2017, Zhu et al., 2018) (Fig. 6.15).

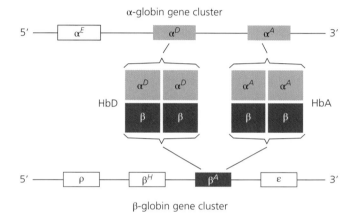

Fig. 6.14. Postnatally expressed Hb isoforms in avian red blood cells. The major isoform, HbA ($\alpha^A_2\beta_2$), has α-chain subunits encoded by the α^A-globin gene, and the minor isoform, HbD ($\alpha^D_2\beta_2$) has α-chain subunits encoded by the α^D-globin gene. Both isoforms share identical β-type subunits encoded by the β^A-globin gene. The remaining members of the avian α- and β-globin gene families (α^E, ρ-, β^H-, and ϵ-globin) are not expressed at appreciable levels in the definitive red cells of adult birds. Within each gene cluster, the intergenic spacing is not drawn to scale.

Reproduced from Opazo et al. (2015).

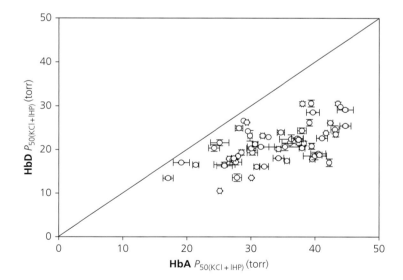

Fig. 6.15. Relative O_2 affinities of the HbA and HbD isoHbs in a phylogenetically diverse set of sixty avian taxa. Data points represent paired P_{50} values of both isoHbs from the same species or subspecies. The fact that all data points fall below the line of equality indicates that HbD has a higher O_2 affinity (lower P_{50}) than HbA in each of the sixty examined taxa. P_{50} values were derived from O_2-equilibria measured in the presence of IHP and Cl⁻ ions. O_2-equilibria were measured in 0.1 mM HEPES buffer at 37°, pH 7.4, in the presence and absence of Cl⁻ ions (0.1 M) and IHP (IHP/Hb ratio = 2.0; see Grispo et al. (2012) for exceptions). Plotted data are from Grispo et al. (2012), Natarajan et al. (2015), Natarajan et al. (2016), and Zhu et al. (2018).

A comprehensive analysis of HbA and HbD in the pheasant (*Phasianus colchicus*) revealed differences in the O_2-binding properties of the two isoHbs (Grispo et al., 2012). The O_2 affinities of pheasant HbA and HbD are modulated by pH in a similar fashion, as Bohr coefficients for both isoHbs were virtually identical (ϕ = –0.43 at 25°C and pH 7.0–7.5). Extended Hill plots and estimates of the two-state MWC parameters revealed the nature of allosteric control mechanisms for both isoHbs (Fig. 6.16). When measured at the same pH, extended Hill plots for HbA and HbD are superimposable, indicating nearly identical association constants in the T- and R-states (K_T and K_R, which can be interpolated from the intercepts of the lower and upper asymptotes, respectively; Fig. 6.16A). Thus, both isoHbs are characterized by similar free energies of cooperativity (ΔG = 8.4 and 8.0 for HbA and

HbD, respectively, at pH ~7.5). Whereas increased proton-binding at low pH reduces Hb-O_2 affinity by lowering K_T without affecting K_R, both association constants are lowered in the presence of IHP (Fig. 6.16B), indicating that this highly charged polyphosphate molecule binds to both ^{T}Hb and ^{R}Hb. Further insights into the allosteric T→R transition of pheasant HbA and HbD are revealed by estimates of Adair association constants (k_{1-4}) for the four successive oxygenation steps. In stripped HbA and HbD at pH ~7.5, the similar k_1 and k_2 values and the markedly increased k_3 and k_4 values indicate that the allosteric transition occurs only after binding the second and third O_2 molecules (Fig. 6.17). This also applies at low pH (~7.0), where lower k_1 and k_2 values show that proton-binding reduces the affinities for binding the first and second O_2 molecules. In the presence of IHP, the even

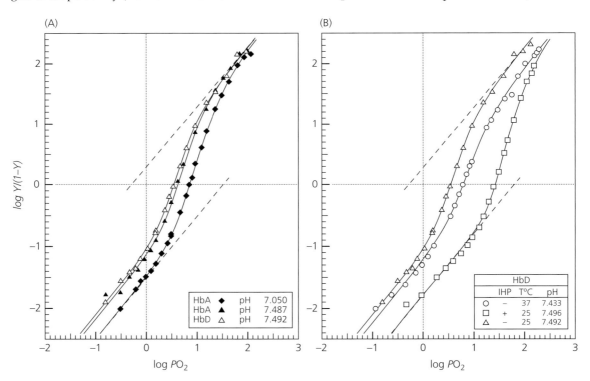

Fig. 6.16. Extended Hill plots of O_2 equilibria for HbA and HbD of pheasant (*Phasianus colchicus*). (A) HbA and HbD at 25°C. (B) HbD at 25°C and 37°C, in the presence and absence of IHP at saturating concentration (IHP/Hb ratio = 23.5). In each plot, the intercept of the lower asymptote with the horizontal line at log $Y/(Y–1)$ = 0 (where Y = fractional O_2 saturation) provides an estimate of K_T, the O_2 association constant of T-state deoxyHb, and the intercept of the upper asymptote with the same line provides an estimate of K_R, the O_2 association constant of R-state oxyHb. O_2-equilibria were measured in 0.1 M NaHEPES buffer containing 0.1 M KCl. Heme concentration = 0.60.

Reproduced from Grispo et al. (2012).

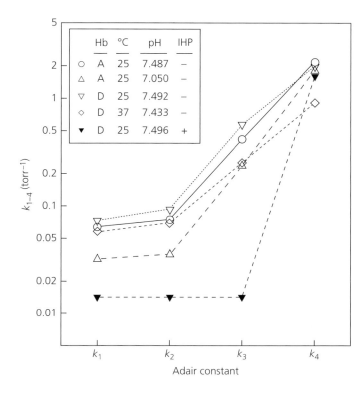

Fig. 6.17. Adair constants (k_1, k_2, k_3, and k_4) for pheasant HbA and HbD as a function of temperature, pH, and the presence and absence of IHP.

Reproduced from Grispo et al. (2012).

lower values of k_1, k_2, and k_2, combined with a markedly increased k_4 (Fig. 6.17), indicate that IHP suppresses the affinities of unliganded hemes for the first, second, and third O_2 molecules but has little effect on the affinity of the remaining unliganded heme. This indicates that IHP binding delays the T→R transition in quaternary structure until the final oxygenation step.

Similar to the avian pattern of isoHb differentiation, turtles, lizards, and snakes also express homologous HbA and HbD isoHbs in definitive erythrocytes (however, it is important to note that although the α-type globin genes are orthologous to those in birds, the β-type globins are not necessarily 1:1 orthologs) (Hoffmann et al., 2010, Storz et al., 2011, Hoffmann et al., 2018). Surprisingly, the reptilian pattern of isoHb differentiation is reversed relative to the avian pattern: In the few reptiles that have been investigated, HbD is the major isoHb and (at least in turtles and snakes) it has a lower O_2 affinity than other isoHbs that incorporate products of the α^A-globin gene (Storz et al., 2011, Damsgaard et al., 2013, Storz et al., 2015, Hoffmann et al., 2018) (Fig. 6.18).

6.6.1 Regulatory changes in isoHb composition in amniote vertebrates

Since the HbA and HbD isoHbs of birds exhibit appreciable differences in O_2-binding properties, regulatory changes in the HbA/HbD ratio could be expected to contribute to inter- or intraspecific differences in blood-O_2 affinity. In the adult red cells of most bird species, the minor HbD isoHb typically accounts for 20–40 percent of total Hb (Grispo et al., 2012, Opazo et al., 2015, Natarajan et al., 2016), but this isoHb is altogether absent in representatives of five major avian clades: Aequornithia (waterbirds), Columbiformes (doves), Coraciformes (rollers and kingfishers), Cuculiformes (cuckoos), and Psittaciformes (parrots). A comparative genomic analysis revealed that this is attributable to multiple independent deletions or inactivations of the α^D-globin gene (Opazo et al., 2015) (Fig. 6.19, Plate 10). In general, HbD expression is also low among representatives of Apodiformes (swifts and hummingbirds), generally not exceeding 25 percent of total Hb (Opazo et al., 2015, Natarajan et al., 2016). A systematic

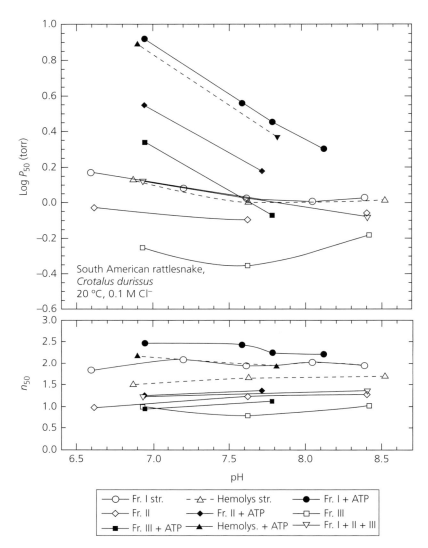

Fig. 6.18. pH dependence of P_{50} and n_{50} values for composite hemolysate and isolated isoHbs of the South American rattlesnake (*Crotalus durissus*). O_2 affinities and cooperativities are shown for stripped (str) hemolysate (triangles), Hb fraction (Fr) I (circles), fraction II (diamonds), and fraction III (squares), and the three fractions combined in a 1:1:1 ratio (inverted triangles), in the absence (open symbols) and presence (solid symbols) of ATP. In contrast to the pattern of isoHb differentiation in birds, where the α^D-containing isoHb has a consistently higher O_2 affinity relative to the α^A-containing isoHb, the α^D-containing isoHbs of rattlesnake (Fr. I and II) have a *lower* O_2 affinity relative to the α^A-containing isoHb (Fr. III). O_2-equilibria were measured at 20°C in the presence Cl⁻ (0.1 M KCl) in 0.1 M HEPES buffer; heme concentration 0.04 mM.

Reproduced from Storz et al. (2015).

analysis of isoHb abundance in avian red cells revealed that relative expression levels of HbD exhibited a significant degree of phylogenetic signal (Opazo et al., 2015) (Plate 11), meaning that phylogenetic relationships among species explain much of the observed variation in trait values. Since HbD (which incorporates products of α^D-globin) has a

consistently higher O_2 affinity than HbA (which incorporates products of α^A-globin), repeated inactivations of the α^D-globin gene has likely contributed to among-species variation in blood-O_2 affinity. This is because the O_2 affinity of blood reflects O_2-binding affinities of the composite mixture of HbA and HbD. For example, in passerine birds, in

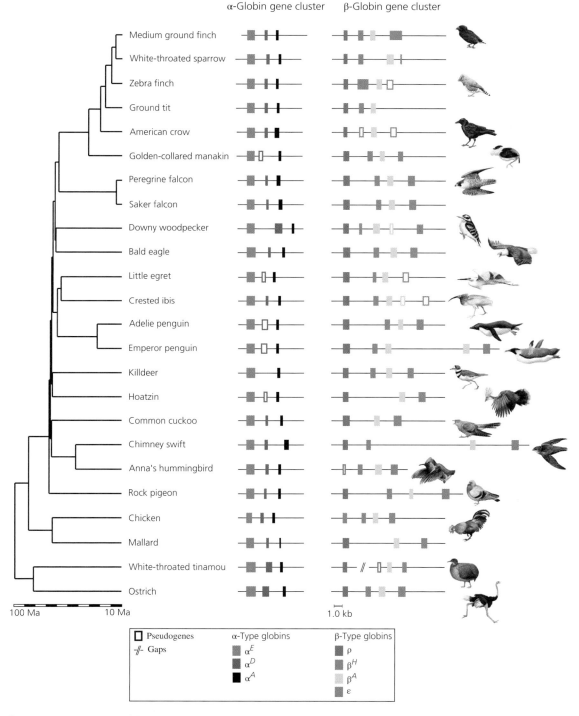

Fig. 6.19. Genomic structure of the avian α- and β-globin gene clusters. The phylogeny depicts a time-calibrated supertree of twenty-four species representing all the major avian lineages.

Reproduced from Opazo et al. (2015). See Plate 10.

which HbD typically accounts for ~40 percent of total Hb, inactivation of α^D-globin (and concomitant loss of HbD) would be expected to reduce blood-O_2 affinity by ~15–20 percent (Opazo et al., 2015). Recurrent deletions and inactivations of α^D-globin must have entailed significant reductions in blood-O_2 affinity in the lineages in which they occurred. It is an open question as to whether such changes were compensated by changes in O_2-binding properties of HbA, the sole remaining adult isoHb, which would require changes in the α^A- and/or β^A-globin genes.

Because HbD is also expressed at high levels during prenatal development (Cirotto et al., 1987, Alev et al., 2008, Alev et al., 2009), inactivation or deletion of α^D-globin could affect O_2-delivery to the developing embryo. Intriguingly, representatives of several groups (families Cuculidae [cuckoos] and Columbidae [pigeons and doves] and the superfamily Psittacoidea [parrots]) do not express HbD in adult red cells, but they have retained α^D-globin genes with intact reading frames. In such taxa, HbD may still be expressed during embryonic development, as has been documented in domestic pigeon (Ikehara et al., 1997).

Within the lifetimes of individual animals, and perhaps on a seasonal basis, regulatory changes in the HbA/HbD ratio could conceivably provide an effective mechanism for reversibly modulating blood-O_2 affinity in response to changes in environmental O_2 availability or changes in internal metabolic demands (Hiebl et al. 1988, Weber et al. 1988, Grispo et al. 2012). Among sauropsid vertebrates, however, there is currently no evidence to suggest that isoHb switching plays an important role in acclimatization to environmental hypoxia. Birds that are native to different elevations exhibit consistent differences in Hb-O_2 affinity due to genetically based increases in the O_2 affinities of HbA and HbD in highland taxa, but there are no detectable elevational differences in HbA/HbD ratios (Projecto-Garcia et al., 2013, Cheviron et al., 2014, Galen et al., 2015, Natarajan et al., 2015b, Natarajan et al., 2016, Kumar et al., 2017, Zhu et al., 2018). Likewise, in turtles, the red cell isoHb composition does not change during short-term acclimation to hypoxia (Damsgaard et al., 2013).

6.6.2 The Root effect and isoHb differentiation in teleost fishes

To assess the possible physiological significance of Hb multiplicity, teleost fishes are an ideal group to study. First, teleosts exhibit the highest levels of functional isoHb diversity among vertebrates (Fyhn et al., 1979, Weber, 1996, Ingermann, 1997, Jensen et al., 1998, Weber, 2000, Wells, 2009). The extensive repertoire of α- and β-type globin genes in this group is partly attributable to a teleost-specific whole-genome duplication event (Opazo et al., 2013). Second, teleosts inhabit aquatic environments that span an extraordinarily broad range of variation in O_2 availability, salinity, ionic composition, pH, and temperature. In principle, the expression of multiple isoHbs with graded O_2 affinities and allosteric regulatory capacities could broaden the permissible range of PO_2 for efficient O_2 uptake and delivery (Weber, 1990, Weber, 2000). The isoHb differentiation in some groups may be adaptive in this regard. Most notably, several taxa including eels, catfish, and salmonids express two electrophoretically distinct isoHb classes (designated as "anodic" and "cathodic") that exhibit pronounced differences in intrinsic O_2 affinity, buffer capacity, and sensitivity to pH, temperature, and organic phosphates (Weber et al., 1976a, Weber et al., 1976b, Weber, 1990, Weber, 1996, Brauner and Weber, 1998, Jensen et al., 1998, Weber, 2000, Weber et al., 2000, Wells, 2009).

As discussed at length in Chapter 7, an important functional specialization of some anodic isoHbs involves an extreme form of pH sensitivity known as the Root effect, whereby the low-affinity T-state conformation of deoxyHb is strongly stabilized at low pH (Brittain, 1987, Pelster and Weber, 1991, Pelster and Randall, 1998, Berenbrink et al., 2005, Brittain, 2005, Berenbrink, 2007). In the glandular epithelium of the swim bladder, reductions of blood pH in dense, countercurrent capillary networks (*rete*) trigger the release of Hb-bound O_2 via the Root effect, thereby promoting O_2 secretion at high PO_2. Swim bladder O_2 secretion provides a mechanism of buoyancy regulation, and therefore enables teleosts to exploit a wide range of depths in the water column. A similar mode of O_2 secretion in the ocular system increases O_2 delivery to highly aerobic cells in the avascular retina of the fish eye, thereby enhancing high-acuity vision. In addition to these well-known functional specializations, recent *in vitro* and *in vivo* studies have demonstrated that Root effect Hbs—in conjunction with mechanisms for maintaining an arterial-venous pH difference—also play a significant role in general tissue O_2 delivery (Rummer and Brauner, 2011, Rummer et al., 2013, Rummer and Brauner, 2015).

The differentiation between (anodic) Root effect Hbs and cathodic isoHbs that have low-to-normal pH sensitivities may represent a physiologically significant division of labor for tissue O_2 delivery, especially under hypoxic and/or hypercapnic stress (Weber, 1990, Fago et al., 1995, Weber, 1996, Jensen et al., 1998, Weber, 2000, Weber et al., 2000, Tamburrini et al., 2001, Wells, 2009). Since the cathodic isoHbs typically have higher O_2 affinities than the Root effect Hbs, they may help secure arterial O_2 loading under conditions of hypoxic hypercapnia where the Root effect Hbs would not be fully saturated. Likewise, since the cathodic isoHbs typically have far lower pH sensitivities, they may help secure tissue O_2 delivery during stress-induced acidosis if anion exchange across the red cell membrane is not sufficient to safeguard intraerythrocytic pH.

Marine and freshwater fishes must often contend with extreme vicissitudes of O_2 availability on a daily or seasonal basis. Given that adult-expressed isoHbs of teleost fish often exhibit physiologically significant differences in oxygenation properties, it seems plausible that regulatory adjustments in red cell isoHb composition could represent an important mechanism of phenotypic plasticity in blood-O_2 transport. Experiments involving the African cichlid *Haplochromis ishmaeli* revealed that exposure to chronic hypoxia during postnatal development induced changes in the relative expression of functionally distinct isoHbs that increased blood-O_2 affinity (Rutjes et al., 2007). A study of red drum (*Sciaenops ocellatus*) documented that a three-week acclimation to hypoxia altered expression levels of genes encoding the α- and β-type subunits of the various adult isoHbs and the expression changes were associated with an increased blood-O_2 affinity (Pan et al., 2017). However, this study did not document a causal link between red cell isoHb composition and blood-O_2 affinity, as it did not control for expected changes in red cell concentrations of

ATP and GTP that are known to have major effects on Hb-O_2 binding. As a mechanism of physiological plasticity during postnatal life, there is currently little evidence to suggest that regulatory changes in red cell isoHb composition make significant contributions to the acclimatization response to hypoxia (Ingermann, 1997, Wells, 2009). In fishes, reversible changes in red cell pH and concentrations of organic phosphates appear to represent far more important mechanisms for modulating blood-O_2 affinity in response to changes in O_2 availability or internal metabolic demand (Weber and Jensen, 1988, Weber, 1996, Val, 2000, Weber, 2000, Nikinmaa, 2001, Brauner and Val, 2006, Wells, 2009, Fago, 2017).

6.6.3 The possible functional significance of Hb heterogeneity

Aside from isoHbs with unique specializations of function such as the Root effect Hbs of teleost fishes, the adaptive significance of Hb multiplicity in the definitive erythrocytes of vertebrates is generally unclear. In teleost fishes, there is no evidence that overall levels of isoHb diversity are generally correlated with physiological capacities or ecological niche breadth (Fyhn et al., 1979, Ingermann, 1997, Wells et al., 1997, Wells, 2009). Salmonid fishes have among the most diverse repertoires of adult-expressed isoHbs (some species express ≥ 9 structurally distinct isoHbs) (Fago et al., 2001), and yet they do not tolerate an especially broad range of water temperatures or O_2 partial pressures in comparison with many other teleost groups. By contrast, cypriniform fishes like the common carp (*Cyprinus carpio*) express only 4 isoHbs (Weber and Lykkeboe, 1978) and yet they are renowned for their ability to tolerate extreme fluctuations in PO_2.

It is also possible that Hb multiplicity confers physiological benefits that are not directly related to inherent oxygenation properties of the proteins. For example, the presence of multiple isoHbs with different net charges may increase Hb solubility in the red blood cell, thereby increasing the upper limit of intracellular Hb concentration (Perutz et al., 1959, Weber, 1990, Ingermann, 1997). Hb heterogeneity could confer this advantage only if the multiple components co-occur in the same red cell. Single-cell spectroscopy data indicate that this is true for the Hbs of teleost fishes and turtles (Brunori et al., 1974, Frische et al., 2001). The presence of multiple isoHbs with different net charges could also affect the distribution of protons across the red cell membrane thereby affecting intracellular pH, which in turn could influence red cell metabolism and O_2 affinity even if the isoHbs in question have identical oxygenation properties. Finally, it is possible that coexpressed isoHbs could also be differentiated along some other axis of functional variation that is not directly related to blood-O_2 transport.

The differentiation in oxygenation properties among developmentally regulated isoHbs has clear adaptive significance in viviparous and oviparous vertebrates alike. In some cases, a physiological division of labor between coexpressed isoHbs may also contribute to the adaptive enhancement of tissue O_2 delivery. In Chapter 7 we will explore the evolution of novel Hb functions and physiological innovations in respiratory gas transport.

References

Alev, C., McIntyre, B. A. S., Nagai, H., et al. (2008). β^A, the major β-globin in definitive red blood cells, is present from the onset of primitive erythropoiesis in chicken. *Developmental Dynamics*, **237**, 1193–7.

Alev, C., Shinmyozu, K., McIntyre, B. A. S., and Sheng, G. (2009). Genomic organization of zebra finch α and β globin genes and their expression in primitive and definitive blood in comparison with globins in chicken. *Development Genes and Evolution*, **219**, 353–60.

Bauer, C. and Jung, H. D. (1975). Comparison of respiratory properties of sheep hemoglobin A and hemoglobin B. *Journal of Comparative Physiology*, **102**, 167–72.

Bauer, C., Ludwig, I., and Ludwig, M. (1968). Different effects of 2,3-diphosphoglycerate and adenosine triphosphate on oxygen affinity of adult and foetal human haemoglobin. *Life Sciences*, **7**, 1339–43.

Bauer, C., Tamm, R., Petschow, D., Bartels, R., and Bartels, H. (1975). Oxygen affinity and allosteric effects of embryonic mouse hemoglobins. *Nature*, **257**, 333–4.

Baumann, R., Bauer, C., and Rathschl. A. M (1972). Causes of postnatal decrease of blood oxygen affinity in lambs. *Respiration Physiology*, **15**, 151–8.

Berenbrink, M. (2007). Historical reconstructions of evolving physiological complexity: O_2 secretion in the eye and swimbladder of fishes. *Journal of Experimental Biology*, **210**, 1641–52.

Berenbrink, M., Koldkjaer, P., Kepp, O., and Cossins, A. R. (2005). Evolution of oxygen secretion in fishes and the emergence of a complex physiological system. *Science*, **307**, 1752–7.

Berner, N. J. and Ingermann, R. L. (1988). Molecular basis of the difference in oxygen-affinity between maternal and fetal red blood cells in the viviparous garter snake, *Thamnophis elegans*. *Journal of Experimental Biology*, **140**, 437–53.

Birchard, G. F., Black, C. P., Schuett, G. W., and Black, V. (1984). Fetal-maternal blood respiratory properties of an ovoviparous snake—the cottonmouth, *Agistrodon piscivorous*. *Journal of Experimental Biology*, **108**, 247–55.

Blunt, M. H., Kitchens, J. L., Mayson, S. M., and Huisman, T. H. J. (1971). Red cell 2,3-diphosphoglycerate and oxygen affinity in newborn goats and sheep. *Proceedings of the Society for Experimental Biology and Medicine*, **138**, 800–3.

Brauner, C. J. and Val, A. L. (2006). Oxygen transfer. In: Val, A. L. and Almeida-Val, V. M. F. (eds.) *Fish Physiology, Vol. 21: The Physiology of Tropical Fishes*, pp. 277–306. Amsterdam, Elsevier.

Brauner, C. J. and Weber, R. E. (1998). Hydrogen ion titrations of the anodic and cathodic haemoglobin components of the European eel *Anguilla anguilla*. *Journal of Experimental Biology*, **201**, 2507–14.

Breepoel, P. M., Kreuzer, F., and Hazevoet, M. (1981a). Interaction of organic phosphates with bovine hemoglobin. 1. Oxylabile and phosphate-labile proton binding. *Pflügers Archiv-European Journal of Physiology*, **389**, 219–25.

Breepoel, P. M., Kreuzer, F., and Hazevoet, M. (1981b). Interaction of organic phosphates with bovine hemoglobin. 2. Oxygen binding equilibria of newborn and adult hemoglobin. *Pflügers Archiv-European Journal of Physiology*, **389**, 227–35.

Brittain, T. (1987). The Root effect. *Comparative Biochemistry and Physiology B-Biochemistry & Molecular Biology*, **86**, 473–81.

Brittain, T. (2002). Molecular aspects of embryonic hemoglobin function. *Molecular Aspects of Medicine*, **23**, 293–342.

Brittain, T. (2005). Root effect hemoglobins. *Journal of Inorganic Biochemistry*, 99, 120–9.

Brittain, T., Hofmann, O. M., Watmough, N. J., Greenwood, C., and Weber, R. E. (1997). A two-state analysis of cooperative oxygen binding in the three human embryonic haemoglobins. *Biochemical Journal*, **326**, 299–303.

Brittain, T., Sutherland, J., and Greenwood, C. (1986). A study of the kinetics of the reaction of ligands with the liganded states of mouse embryonic hemoglobins. *Biochemical Journal*, **234**, 151–5.

Brittain, T. and Wells, R. M. G. (1983). Oxygen transport in early mammalian development: molecular physiology of embryonic hemoglobins. *Development in Mammals*, **5**, 135–54.

Brunori, M., Giardina, B., Antonini, E., Benedetti, P. A., and Bianchini, G. (1974). Distribution of hemoglobin components of trout blood among erythrocytes- observations by single-cell spectroscopy. *Journal of Molecular Biology*, **86**, 165–9.

Bunn, H. F. and Briehl, R. W. (1970). Interaction of 2,3-diphosphoglycerate with various human hemoglobins. *Journal of Clinical Investigation*, **49**, 1088–95.

Bunn, H. F. and Forget, B. G. (1986). *Hemoglobin: Molecular, Genetic and Clinical Aspects*, Philadelphia, PA, W. B. Saunders Company.

Chen, W. H., Dumoulin, A., Li, X. F., et al. 2000. Transposing sequences between fetal and adult hemoglobins indicates which subunits and regulatory molecule interfaces are functionally related. *Biochemistry*, **39**, 3774–81.

Cheviron, Z. A., Natarajan, C., Projecto-Garcia, J., et al. (2014). Integrating evolutionary and functional tests of adaptive hypotheses: a case study of altitudinal differentiation in hemoglobin function in an Andean sparrow, *Zonotrichia capensis*. *Molecular Biology and Evolution*, **31**, 2948–62.

Cirotto, C., Panara, F., and Arangi, I. (1987). The minor hemoglobins of primitive and definitive erythrocytes of the chicken embryo—evidence for Hemoglobin-L. *Development*, **101**, 805–13.

Clementi, M. E., Scatena, R., Mordente, A., et al. (1996). Oxygen transport by fetal bovine hemoglobin. *Journal of Molecular Biology*, **255**, 229–34.

Cooper, S. J. B., Wheeler, D., De Leo, A., et al. (2006). The mammalian α^D-globin gene lineage and a new model for the molecular evolution of α-globin gene clusters at the stem of the mammalian radiation. *Molecular Phylogenetics and Evolution*, **38**, 439–48.

Damsgaard, C., Storz, J. F., Hoffmann, F. G., and Fago, A. (2013). Hemoglobin isoform differentiation and allosteric regulation of oxygen binding in the turtle, *Trachemys scripta*. *American Journal of Physiology-Regulatory Integrative and Comparative Physiology*, **305**, R961–7.

Dawson, T. J. and Evans, J. V. (1966). Effect of hypoxia on oxygen transport in sheep with different hemoglobin types. *American Journal of Physiology*, **210**, 1021–5.

Dhindsa, D. S., Hoversland, A. S., and Templeton, J. W. (1972). Postnatal changes in oxygen affinity and concentrations of 2,3-diphosphoglycerate in dog blood. *Biology of the Neonate*, **20**, 226–35.

Doyle, M. L., Gill, S. J., Decristofaro, R., Castagnola, M., and Dicera, E. (1989). Temperature-dependence and pH dependence of the oxygen-binding reaction of human fetal hemoglobin. *Biochemical Journal*, **260**, 617–19.

Fago, A. (2017). Functional roles of globin proteins in hypoxia-tolerant ectothermic vertebrates. *Journal of Applied Physiology*, **123**, 926–34.

Fago, A., Carratore, V., Diprisco, G., Feuerlein, R. J., Sottrupjensen, L., and Weber, R. E. (1995). The cathodic hemoglobin of *Anguilla anguilla*—amino acid sequence and oxygen equilibria of a reverse Bohr effect hemoglobin with high oxygen-affinity and high phosphate sensitivity. *Journal of Biological Chemistry*, **270**, 18897–902.

Fago, A., Forest, E., and Weber, R. E. (2001). Hemoglobin and subunit multiplicity in the rainbow trout (*Oncorhynchus mykiss*) hemoglobin system. *Fish Physiology and Biochemistry*, **24**, 335–42.

Fantoni, A., Farace, M. G., and Gambari, R. (1981). Embryonic hemoglobins in man and other mammals. *Blood*, **57**, 623–33.

Flavin, M., Blouquit, Y., Duprat, A. M., and Rosa, J. (1978). Biochemical studies of hemoglobin switch during metamorphosis in salamander, *Pleurodeles waltl*. 2. Comparative studies of larval and adult hemoglobins. *Comparative Biochemistry and Physiology B-Biochemistry and Molecular Biology*, **61**, 539–44.

Flavin, M., Thillet, J., and Rosa, J. (1983). Oxygen equilibrium of larval and adult hemoglobins of the salamander, *Pleurodeles waltl*. *Comparative Biochemistry and Physiology A-Physiology*, **75**, 81–5.

Forget, B. G. and Hardison, R. C. (2009). The normal structure and regulation of human globin gene clusters. *In*: Steinberg, M. H., Forget, B. G., Higgs, D. R., and Weatherall, D. J. (eds.) *Disorders of Hemoglobin: Genetics, Pathophysiology, and Clinical Management*, 2nd edition, pp. 46–61. Cambridge, Cambridge University Press.

Franzke, R. and Jelkmann, W. (1982). Characterization of the pyruvate kinase which induces the low 2,3-DPG level of fetal rabbit red cells. *Pflügers Archiv-European Journal of Physiology*, **394**, 21–5.

Frier, J. A. and Perutz, M. F. (1977). Structure of human fetal deoxyhemoglobin. *Journal of Molecular Biology*, **112**, 97–112.

Frische, S., Bruno, S., Fago, A., Weber, R. E., and Mozzarelli, A. (2001). Oxygen binding by single red blood cells from the red-eared turtle *Trachemys scripta*. *Journal of Applied Physiology*, **90**, 1679–84.

Fyhn, U. E. H., Fyhn, H. J., Davis, B. J., et al. (1979). Hemoglobin heterogeneity in Amazonian fishes. *Comparative Biochemistry and Physiology A-Physiology*, **62**, 39–66.

Galen, S. C., Natarajan, C., Moriyama, H., et al. (2015). Contribution of a mutational hotspot to adaptive changes in hemoglobin function in high-altitude Andean house wrens. *Proceedings of the National Academy of Sciences of the United States of America*, **112**, 13958–63.

Gaudry, M. J., Storz, J. F., Butts, G. T., Campbell, K. L., and Hoffman, F. G. (2014). Repeated evolution of chimeric fusion genes in the β-globin gene family of laurasiatherian mammals. *Genome Biology and Evolution*, **6**, 1219–33.

Giardina, B., Scatena, R., Clementi, M. E., et al. (1993). Physiological relevance of the overall delta H of oxygen binding to fetal human hemoglobin. *Journal of Molecular Biology*, **229**, 512–16.

Goodman, M., Czelusniak, J., Koop, B. F., Tagle, D. A., and Slightom, J. L. (1987). Globins—a case-study in molecular phylogeny. *Cold Spring Harbor Symposia on Quantitative Biology*, **52**, 875–90.

Grigg, G. C. and Harlow, P. (1981). A fetal-maternal shift of blood-oxygen affinity in an Australian viviparous lizard, *Sphenomorphus quoyii* (Reptilia, Scinidae). *Journal of Comparative Physiology*, **142**, 495–9.

Grispo, M. T., Natarajan, C., Projecto-Garcia, J., et al. (2012). Gene duplication and the evolution of hemoglobin isoform differentiation in birds. *Journal of Biological Chemistry*, **287**, 37647–58.

Hardison, R. C. (2012). Evolution of hemoglobin and its genes. *Cold Spring Harbor Perspectives in Medicine*, **2**, a011627.

Hellegers, A. and Schruefer, J. J. (1961). Nomograms and empirical equations relating oxygen tension, percentage saturation, and pH in maternal and fetal blood. *American Journal of Obstetrics and Gynecology*, **81**, 377–84.

Hiebl, I., Weber, R. E., Schneeganss, D., Kosters, J., and Braunitzer, G. (1988). High-altitude respiration of birds - structural adaptations in the major and minor hemoglobin components of adult Ruppells griffon (Gryps rueppellii)—a new molecular pattern for hypoxia tolerance. *Biological Chemistry Hoppe-Seyler*, **369**, 217–232.

Hoffmann, F. G., Opazo, J. C., and Storz, J. F. (2008a). New genes originated via multiple recombinational pathways in the β-globin gene family of rodents. *Molecular Biology and Evolution*, **25**, 2589–600.

Hoffmann, F. G., Opazo, J. C., and Storz, J. F. (2008b). Rapid rates of lineage-specific gene duplication and deletion in the α-globin gene family. *Molecular Biology and Evolution*, **25**, 591–602.

Hoffmann, F. G. and Storz, J. F. (2007). The α^D-globin gene originated via duplication of an embryonic α-like globin gene in the ancestor of tetrapod vertebrates. *Molecular Biology and Evolution*, **24**, 1982–90.

Hoffmann, F. G., Storz, J. F., Gorr, T. A., and Opazo, J. C. (2010). Lineage-specific patterns of functional diversification in the α- and β-globin gene families of tetrapod vertebrates. *Molecular Biology and Evolution*, **27**, 1126–38.

Hoffmann, F. G., Vandewege, M. W., Storz, J. F., and Opazo, J. C. (2018). Gene turnover and diversification of the α- and β-globin gene families in sauropsid vertebrates. *Genome Biology and Evolution*, **10**, 344–58.

Hofmann, O., Carrucan, G., Robson, N., and Brittain, T. (1995a). The chloride effect in the human embryonic hemoglobins. *Biochemical Journal*, **309**, 959–62.

Hofmann, O., Mould, R., and Brittain, T. (1995b). Allosteric modulation of oxygen-binding to the three human embryonic hemoglobins. *Biochemical Journal*, **306**, 367–70.

Huehns, E. R., Keil, J. V., Dance, N., et al. (1964). Human embryonic haemoglobins. *Nature*, **201**, 1095–7.

Huggett, A. S. (1927). Foetal blood-gas tensions and gas trans-fusion through the placenta of the goat. *Journal of Physiology-London*, **52**, 373–84.

Ikehara, T., Eguchi, Y., Kayo, S., and Takei, H. (1997). Isolation and sequencing of two α-globin genes α(A) and α(D) in pigeon and evidence for embryo-specific expression of the α(D)-globin gene. *Biochemical and Biophysical Research Communications*, **234**, 450–3.

Ingermann, R. L. (1997). Vertebrate hemoglobins. *Handbook of Physiology, Comparative Physiology*, **30**, 357–408.

Ingermann, R. L. and Terwilliger, R. C. (1981a). Intraerythrocytic organic phosphates of fetal and adult seaperch (Embiotoca lateralis)—their role in maternal-fetal oxygen transport. *Journal of Comparative Physiology*, **144**, 253–9.

Ingermann, R. L. and Terwilliger, R. C. (1981b). Oxygen affinities of maternal and fetal hemoglobins of the viviparous seaperch, *Embiotoca lateralis*. *Journal of Comparative Physiology*, **142**, 523–31.

Iuchi, I. (1973). Chemical and physiological properties of larval and adult hemoglobins in rainbow trout, *Salmo gairdnerii irideus*. *Comparative Biochemistry and Physiology*, **44**, 1087–101.

Janecka, J. E., Nielsen, S. S. E., Andersen, S. D., et al. (2015). Genetically based low oxygen affinities of felid hemoglobins: lack of biochemical adaptation to high-altitude hypoxia in the snow leopard. *Journal of Experimental Biology*, **218**, 2402–9.

Jelkmann, W. and Bauer, C. (1977). Oxygen-affinity and phosphate compounds of red blood cells during intrauterine development of rabbits. *Pflügers Archiv-European Journal of Physiology*, **372**, 149–56.

Jelkmann, W. and Bauer, C. (1980). 2,3-DPG levels in relation to red cell enzyme activities in rat fetuses and hypoxic newborns. *Pflügers Archiv-European Journal of Physiology*, **389**, 61–8.

Jensen, F. B., Fago, A., and Weber, R. E. (1998). Hemoglobin structure and function. *In*: Perry, S. F. and Tufts, B. L. (eds.) *Fish Physiology, Vol. 17: Fish Respiration*, pp. 1–40. New York, NY, Academic Press.

Johnson, R. M., Buck, S., Chiu, C. H., et al. (1996). Fetal globin expression in new world monkeys. *Journal of Biological Chemistry*, **271**, 14684–91.

Johnson, R. M., Buck, S., Chiu, C. H., et al. (2000). Humans and old-world monkeys have similar patterns of fetal globin expression. *Journal of Experimental Zoology*, **288**, 318–26.

King, L. A. (1994). Adult and fetal hemoglobins in the oviparous swell shark, *Cephaloscyllium ventriosum*. *Comparative Biochemistry and Physiology B-Biochemistry and Molecular Biology*, **109**, 237–43.

Kleinschmidt, T., Rucknagel, K. P., Weber, R. E., Koop, B. F., and Braunitzer, G. (1987). Primary structure and functional properties of the hemoglobin from the free-tailed bat *Tadarida brasiliensis* (Chiroptera). Small effect of carbon dioxide on oxygen affinity. *Biological Chemistry Hoppe-Seyler*, **368**, 681–90.

Kumar, A., Natarajan, C., Moriyama, H., et al. (2017). Stability-mediated epistasis restricts accessible mutational pathways in the functional evolution of avian hemoglobin. *Molecular Biology and Evolution*, **34**, 1240–51.

Manwell, C. (1958). Ontogeny of hemoglobin in the skate *Raja binoculata*. *Science*, **128**, 419–20.

Manwell, C. (1963). Fetal and adult hemoglobins of spiny dogfish *Squalus suckleyi*. *Archives of Biochemistry and Biophysics*, **101**, 504–11.

McCutcheon, F. H. (1936). Hemoglobin function during the life history of the bullfrog. *Journal of Cellular and Comparative Physiology*, **8**, 63–81.

Natarajan, C., Hoffman, F. G., Lanier, H. C., et al. (2015a). Intraspecific polymorphism, interspecific divergence, and the origins of function-altering mutations in deer mouse hemoglobin. *Molecular Biology and Evolution*, **32**, 978–97.

Natarajan, C., Hoffmann, F. G., Weber, R. E., et al. (2016). Predictable convergence in hemoglobin function has unpredictable molecular underpinnings. *Science*, **354**, 336–40.

Natarajan, C., Projecto-Garcia, J., Moriyama, H., et al. (2015b). Convergent evolution of hemoglobin function in high-altitude Andean waterfowl involves limited parallelism at the molecular sequence level. *PLoS Genetics*, **11**, e1005681.

Nikinmaa, M. (2001). Haemoglobin function in vertebrates: evolutionary changes in cellular regulation in hypoxia. *Respiration Physiology*, **128**, 317–29.

Novy, M. J. and Parer, J. T. (1969). Absence of high blood oxygen affinity in fetal cat. *Respiration Physiology*, **6**, 144–50.

Opazo, J. C., Butts, G. T., Nery, M. F., Storz, J. F., and Hoffmann, F. G. (2013). Whole-genome duplication and the functional diversification of teleost fish hemoglobins. *Molecular Biology and Evolution*, **30**, 140–53.

Opazo, J. C., Hoffman, F. G., Natarajan, C., et al. (2015). Gene turnover in the avian globin gene family and evolutionary changes in hemoglobin isoform expression. *Molecular Biology and Evolution*, **32**, 871–87.

Opazo, J. C., Hoffmann, F. G., and Storz, J. F. (2008a). Differential loss of embryonic globin genes during the radiation of placental mammals. *Proceedings of the National Academy of Sciences of the United States of America*, **105**, 12950–5.

Opazo, J. C., Hoffmann, F. G., and Storz, J. F. (2008b). Genomic evidence for independent origins of β-like globin genes in monotremes and therian mammals. *Proceedings of the National Academy of Sciences of the United States of America*, **105**, 1590–5.

Opazo, J. C., Sloan, A. M., Campbell, K. L., and Storz, J. F. (2009). Origin and ascendancy of a chimeric fusion gene: the β/δ-globin gene of paenungulate mammals. *Molecular Biology and Evolution*, **26**, 1469–78.

Pan, Y. H. K., Ern, R., Morrison, P. R., Brauner, C. J., and Esbaugh, A. J. (2017). Acclimation to prolonged hypoxia alters hemoglobin isoform expression and increases hemoglobin oxygen affinity and aerobic performance in a marine fish. *Scientific Reports*, **7**, 7834.

Pelster, B. and Randall, D. J. (1998). The physiology of the Root effect. *In:* Perry, S. F. and Tufts, B. L. (eds.) *Fish Physiology, Vol. 17: Fish Respiration.*, pp. 113–39. New York, NY, Academic Press.

Pelster, B. and Weber, R. E. (1991). The physiology of the Root effect. *Advances in Comparative and Environmental Physiology*, **8**, 51–77.

Perutz, M. F., Steinrauf, L. K., Stockell, A., and Bangham, A. D. (1959). Chemical and crystallographic study of the two fractions of adult horse haemoglobin. *Journal of Molecular Biology*, **1**, 402–4.

Petschow, R., Petschow, D., Bartels, R., Baumann, R., and Bartels, H. (1978). Regulation of oxygen-affinity in blood of fetal, newborn and adult mouse. *Respiration Physiology*, **35**, 271–82.

Philipsen, S. and Hardison, R. C. (2018). Evolution of hemoglobin loci and their regulatory elements. *Blood Cells, Molecules, and Disease*, **70**, 2–12.

Pough, F. H. (1977). Ontogenetic change in molecular and functional properties of blood of garter snakes, *Thamnophis sirtalis*. *Journal of Experimental Zoology*, **201**, 47–55.

Projecto-Garcia, J., Natarajan, C., Moriyama, H., et al. (2013). Repeated elevational transitions in hemoglobin function during the evolution of Andean hummingbirds. *Proceedings of the National Academy of Sciences of the United States of America*, **110**, 20669–74.

Purdie, A., Wells, R. M. G., and Brittain, T. (1983). Molecular aspects of embryonic mouse hemoglobin ontogeny. *Biochemical Journal*, **215**, 377–83.

Qvist, J., Weber, R. E., and Zapol, W. M. (1981). Oxygen equilibrium properties of blood and hemoglobin of fetal and adult Weddell seals. *Journal of Applied Physiology*, **50**, 999–1005.

Ragsdale, F. R., Herman, J. K., and Ingermann, R. L. (1995). Nucleotide triphosphate levels versus oxygen affinity of rattlesnake red cells. *Respiration Physiology*, **102**, 63–9.

Ragsdale, F. R. and Ingermann, R. L. (1991). Influence of pregnancy on the oxygen affinity of red cells from the northern Pacific rattlesnake *Crotalus viridis oreganus*. *Journal of Experimental Biology*, **159**, 501–5.

Ragsdale, F. R. and Ingermann, R. L. (1993). Biochemical bases for difference in oxygen-affinity of maternal and fetal red blood cells of rattlesnakes. *American Journal of Physiology*, **264**, R481–6.

Riggs, A. (1951). The metamorphosis of hemoglobin in the bullfrog. *Journal of General Physiology*, **35**, 23–40.

Rummer, J. L. and Brauner, C. J. (2011). Plasma-accessible carbonic anhydrase at the tissue of a teleost fish may greatly enhance oxygen delivery: *in vitro* evidence in a rainbow trout, *Oncorynchus mykiss*. *Journal of Experimental Biology*, **214**, 2319–28.

Rummer, J. L. and Brauner, C. J. (2015). Root effect haemoglobins in fish may greatly enhance general oxygen delivery relative to other vertebrates. *PloS One*, **10**, e0139477.

Rummer, J. L., McKenzie, D. J., Innocenti, A., Supuran, C. T., and Brauner, C. J. (2013). Root effect hemoglobin may have evolved to enhance general tissue oxygen delivery. *Science*, **340**, 1327–9.

Runck, A. M., Moriyama, H., and Storz, J. F. (2009). Evolution of duplicated β-globin genes and the structural basis of hemoglobin isoform differentiation in *Mus*. *Molecular Biology and Evolution*, **26**, 2521–32.

Runck, A. M., Weber, R. E., Fago, A., and Storz, J. F. (2010). Evolutionary and functional properties of a two-locus β-globin polymorphism in Indian house mice. *Genetics*, **184**, 1121–31.

Rutjes, H. A., Nieveen, M. C., Weber, R. E., Witte, F., and Van Den Thillart, G. E. E. J. M. (2007). Multiple strategies of Lake Victoria cichlids to cope with lifelong hypoxia include hemoglobin switching. *American Journal of Physiology-Regulatory Integrative and Comparative Physiology*, **293**, R1376–83.

Sankaran, V. G., Xu, J., and Orkin, S. H. (2010). Advances in the understanding of haemoglobin switching. *British Journal of Haematology*, **149**, 181–94.

Scheepens, A., Mould, R., Hofmann, O., and Brittain, T. (1995). Some effects of posttranslational N-terminal acetylation of the embryonic ζ-globin protein. *Biochemical Journal*, **310**, 597–600.

Schimenti, J. C. and Duncan, C. H. (1985). Structure and organization of the bovine β-globin genes. *Molecular Biology and Evolution*, **2**, 514–25.

Storz, J. F. 2016. Gene duplication and evolutionary innovations in hemoglobin-oxygen transport. *Physiology*, **31**, 223–32.

Storz, J. F., Hoffmann, F. G., Opazo, J. C., Sanger, T. J., and Moriyama, H. (2011). Developmental regulation of hemoglobin synthesis in the green anole lizard *Anolis carolinensis*. *Journal of Experimental Biology*, **214**, 575–81.

Storz, J. F., Natarajan, C., Moriyama, H., et al. (2015). Oxygenation properties and isoform diversity of snake hemoglobins. *American Journal of Physiology-Regulatory Integrative and Comparative Physiology*, **309**, R1178–91.

Storz, J. F., Weber, R. E., and Fago, A. (2012). Oxygenation properties and oxidation rates of mouse hemoglobins that differ in reactive cysteine content. *Comparative Biochemistry and Physiology A-Molecular and Integrative Physiology*, **161**, 265–70.

Tamburrini, M., Verde, C., Olianas, A., et al. (2001). The hemoglobin system of the brown moray *Gymnothorax unicolor*—structure/function relationships. *European Journal of Biochemistry*, **268**, 4104–11.

Tomita, S. (1981). Modulation of the oxygen equilibria of human-fetal and adult hemoglobins by 2,3-diphosphoglyceric acid. *Journal of Biological Chemistry*, **256**, 9495–500.

Townes, T. M., Fitzgerald, M. C., and Lingrel, J. B. (1984). Triplication of a four-gene set during evolution of the goat β-globin locus produced three genes now expressed differentially during development. *Proceedings of the National Academy of Sciences of the United States of America*, **81**, 6589–93.

Tuchinda, S., Nagai, K., and Lehmann, H. (1975). Oxygen dissociation curve of hemoglobin Portland. *FEBS Letters*, **49**, 390–1.

Tweeddale, P. M. (1973). DPG and oxygen affinity of maternal and fetal pig blood and hemoglobins. *Respiration Physiology*, **19**, 12–18.

Tyuma, I. and Shimizu, K. (1969). Different response to organic phosphates of human fetal and adult hemoglobins. *Archives of Biochemistry and Biophysics*, **129**, 404–5.

Val, A. L. (2000). Organic phosphates in the red blood cells of fish. *Comparative Biochemistry and Physiology A-Molecular and Integrative Physiology*, **125**, 417–35.

Watt, K. W. K. and Riggs, A. (1975). Hemoglobins of the tadpole of the bullfrog, *Rana catesbeiana*. Structure and function of isolated components. *Journal of Biological Chemistry*, **250**, 5934–44.

Weber, R. E. (1990). Functional significance and structural basis of multiple hemoglobins with special reference to ectothermic vertebrates. *In:* Truchot, J.-P. and Lahlou, B. (eds.) *Animal Nutrition and Transport Processes*, pp. 58–75. Basel, Switzerland, S. Karger.

Weber, R. E. (1994). Hemoglobin-based O$_2$ transfer in viviparous animals. *Israel Journal of Zoology*, **40**, 541–50.

Weber, R. E. (1995). Hemoglobin adaptations to hypoxia and altitude—the phylogenetic perspective. *In:* Sutton, J. R., Houston, C. S., and Coates, G. (eds.) *Hypoxia and the Brain*, pp. 31–44. Burlington, VT, Queen City Printers.

Weber, R. E. (1996). Hemoglobin adaptations in Amazonian and temperate fish with special reference to hypoxia, allosteric effectors and functional heterogeneity. *In:* Val,

A. L., Almeida-Val, V. M. F. and Randall, D. J. (eds.) *Physiology and Biochemistry of the Fishes of the Amazon*, pp. 75–90. Manaus, INPA.

Weber, R. E. (2000). Adaptations for oxygen transport: Lessons from fish hemoglobins. *In*: Di Prisco, G., Giardina, B., and Weber, R. E. (eds.) *Hemoglobin Function in Vertebrates. Molecular Adaptation to Extreme and Temperate Environments*. Berlin, Springer-Verlag.

Weber, R. E., Fago, A., Malte, H., Storz, J. F., and Gorr, T. A. (2013). Lack of conventional oxygen-linked proton and anion binding sites does not impair allosteric regulation of oxygen binding in dwarf caiman hemoglobin. *American Journal of Physiology-Regulatory Integrative and Comparative Physiology*, **305**, R300–12.

Weber, R. E., Fago, A., Val, A. L., et al. (2000). Isohemoglobin differentiation in the bimodal-breathing amazon catfish *Hoplosternum littorale*. *Journal of Biological Chemistry*, **275**, 17297–305.

Weber, R. E. and Hartvig, M. (1984). Specific fetal hemoglobin underlies the fetal-maternal shift in blood-oxygen affinity in a viviparous teleost. *Molecular Physiology*, **6**, 27–32.

Weber, R. E., Hiebl, I., and Braunitzer, G. (1988). High-altitude and hemoglobin function in the vultures Gyps rueppellii and Aegypius monachus. *Biological Chemistry Hoppe-Seyler*, **369**, 233–240.

Weber, R. E. and Jensen, F. B. 1988. Functional adaptations in hemoglobins from ectothermic vertebrates. *Annual Review of Physiology*, **50**, 161–79.

Weber, R. E., Kleinschmidt, T., and Braunitzer, G. (1987). Embryonic pig hemoglobins Gower-I ($\zeta_2\epsilon_2$), Gower-II ($\alpha_2\epsilon_2$), Heide-I($\zeta_2\theta_2$) and Heide-II($\alpha_2\theta_2$)—oxygen-binding functions related to structure and embryonic oxygen-supply. *Respiration Physiology*, **69**, 347–57.

Weber, R. E., Lalthantluanga, R., and Braunitzer, G. (1988). Functional characterization of fetal and adult yak hemoglobins—an oxygen binding cascade and its molecular basis. *Archives of Biochemistry and Biophysics*, **263**, 199–203.

Weber, R. E. and Lykkeboe, G. (1978). Respiratory adaptations in carp blood. Influences of hypoxia, red cell organic phosphates, divalent cations and CO_2 on hemoglobin-oxygen affinity. *Journal of Comparative Physiology B*, **128**, 127–37.

Weber, R. E., Lykkeboe, G., and Johansen, K. (1976a). Physiological properties of eel hemoglobin—hypoxic acclimation, phosphate effects and multiplicity. *Journal of Experimental Biology*, **64**, 75–88.

Weber, R. E. and White, F. N. (1986). Oxygen binding in alligator blood related to temperature, diving, and alkaline tide. *American Journal of Physiology*, **251**, R901– 8.

Weber, R. E. and White, F. N. (1994). Chloride-dependent organic phosphate sensitivity of the oxygenation reaction in crocodilian hemoglobins. *Journal of Experimental Biology*, **192**, 1–11.

Weber, R. E., Wood, S. C., and Lomholt, J. P. (1976b). Temperature acclimation and oxygen-binding properties of blood and multiple hemoglobins of rainbow trout. *Journal of Experimental Biology*, **65**, 333–45.

Wells, R. M. G. (1979). Hemoglobin-oxygen affinity in developing embryonic erythroid cells of the mouse. *Journal of Comparative Physiology*, **129**, 333–8.

Wells, R. M. G. (2009). Blood-gas transport and hemoglobin function: adaptations for functional and environmental hypoxia. *In*: Richards, J. G., Farrell, A. P., and Brauner, C. J. (eds.) *Fish Physiology, Vol. 27: Hypoxia*, pp. 255–99. Amsterdam, Elsevier.

Wells, R. M. G., Baldwin, J., Seymour, R. S., and Weber, R. E. (1997). Blood oxygen transport and hemoglobin function in three tropical fish species from northern Australian freshwater billabongs. *Fish Physiology and Biochemistry*, **16**, 247–58.

Wells, R. M. G. and Brittain, T. (1981). Transition to cooperative oxygen-binding by embryonic hemoglobin in mice. *Journal of Experimental Biology*, **90**, 351–5.

Wells, R. M. G. and Brittain, T. (1983). Non-cooperative oxygen binding in the erythrocytes of preimplanted sheep embryos. *Comparative Biochemistry and Physiology A-Physiology*, **76**, 387–8.

Wood, S. C. (1971). Effects of metamorphosis on blood respiratory properties and erythrocyte adenosine triphosphate level of salamander *Dicamptodon ensatus* (Eschscholtz). *Respiration Physiology*, **12**, 53–65.

Wood, W. G. (1976). Hemoglobin synthesis during human fetal development. *British Medical Bulletin*, **32**, 282–7.

Zheng, T., Brittain, T., Watmough, N. J., and Weber, R. E. (1999). The role of amino acid alpha(38) in the control of oxygen binding to human adult and embryonic haemoglobin Portland. *Biochemical Journal*, **343**, 681–5.

Zhu, X., Guan, Y., Signore, A. V., et al. (2018). Divergent and parallel routes of biochemical adaptation in high-altitude passerine birds from the Qinghai-Tibet Plateau. *Proceedings of National Academy of Sciences USA*, **115**, 1865–70.

CHAPTER 7

The evolution of novel hemoglobin functions and physiological innovation

7.1 The evolution of novel Hb functions

In the epic sweep of life's history on Earth, globin proteins such as vertebrate Hb were only recently co-opted for a respiratory function in circulatory gas transport. Even after blood-O_2 transport became an entrenched feature of vertebrate physiology, red blood cell Hbs evolved additional specializations of function in particular lineages. In some cases, like the Root effect of fish Hbs, these new functions represent key physiological innovations that have contributed to adaptive radiation. Here we will explore several case studies of how the evolution of novel allosteric properties have enhanced and expanded the physiological capacities of particular vertebrate groups.

7.2 Root effect Hb and the evolution of ocular and swimbladder O_2 secretion in teleost fishes

As we saw in Chapter 6, the pH-sensitive Root effect Hbs of some teleost fishes play a specialized role in "O_2 secretion"—the localized increase in PO_2 in a particular organ or tissue at a level exceeding the prevailing PO_2 in arterial blood (which, since it comes from the gills, has gas partial pressures in equilibrium with the water) (Biot, 1807). In teleost fishes, O_2 secretion into the compressible, gas-filled swimbladder provides an internal means of regulating buoyancy (Scholander and van Dam, 1954, Scholander,

1954, Pelster and Weber, 1991, Pelster and Scheid, 1992, Pelster and Randall, 1998, Berenbrink et al., 2005, Berenbrink, 2007). The ability to inflate the swimbladder with almost pure O_2 against extreme hydrostatic pressures facilitated the colonization of the deep sea and opened up many new ecological opportunities for the ancestors of today's teleosts, likely spurring the adaptive radiation of a taxon that accounts for roughly half of all living vertebrate species. As a mechanism for achieving neutral buoyancy, swimbladder O_2 secretion is also thought to have removed numerous constraints on form and function, especially in relation to the structure of pectoral and caudal fins (Lauder and Liem, 1983). A similar mode of O_2 secretion in the fish eye provides a means of oxygenating the metabolically active but largely avascular retina, thereby enhancing high-acuity vision (Wittenberg and Wittenberg, 1962, Wittenberg and Wittenberg, 1974, Berenbrink et al., 2005, Berenbrink, 2007, Verde et al., 2008). For these reasons, O_2 secretion represents a key physiological innovation in the evolution of teleost fishes.

Let us now explore the role of Root effect Hbs in the functioning of this unique and complex physiological system. In teleost fishes, arterial blood that perfuses the glandular epithelium of the swimbladder or the retinal pigment epithelium of the eye (which composes the blood-retinal barrier) is acidified by the release of CO_2 and lactic acid. The drop in pH promotes the release of Hb-bound O_2 in spite of the high PO_2—this is the Root effect. At low pH,

Hemoglobin: Insights into Protein Structure, Function, and Evolution. Jay F. Storz, Oxford University Press (2019).
© Jay F. Storz 2019. DOI: 10.1093/oso/9780198810681.001.0001

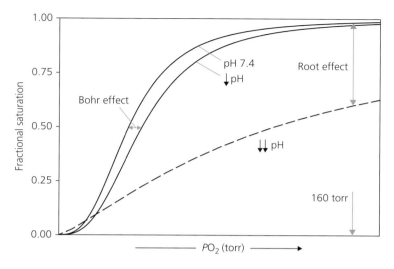

Fig. 7.1. The effect of pH on the Hb-O_2 equilibrium curve. At slightly acid pH, the O_2 equilibrium curve for most vertebrate Hbs is shifted to the right (the Bohr effect). At very low pH, the O_2 equilibrium curve for a typical "Root effect" Hb is dramatically right-shifted, and complete O_2 saturation is not achieved even at several hundred atmospheres of pressure. A conventional means of quantifying the magnitude of the Root effect is to measure the acid-induced reduction in Hb-O_2 saturation in air-equilibrated hemolysates or Hb solutions (Farmer et al., 1979, Berenbrink et al., 2005). For this reason, the PO_2 of dry air at sea level (~160 torr) is indicated on the X-axis.

Root effect Hbs exhibit a large decrease in O_2 affinity (i.e., a large Bohr effect) and a corresponding decrease in cooperativity, which reduces the Hb-O_2 carrying capacity (Root, 1931, Scholander and van Dam, 1954, Brittain, 1987, Pelster and Weber, 1991, Pelster and Scheid, 1992, Pelster and Randall, 1998, Brittain, 2005, Berenbrink, 2007, Waser, 2011). It is the acid-induced reduction in Hb-O_2 carrying capacity that distinguishes the Root effect from the normal Bohr effect (Fig. 7.1); moreover, acid-induced O_2 unloading from Root effect Hb occurs even at O_2 partial pressures in excess of atmospheric levels. The extreme pH sensitivity of Root effect Hbs is illustrated by data for "Hb IV" of rainbow trout, *Oncorhynchus mykiss* (formerly known as *Salmo irideus*) (Fig. 7.2). O_2 equilibrium curves for Trout IV reveal two key features of Root effect Hbs. First, the curves depicted in Fig. 7.2 reveal the extraordinarily broad range of O_2 affinities that are expressed by a single isoHb, as P_{50} values span approximately three orders of magnitude over a range of less than two pH units. The second key feature is the pH-dependence of cooperativity (reflected by differences in the shapes of the curves) as Hill coefficients range from normal values of n_{50} greater than 2 at pH 6.7, to a value of less than 1.0 at pH 6.1.

The Root effect is necessary but not sufficient for O_2 secretion. In arterial capillary blood that perfuses the swimbladder epithelium, Hb-bound O_2 that is released into physical solution is further concentrated by countercurrent circulation in a vascular exchange system known as the *rete mirable* (in Latin, "wonderful network"; plural, *retia mirabilia*). The *rete* is a dense network of interdigitated arterial and venous capillaries that—in most teleost species—are laced around the acid-producing gas glands in the swimbladder wall. The high density of the capillary network provides an increased surface area for countercurrent exchange whereby dissolved O_2 and CO_2 in venous capillaries leaving the swimbladder diffuse back into the less acidic blood of arterial capillaries entering the swimbladder (Steen, 1963, Kobayashi et al., 1990, Pelster and Scheid, 1992, Berenbrink, 2007) (Fig. 7.3). The back-diffusion of dissolved O_2 from venous to arterial capillaries in the *rete* is responsible for concentrating O_2 in the swimbladder epithelium at a level that vastly exceeds what could be produced by normal arterial Hb-O_2 delivery. The localized increase in PO_2 then fills the swimbladder via diffusion. Countercurrent gas exchange in the *rete mirabile*—in combination with the Root effect—can produce O_2 partial pressures

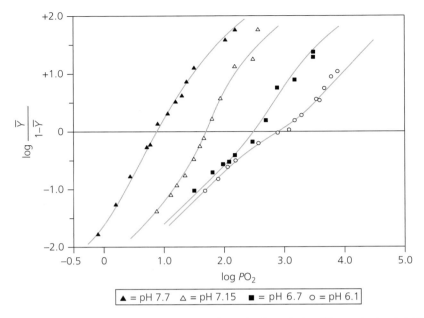

Fig. 7.2. O_2-binding curves of purified stripped Hb IV from rainbow trout (*Oncorhynchus mykiss*) at different pH values in 0.05 Bis-tris [bis(2-hydroxyethyl)imino-tris(hydroxymethyl)methane] buffer at 14°C. The binding isotherms were estimated spectrophotometrically on the basis of absorbance changes between 650 and 450 nm.

Modified from Brunori et al. (1978), with permission from the National Academy of Sciences.

within the swimbladder exceeding 50 atmospheres (38,000 torr) (Scholander and van Dam, 1954)!

Far less is known about the mechanism of O_2 secretion in the eye, although it involves a similar vascular countercurrent system called the choroid *rete mirabile* (Wittenberg and Wittenberg 1962, 1974). As with the swimbladder mechanism, it appears that retinal pigment epithelial cells produce lactic acid, thereby acidifying blood in the dense network of capillaries underlying the retina. PO_2 is increased via the Root effect, and is further amplified by countercurrent gas exchange in the choroid *rete* (Wittenberg and Wittenberg, 1962, Wittenberg and Wittenberg, 1974, Pelster and Weber, 1991, Pelster and Randall, 1998, Waser and Heisler, 2005, Berenbrink, 2007).

7.2.1 Stepwise evolution of a complex physiological system

Root effect Hbs contribute to swimbladder and ocular O_2 secretion because they can unload O_2 upon acidification of capillary blood, even at very high PO_2. However, O_2 secretion also requires a specialized

tissue-specific metabolism geared toward the production and release of acid metabolites (to induce the Root effect) and a vascular countercurrent exchange in the *rete mirabile* (to localize and amplify the O_2 concentration). Each of these physiological and anatomical components must be in place for O_2 secretion to occur. As with any multifaceted, complex trait, the question arises as to how the system evolved in an incremental, stepwise fashion. For reasons just explained, the physiological capacity for O_2 secretion has clear adaptive value for teleost fishes, but this capacity depends on the integrated functioning of multiple components, none of which have obvious adaptive value in isolation. Did the individual components have to evolve in a particular order? Which came first, the Root effect or the *rete mirabile*? Were some features of the system co-opted for a role in O_2 secretion even though they originally evolved to perform a different function?

There are good physiological reasons to expect that the Root effect originated prior to either of the two types of *rete mirabile*. In the absence of acid-induced O_2 unloading by Root effect Hb, the low

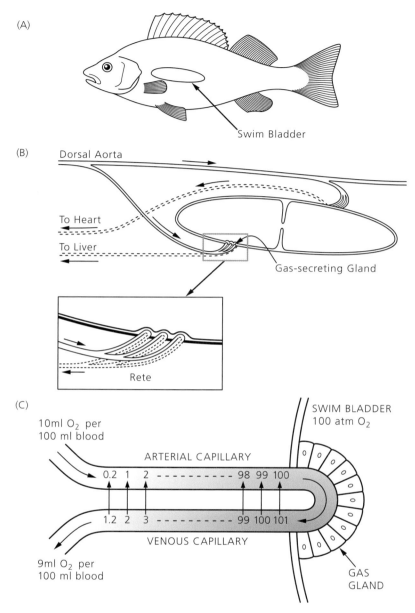

Fig. 7.3. Swimbladder O_2 secretion in teleost fishes. (A) The swimbladder of teleosts is located in the abdominal cavity, just below the vertebral column. (B) Diagram of circulation to the swimbladder. Blood perfuses the glandular epithelium of the swimbladder via the *rete mirabile*, a dense network of capillaries that serves as a countercurrent gas exchange system. As shown in the inset, countercurrent gas exchange in the *rete* depends on passive diffusion between interdigitated arterial and venous capillaries that run in opposite directions. (C) Diagrammatic representation of the countercurrent multiplier system of the fish swimbladder. For simplicity, the numerous capillaries of the *rete* are depicted as a single loop. In this example, arterial blood entering the *rete* contains 10 ml O_2 per 100 ml blood and venous blood exiting the *rete* contains 9 ml per 100 ml. The difference in O_2 content between the arterial supply and venous drainage (1 ml per 100 ml blood in this example) is the quantity of O_2 that is unloaded to the swimbladder. Production of CO_2 and lactic acid by the gas gland induces Hb to unload O_2 (the Root effect) during capillary transit, resulting in an increased PO_2 in the venous capillary. Thus, blood in the venous capillaries has a lower O_2 *content* but a higher O_2 *tension* (PO_2) relative to adjacent arterial capillaries. The resultant arteriovenous pressure gradient causes dissolved O_2 and CO_2 in the venous capillary blood to diffuse back into the arterial capillaries (countercurrent exchange), producing a localized increase in PO_2 that fills the swimbladder via diffusion.

Modified from Schmidt-Nielsen (Schmidt-Nielsen, 1997) with permission from Cambridge University Press.

blood PO_2 in the venous capillaries of the *rete mirabile* would create a large pressure gradient with the high blood PO_2 in the interdigitated arterial capillaries. This would result in the diffusive shunting of O_2 from the arterial supply to the venous drainage of the tissue capillary bed, thereby short-circuiting tissue O_2 delivery (Berenbrink et al., 2005, Berenbrink, 2007). In the context of swimbladder inflation or retinal oxygenation, the Root effect has clear adaptive significance in the presence of *retia mirabilia*. However, the extreme pH-dependent reduction in Hb-O_2 carrying capacity has potentially significant downsides with regard to blood gas transport in the general circulation. For example, exercise- or hypoxia-induced acidosis of systemic blood could greatly hinder O_2 saturation of Root effect Hb in the gills, and would therefore impair tissue O_2 delivery (Berenbrink et al., 2005, Berenbrink, 2007). This could be especially problematic for teleost fishes because their red cells have low buffer capacities—largely due to the low specific buffer value of Hb—so their red blood cells are more readily acidified by even small changes in CO_2 or lactic acid production (Jensen, 1989, Berenbrink, 2006). Many extant species of teleosts have a means of safeguarding intracellular pH (and, hence, Hb-O_2 saturation) through the β-adrenergic activation of a Na^+/H^+-exchanger (βNHE) in the red cell membrane (Motais et al., 1992, Nikinmaa, 1992, Berenbrink and Bridges, 1994). Thus, to evaluate the significance of the Root effect when it first evolved, it is important to understand the physiological context of its historical origin with regard to both red cell buffer capacity and βNHE activity.

The evolution of O_2 secretion and its various subsidiary components was investigated in a *tour de force* study by Berenbrink et al. (2005)—a study that illustrates the value of integrating mechanistic and evolutionary approaches. These authors used phylogenetic comparative methods to reconstruct the evolution of swimbladder and choroid *retia mirabilia*, the Bohr effect, the Root effect, red cell βNHE, and Hb buffer values during the diversification of jawed vertebrates. Starting with trait measurements for living species (representing the branch-tips of the phylogeny), the authors used model-based statistical methods to reconstruct character states of inferred ancestors (representing internal nodes of the phylogeny). These ancestral reconstructions

enabled the authors to identify when particular features first originated and when (in some lineages) they were secondarily lost. They were therefore able to infer the sequential order in which components of the O_2 secretion system were acquired, and by identifying correlated changes in different traits, they were able to identify constraints and propensities in the evolution of this complex physiological system.

The analysis of Berenbrink et al. (2005) revealed that the choroid *rete mirabile* originated roughly 250 million years ago in the most recent common ancestor of teleost fishes and the bowfin, *Amia calva* (Fig. 7.4). Subsequently, swimbladder *retia mirabilia* evolved independently in four separate teleost lineages that already possessed choroid *retia* (Groups A, B, C, and D in Fig. 7.4). It therefore appears that features associated with ocular O_2 secretion set the stage for the later evolution of swimbladder O_2 secretion. As stated by Berenbrink (2007): "[T]he widespread and extraordinary physiological capacity for O_2 secretion, which has been studied in relation to the swimbladder and buoyancy regulation for 200 years, originally evolved in the eye of fishes and for an entirely different purpose." Although most living teleosts possess both a choroid *rete* and a swimbladder *rete*, both structures have been secondarily lost—individually and in tandem—in several different lineages (Fig. 7.4).

Reconstruction of ancestral character states indicated that a pronounced Root effect of more than 40 percent was already present when the choroid *rete mirabile* originated in the common ancestor of teleosts and bowfins (Berenbrink et al., 2005) (Fig. 7.5, Plate 12A). This is consistent with the idea that possession of a gas-permeable *rete mirabilile* without a Root effect would be detrimental due to the short-circuiting of tissue O_2 delivery (Berenbrink et al., 2005; Berenbrink, 2007). Reciprocally, secondary reductions in the Root effect below 40 percent occurred multiple times independently within teleosts, but only in lineages in which the choroid *rete mirabile* had already been lost (Fig. 7.5, Plate 12B). This nonrandom pattern of associated change suggests that the Root effect is selectively maintained in species that possess a choroid *rete mirabile* and that it becomes dispensable in the absence of a *rete*.

A sizable Root effect was already present when the choroid *rete* evolved, and it was also present in each

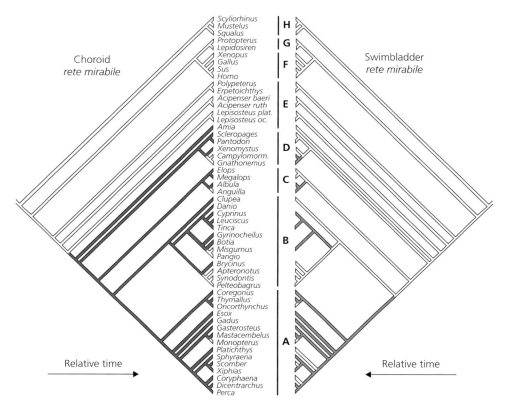

Fig. 7.4. Phylogenetic inferences regarding the evolutionary origins of the choroid and swimbladder *retia* in ray-finned fish. The inferred presence of both types of *retia* are indicated by shaded branches of the mirror-image cladograms. A, Euteleostei; B, Otocephala; C, Elopomorpha; D, Osteoglossomorpha; E, early-branching ray-finned fish (gars, sturgeon, bichir, and reedfish); F, Tetrapods (amphibians, reptiles, birds, and mammals); G, Dipnoi (lungfish); Chondrichthyes (cartilaginous fish). In the phylogeny of ray-finned fish, there is uncertainty about the order in which some lineages branched off the line of descent leading to Euteleostei and Otocephala (clades A and B). Contrary to the tree topology shown here, recent evidence suggests that bowfins (*Amia*) may share a more recent common ancestor with gars (represented by *Lepisosteus* in clade E) than with teleosts (clades A–D) (Near et al., 2012, Betancur-R et al., 2017). This alternative topology is consistent with two alternative scenarios: (1) the choroid *rete* evolved independently in *Amia* and in the most recent common ancestor of teleosts, or (2) it originated in the common ancestor of teleosts + (bowfins, gars) and was secondarily lost in the gar lineage. Evidence also suggests that Elopomorpha (eels and tarpons, clade C) rather than Osteoglossomorpha (mooneyes and boneytongues, clade D) may represent the earliest-branching lineage of teleosts (Near et al., 2012, Betancur-R et al., 2017). This alternative branching order does not affect inferences about the origins of *retia mirabilia*; in either topology, the most parsimonious inference is that a choroid *rete* was already present in the common ancestor of teleosts, and that swimbladder *retia* evolved independently in representatives of both Elopomorpha and Osteoglossomorpha.

Reproduced from Berenbrink et al. (2005) with permission from the Association for the Advancement of Science.

of the four descendent lineages in which a swim-bladder *rete* evolved subsequently (Fig. 7.5). Given that the Root effect could not have played a role in O_2 secretion at its inception, what was its original *raison d'être*, and how did ancestral teleosts compensate for the impaired Hb-O_2 saturation under general blood acidosis? With regard to the latter question, ancestral state reconstructions indicated that the Root effect evolved in the presence of a relatively high specific

Hb buffer value (Berenbrink et al. 2005). Thus, in the ancestors of today's ray-finned fishes, it may be that pH levels in the red cell never fell low enough to elicit the Root effect. Moreover, in extant representatives of several early-branching lineages of "primitive" ray-finned fishes that possess a large Root effect but which lack βNHE, the pH at which the Root effect is triggered is well below the range of blood pH values that is expected during generalized blood

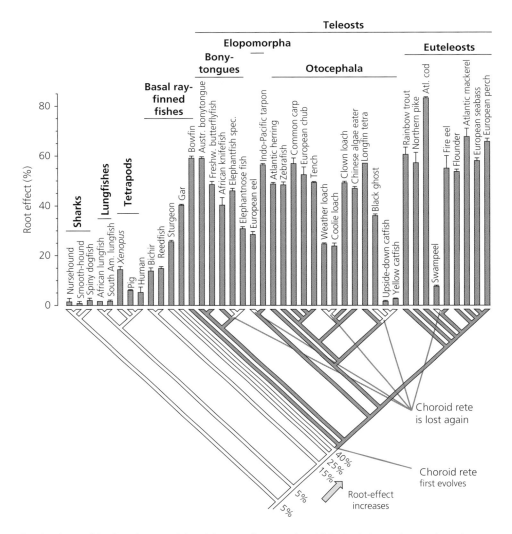

Fig. 7.5. Inferred evolution of the choroid *rete mirabile* and the Root effect in ray-finned fish. The cladogram depicts relationships among jawed vertebrates, along with the inferred presence or absence of a choroid *rete* (denoted by filled or unfilled branches, respectively). The magnitude of the Root effect (mean values and SEM) was estimated as the maximum acid-induced deoxygenation of air-equilibrated hemolysates at physiological body temperatures. The evolutionary reconstruction suggests that the Root effect had increased to a value of ~40 percent prior to the origin of the choroid *rete* in the most recent common ancestor of bowfins (*Amia*) and teleosts. Note that this association between the Root effect and choroid *rete* remains largely unaffected if bowfins share a more recent common ancestor with gars than with teleosts, as suggested by Near et al. (2012) and Betancur-R et al. (2017). In living teleosts, Root effects lower than 40 percent are not observed in taxa that have secondarily lost the choroid rete.

Reproduced from Berenbrink (2011) with permission from Elsevier.

acidosis (Regan and Brauner, 2010a). Thus, the combination of a high buffer value and low Root effect onset pH should mitigate any impairment of Hb-O$_2$ loading in the gills during acidotic stress. This applies to the ancestors of today's ray-finned fishes and in contemporary species that have retained the ances-

tral condition (Brauner and Berenbrink, 2007, Regan and Brauner, 2010a, Regan and Brauner, 2010b).

The ancestral state reconstructions of Berenbrink et al. (2005) indicated that Hb buffer values declined steadily in the line of descent leading to modern teleosts. A reduced Hb buffer value can increase the efficiency of tissue O_2 delivery because a smaller increase in CO_2 and/or lactic acid production (i.e., a smaller decrease in pH) is required to elicit the Bohr effect (Berenbrink, 2006). Since teleosts have evolved a reduction in Hb buffer value in combination with a Root effect onset pH in the physiological range (Pelster and Weber, 1991, Berenbrink et al., 2005, Waser and Heisler, 2005, Berenbrink, 2007, Harter and Brauner, 2017), the increased vulnerability to impaired Hb-O_2 saturation under general blood acidosis may have been the selection pressure that originally favored βNHE expression in the red cell membrane as a means of regulating intracellular pH. βNHE activities underwent secondary reductions in several different teleost

lineages—the very same lineages that secondarily lost the choroid *retia* (Fig. 7.6). This pattern of associated change suggests that βNHE activity is selectively maintained to safeguard retinal oxygenation.

To summarize, the evolutionary reconstruction of Berenbrink et al. (2005) suggests that the common ancestor of jawed vertebrates possessed Hb with a high buffer value and a low level of pH sensitivity (small Bohr effect). This ancestral Hb functioned in a red cell environment in which anion exchange equilibrated bicarbonate across the cell membrane such that changes in plasma pH produced concomitant changes in red cell pH (Nikinmaa, 1997). In the line of descent leading to modern teleosts, the Bohr effect gradually increased (Plate 13). Within tetrapods,

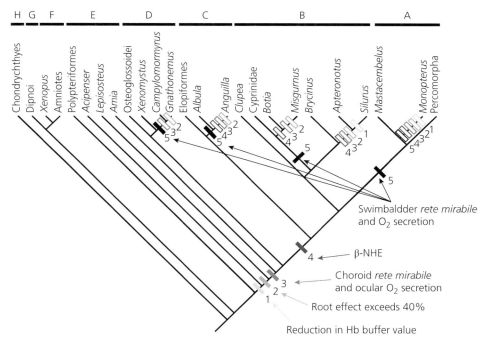

Fig. 7.6. Inferred evolution of O_2 secretion in jawed vertebrates. The most recent common ancestor of jawed vertebrates did not possess red cell βNHE, *retia mirabilia*, or Root effect Hb. Evolutionary reconstructions also suggest that—relative to most modern-day teleosts—the ancestral Hb buffer value was high and the Bohr effect was small (Berenbrink et al., 2005, Berenbrink, 2007). Numbered, filled hashmarks on branches of the tree denote evolutionary gains of particular biochemical/physiological/anatomical features involved in O_2 secretion. Secondary losses of the same features are denoted by numbered, unfilled hashmarks. The taxonomic groupings A–H are the same as in Fig. 7.4, with teleosts represented by clades A–D. Contrary to the phylogenetic relationships shown here for the four main teleost lineages, ((A,B)C,D), recent evidence suggests that Osteoglossomorpha (clade D) branched off before Elopomorpha (clade C) in the diversification of teleosts (Near et al., 2012, Betancur-R et al., 2017). This alternative topology, ((A,B)D,C), implies an earlier origin of βNHE in the stem lineage of teleosts and an additional secondary loss of this feature in osteoglossomorphs. For an explanation of how this alternative tree topology affects inferences about the origins of the choroid *rete*, see the legend for Fig. 7.4.

Modified from Berenbrink et al. (2005) with permission from the Association for the Advancement of Science.

a pronounced Bohr effect also evolved independently in two different amniote lineages (birds and mammals), but it was only in teleosts that the increased pH sensitivity was further elaborated into a bonafide Root effect. As stated by Berenbrink et al. (2005): "the Root effect initially did not evolve as an adaptation for O_2 secretion but may merely have been a by-product of a specific mechanism by which ray-finned fishes increased their Bohr effect." Subsequently, Hb buffer values started to decrease in the common ancestor of teleosts after divergence from the ancestors of bichir and reedfish (order Polypteriformes) (Fig. 7.6). The decline in Hb buffer value would have permitted a more efficient exploitation of the Bohr effect for tissue O_2 delivery. In the ancestor of bowfin + teleosts, the Root effect increased in parallel with the Bohr effect and provided a mechanism for ocular O_2 secretion once the choroid *rete mirabile* evolved. Subsequently, a paralog of the vertebrate-specific family of Na^+/H^+ exchangers (βNHE) was co-opted for expression in the red cell membrane (Fig. 7.6). βNHE activity uncoupled red cell pH from plasma pH and therefore provided a means of safeguarding Hb-O_2 saturation during general circulatory acidosis. Over 100 million years after the origin of the choroid *rete mirabilile*, the same cellular/developmental machinery was recruited for swimbladder O_2 secretion in four separate teleost lineages (Figs. 7.4, 7.6). Thus, in the ancient ancestry of teleosts, anatomical and physiological changes associated with ocular O_2 secretion left a deep imprint on seemingly unrelated aspects of respiratory gas transport in all descendent species.

7.2.2 The Role of the Root effect in general tissue O_2 delivery

Although the Root effect has attracted the attention of physiologists because of its role in O_2 secretion, recent experimental evidence demonstrates that Root effect Hbs can also contribute to the enhancement of general tissue O_2 delivery (Rummer and Brauner, 2011, Rummer et al., 2013, Rummer and Brauner, 2015). In the tissue capillaries, metabolically produced CO_2 diffuses into the blood and—once inside the red cell—the enzyme carbonic anhydrase catalyzes the conversion of CO_2 to HCO_3^-

and H^+. As mentioned in section 7.2.1, βNHE activity prevents red cell pH from falling too low by removing H^+ from the red cell in exchange for Na^+. Since there is no plasma-accessible carbonic anhydrase in the gills of teleosts (Harter and Brauner, 2017), hydration/dehydration of CO_2 occurs predominantly within the red cell and this provides a means of regulating intracellular pH independently of plasma pH, thereby protecting arterial O_2 saturation. However, the presence of plasma-accessible carbonic anhydrase in the endothelium of arterial capillaries can short-circuit Na^+/H^+ exchange across the red cell membrane, which results in a rapid acidification of the red cell during capillary transit (Wang et al., 1998, Henry and Swenson, 2000, Randall et al., 2014, Harter and Brauner, 2017). The transient drop in intracellular pH triggers O_2 unloading from the Root effect Hb (Fig. 7.7) and increases the oxygenation of muscle tissue beyond what would be possible in the absence of plasma-accessible carbonic anhydrase (Rummer et al., 2013). In principle, the absence of plasma-accessible carbonic anhydrase in the venous circulation should allow red cell pH to recover before reaching the gills due to the resumption of Na^+/H^+ exchange across the red cell membrane.

In vitro experiments demonstrated that addition of carbonic anhydrase to acidified, adrenergically stimulated blood from rainbow trout increased PO_2 by greater than 30 torr (Rummer and Brauner, 2011). Subsequent *in vivo* experiments demonstrated that exposure of intact rainbow trout to hypercarbia (1.5 percent CO_2) caused a reduction in plasma pH and red cell pH and increased red muscle PO_2 by 30 torr without any change in perfusion. This change in tissue oxygenation was measured using fiber optic O_2 sensors implanted directly in the muscle tissue. The causal link between plasma-accessible carbonic anhydrase and muscle PO_2 was further confirmed by experiments in which arterial injection of a membrane-impermeant inhibitor of carbonic anhydrase completely abolished the increase in muscle PO_2 (Rummer et al., 2013). These results demonstrate that Root effect Hb—in combination with a mechanism to increase the arteriovenous difference in pH (caused by the presence of plasma-accessible carbonic anhydrase in the tissue capillaries but not in the venous circulation or in the gills)—can greatly enhance Hb-O_2 unloading to tissues other

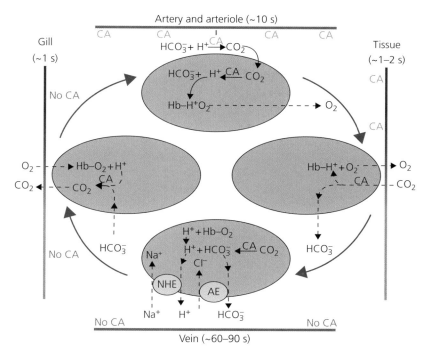

Fig. 7.7. Diagrammatic representation of red blood cell gas exchange throughout the circulatory system of teleost fishes. As blood flows through the gills, arteries, tissue capillaries, and venous circulation, the accessibility of carbonic anhydrase (CA) in the plasma plays an important role in enhancing the exchange of O_2, CO_2, and bicarbonate (HCO_3^-) between the plasma and red blood cells. The approximate transit times of blood through different parts of the circulation are shown in parentheses. Plasma-accessible CA in the arteries and arterioles short-circuits Na^+/H^+ exchange across the red cell membrane, resulting in the rapid acidification of the red cell during capillary transit. This transient drop in red cell pH promotes O_2 unloading via the Bohr effect. Plasma-accessible CA is not present in the gills or in the venous circulation, which allows red blood cell pH to recover prior to gas exchange at the gill.

NHE, sodium-proton exchange; AE, anion exchange.

Reproduced from Randall et al. (2014) with permission from the Journal of Experimental Biology.

than the swimbladder and retina (Rummer and Brauner, 2011, Rummer et al., 2013, Rummer and Brauner, 2015). This unique mode of enhancing tissue O_2 delivery in teleosts may have played a key role in the physiological versatility and ecological success of this exceedingly speciose group (Nikinmaa, 1997, Brauner and Randall, 1998, McKenzie et al., 2004, Berenbrink et al., 2005, Berenbrink, 2007, Randall et al., 2014, Harter and Brauner, 2017).

7.2.3 Structural basis of the Root effect

Interpreted in the context of the two-state MWC allosteric model, the mechanism of the Root effect in teleost Hb was originally thought to involve a strong stabilization of the T-state at acid pH, which

could account for the reduced O_2 affinity, and a corresponding inhibition of the allosteric T→R transition, which could account for the reduced cooperativity. To explain the stabilization of the T-state at low pH, Perutz and Brunori (Perutz and Brunori, 1982) suggested a key role for an intrasubunit salt bridge between the imidazole of the C-terminal His β146 and either Asp or Glu β94 (numbering as in the human sequence). These authors essentially viewed the Root effect as an exaggerated form of the Bohr effect, involving many of the same ionizable groups (Fig. 7.8A). In the T-state of Root effect Hbs, the intrachain salt bridge between the C-terminal His β146β and Asp/Glu β94 was thought to be stabilized by additional bonding interactions with Ser β93. Although all Root effect Hbs have Ser β93,

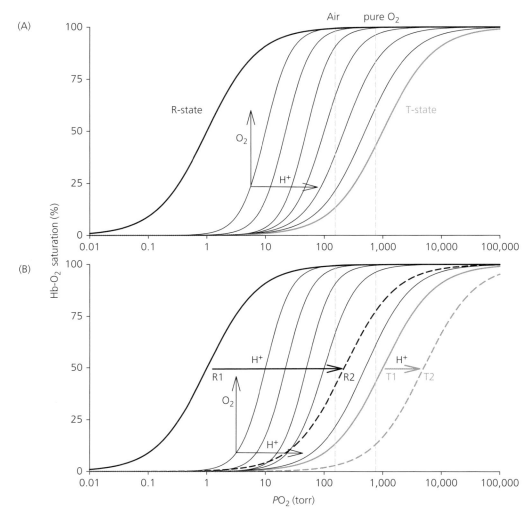

Fig. 7.8. Alternative models to explain the loss of cooperative O_2 binding and incomplete O_2 saturation of Root effect Hbs at low pH, even when equilibrated with air or pure O_2 (dashed vertical lines). (A) According to the conventional two-state MWC model (see Chapter 3), Hb exists in a conformational equilibrium between the R-state (high affinity for O_2, low affinity for H^+) and the T-state (low affinity for O_2, high affinity for H^+). High PO_2 promotes the allosteric transition to the high-affinity R-state (black curve), resulting in cooperative O_2 binding, as indicated by the increased steepness of the curves at intermediate O_2 saturations. Protons (H^+) preferentially bind and stabilize Hb in the T-state, which causes a rightward shift in the O_2 equilibrium curve (decreased O_2 affinity) by delaying the T→R transition. Strong stabilization of the T-state at very low pH abolishes cooperative O_2 binding by preventing the allosteric T→R transition altogether. (B) The multistate model extends the traditional two-state MWC model by postulating a role for pH-mediated tertiary control of O_2 affinity. Specifically, proton-binding stabilizes low-affinity tertiary substates within each quaternary state (substates R1, R2, and T1, T2), resulting in a rightward shift in the O_2 equilibrium curve and reduced cooperativity.

Modified from Berenbrink (2011) with permission from Elsevier.

whereas mammals and most other vertebrates have Cys at this site, the stereochemical mechanism proposed by Perutz and Brunori (1982) clearly does not provide a universal explanation. First, crystallographic analysis of Root effect Hbs have demon-

strated that Ser $\beta93$ forms a hydrogen bond with Asp $\beta94$ in the T-state, but the latter residue is too far away to form a salt bridge with the C-terminal His $\beta146$ (Yokoyama et al., 2004). In fact, crystal structures of Root effect Hbs have demonstrated that the C-terminal His $\beta146$ does not contribute to the formation of an intrachain salt bridge in either the T- or R-state (Camardella et al., 1992, Ito et al., 1995, Yokoyama et al., 2004), indicating that its protonation status would be unaffected by allosteric transitions in quaternary structure. Second, site-directed mutagenesis experiments revealed that key substitutions at $\beta93$ and other candidate sites did not alter pH sensitivity in the way predicted by the model (Nagai et al., 1985, Luisi and Nagai, 1986, Unzai et al., 2008).

According to the original formulation of the two-state MWC model, allosteric effectors modulate Hb-O_2 affinity by shifting the equilibrium between the R and T states and do not affect the intrinsic O_2 binding affinities of subunits within either quaternary state (i.e., K_R and K_T remain constant). It is clear that the traditional MWC framework is not adequate to explain the mechanism of the Root effect, as precise O_2 binding measurements have demonstrated that O_2 affinities of Root effect Hbs are variable within the T-state and are modulated by pH (Brunori et al., 1978, Yokoyama et al., 2004, Bellelli and Brunori, 2011, Berenbrink, 2011). These experimental results indicate that the Root effect cannot be simply explained by stabilization of the T-state or destabilization of the R-state. This is evident in Fig. 7.2, as K_T spans a wide range of values over the physiologically relevant pH range. This suggests a key role for proton-linked changes in tertiary structure (i.e., modulation of O_2 affinity within the T-state) in addition to the T↔R transition in quaternary structure envisioned by the two-state MWC model.

Crystallographic analyses of Root effect Hbs have suggested possible structural mechanisms underlying both tertiary and quaternary control of O_2 affinity. With regard to tertiary control, an analysis of Hb from northern bluefin tuna, *Thunnus thynnus*, revealed the presence of an intrasubunit salt bridge in the T-state between His $\beta69$ and Asp $\beta72$ at a pH of less than 7.5 (Yokoyama et al., 2004). Deprotonation of His $\beta69$ appears to be linked to changes in tertiary structure that are induced by heme ligation of the same β-chain subunit. With regard to the possible coupling of tertiary and quaternary effects, crystallographic studies of Hbs from both bluefin tuna and the emerald rock cod, *Trematomus bernacchii*, revealed an additional proton-binding site that is only formed in the T-state between paired Asp residues on the α- and β-chain subunits of opposing dimers. In T-state Hb of bluefin tuna, the proton is bound between the carboxyl side chains of Asp α_196 and Asp β_2101 (Yokoyama et al., 2004). In T-state Hb of emerald rock cod, the proton is bound between Asp α_195 and Asp β_2101 (Ito et al., 1995). In both cases, binding a shared proton between paired α_1-β_2 Asp residues is thought to stabilize the T-state (Fig. 7.8B). Protonation of the paired Asp residues depends critically on the amino acid state of $\alpha96$ and other structurally remote residues; these indirect, long-range interactions have made it difficult to identify a specific set of amino acid substitutions that are necessary and sufficient for expression of the Root effect (Brittain, 2008, Unzai et al., 2008). As stated by Brittain (2008), the Root effect "requires the permissive actions of a number of distal residues that allow the important functional amino acids to take up the required positions in space."

In at least some Root effect Hbs, the low cooperativity or even "negative cooperativity" ($n_{50} < 1.0$) at low pH is attributable to extreme differences in O_2 affinity between the α- and β-chains (Fig. 7.9) (Decker and Nadja, 2007). In the case of Root effect Hb from bluefin tuna, crystallographic evidence suggests that it is the α-chain that has an especially low O_2 affinity at acid pH, as the distal E7 histidine ($\alpha60$ in tuna Hb) forms a hydrogen bond with one of the heme propionates rather than the free atom of the heme-bound O_2. As discussed in Chapter 4, stabilization of the Fe(II)-O bond by the distal E7 His is a key determinant of O_2 binding affinity. The drastically reduced affinity of one of the two types of subunits at acid pH could explain the biphasic O_2 equilibrium curves observed for some Root effect Hbs (Figs. 7.2, 7.9) (Brunori et al., 1978, Yokoyama et al., 2004).

It has proven remarkably difficult to identify the specific combination of amino acid substitutions that confer the Root effect in teleost Hbs, although it is clear that proton-binding sites are different from those that contribute to the Bohr effect in human Hb (Nagai et al., 1985, Luisi and Nagai, 1986, Camardella et al., 1992, Ito et al., 1995, Jensen et al., 1998a, Mazzarella et al., 1999, Yokoyama et al., 2004, Mazzarella et al., 2006a, Mazzarella et al., 2006b, Unzai et al., 2008, Merlino et al., 2010, Berenbrink, 2011). Yokoyama et al. (2004) articulated the limits of

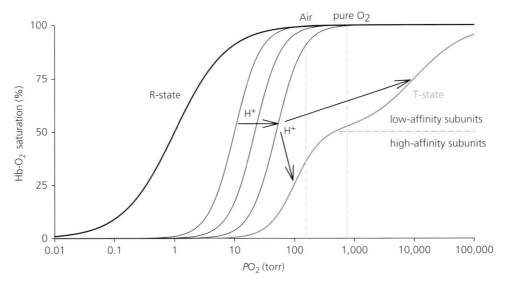

Fig. 7.9. Hypothesized mechanism to explain the reduced cooperativity of Root effect Hbs at low pH. In the Root effect Hbs of some teleost fish species, the α- and β-chain subunits have dramatically different O_2 affinities, which causes "negative cooperativity" ($n_{50}<1.0$), a flattening of the O_2 equilibrium curve at intermediate O_2 saturations. This behavior is not observed in the normal Bohr effect Hbs of other vertebrates, as cooperativity typically remains constant in spite of pH-induced changes in O_2 affinity.

Modified from Berenbrink (2011) with permission from Elsevier.

using our understanding of the Bohr effect of human Hb (HbA) to deduce the structural basis of the Root effect in teleost Hbs: "Despite the strong conservation of the fold and mechanism of tetrameric Hb throughout vertebrate history, a true understanding of each Hb variety with markedly different properties from HbA requires an independent structure determination. Simply decorating the main chain of HbA with the side chains of tuna Hb gives no indication that His-69β and Asp-72β form a ligation state-dependent salt bridge." It may be that there is no unitary structural mechanism that applies to the Root effect Hbs of all species. Just as the Bohr effect has different molecular underpinnings in different vertebrate taxa, the Root effect may also have a somewhat different structural basis in representatives of different teleost lineages.

7.3 Bicarbonate sensitivity of crocodilian Hbs

Crocodilians are able to remain submerged underwater for extraordinarily long periods of time—a physiological capacity that allows some species, such as the Nile crocodile (*Crocodylus niloticus*), to kill large, mammalian prey like wildebeest and zebras by dragging them underwater and drowning them. This capacity for breath-hold diving is aided by a unique allosteric property of crocodilian Hb whereby O_2 affinity is principally governed by the binding of bicarbonate ions (HCO_3^-) rather than organic phosphates (Bauer et al., 1981, Leclercq et al., 1981, Perutz et al., 1981, Perutz, 1983, Brittain and Wells, 1991, Jensen et al., 1998b) (Fig. 7.10). Studies on purified Hb (Bauer et al., 1981) and whole blood (Jensen et al., 1998b) demonstrate that two bicarbonate ions are bound per tetramer upon deoxygenation. During breath-hold submergence, a progressive increase in the red cell concentration of bicarbonate (derived from the hydration of CO_2) reduces Hb-O_2 affinity, thereby promoting O_2 unloading to the cells of metabolizing tissues. This is a highly efficient physiological strategy since the circulatory O_2 store can be almost fully depleted. This mechanism of allosteric regulatory control seems to be unique among tetrapod Hbs. As stated by Perutz (1983): "It is surprising that the crocodilian hemoglobins' simple and direct reciprocating

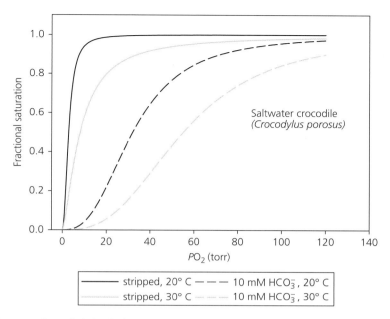

Fig. 7.10. O_2 equilibrium curves for purified Hbs of saltwater crocodile, *Crocodylus porosus* (0.1 M HEPES buffer, pH 7.8, in the presence and absence of 10 mM bicarbonate). Over a range of temperatures, Hb-O_2 affinity is markedly reduced (curves are right-shifted) in the presence of bicarbonate.

Curves were generated using data reported in Brittain and Wells (1991).

action between oxygen and one of the end products of oxidative metabolism has not been adopted by other vertebrates."

In addition to supporting aerobic metabolism during prolonged submergence, the bicarbonate sensitivity of crocodilian Hb may also play a key role in fueling the increase in metabolic rate during digestion (Weber and White, 1986, Busk et al., 2000). Like many reptiles, crocodilians habitually ingest large meals at infrequent intervals, leading to drastic changes in the acid-base status of the blood (Coulson et al., 1950, Busk et al., 2000, Wang et al., 2001). As stated by Coulson and Hernandez (Coulson and Hernandez, 1983): "No other animal is subject by nature to the enormous changes in blood pH and plasma bicarbonate that occur almost daily in Crocodilia. They go into what ordinarily would be considered a grave acidemia every time they exert themselves, and the simple act of eating forces them into an equally pronounced alkalosis." Whereas activity is predominantly fueled by a short-term increase in anaerobic metabolism of skeletal muscle,

resulting in acidosis, digestion is associated with a prolonged increase in aerobic metabolism of the gastrointestinal tract (Wang et al., 2001). The ensuing postprandial alkalosis ("alkaline tide") results from the massive secretion of hydrochloric acid into the lumen of the stomach (which aids the digestion of skeletal material), leading to anion exchange across the gut lining that produces a drastic reduction in the blood concentration of Cl^- and a concomitant increase in the concentration of bicarbonate (Coulson and Hernandez, 1983, Busk et al., 2000). The metabolic alkalosis associated with the digestion of large, bone-rich meals would normally be expected to cause a pronounced increase in blood-O_2 affinity due to the increase in pH. Consequently, O_2 unloading to the tissues would be curtailed during a period of peak metabolic demand.

Weber and White (1986) suggested that the responsiveness of crocodilian Hb to bicarbonate counteracts the pH effect, thereby ensuring continued transport of O_2 to fuel the postprandial increase in aerobic metabolism. To test this hypothesis, Busk et al. (2000) measured *in vivo* arterial blood gases in cannulated alligators (*Alligator mississippiensis*) before and after voluntary feeding, where meal sizes were 7.5 percent of body mass. Over a 24-hour period, digestion was associated with an approximately fourfold increase in metabolic rate. Using blood drawn from cannulae that were implanted in the femoral artery, these researchers demonstrated that the increased metabolic rate was accompanied by an increase in plasma [HCO_3^-] from 24.4 to 36.9 mM (Fig. 7.11) and an equimolar reduction in plasma [Cl^-]. However, the estimated *in vivo* blood-O_2 affinity remained largely unchanged because the bicarbonate effect—in conjunction with ventilatory adjustments that increased PCO_2—was sufficient to offset the effects of increased red cell pH.

At the *in vivo* arterial pH of 7.510 for fasting alligators, the predicted *in vivo* blood P_{50} was 27.7 torr. At 24 hours postfeeding, arterial pH increased from 7.51 to 7.58, a change that would ordinarily

be expected to produce a pronounced increase in blood-O_2 affinity (reduction in P_{50}) via the Bohr effect. For example, in fasting alligators, this increase in pH was predicted to reduce *in vivo* P_{50} to 23.8 torr. In spite of the increased pH, the predicted *in vivo* P_{50} for postprandial alligators was 29.2 torr (Fig. 7.11). This value is ~1.5 torr higher than that predicted under fasting conditions, and is therefore consistent with the hypothesis that the bicarbonate sensitivity of crocodilian Hbs protects tissue O_2 delivery during metabolic alkalosis by preventing a detrimental increase in Hb-O_2 affinity. Moreover, although naturally occurring organic phosphates in red cells of adult crocodiles and alligators (ATP, DPG, and IPP) have no detectable allosteric effect under normal conditions, their inhibitory effects on O_2 binding are manifest when Cl^- concentration is very low (Fig. 7.12) (Weber and White, 1994, Weber et al., 2013), as would be the case during the alkaline tide.

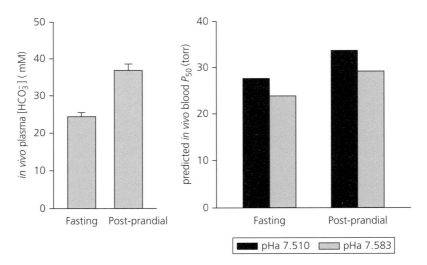

Fig. 7.11. Effects of the postprandial alkaline tide on *in vivo* blood properties of American alligators (*Alligator mississipiensis*). *In vivo* plasma bicarbonate concentrations [HCO_3^-](mean ± SEM) (left) and predicted *in vivo* blood P_{50} values (right) in fasting and postprandial alligators that were chronically cannulated. In the case of the postprandial alligators, measurements of arterial blood gases were taken 24 h after the animals had ingested a meal equal to 7.5 percent of their own body mass (see Busk et al. [2000] for details). Predicted *in vivo* blood P_{50} values are shown for fasting alligators at their *in vivo* arterial pH (7.510) and at the *in vivo* pH of postprandial alligators (7.583). Likewise, predicted *in vivo* P_{50} values are shown for postprandial alligators at their *in vivo* arterial pH and at the *in vivo* pH of fasting alligators. During digestion, the increase in plasma bicarbonate concentration causes a concomitant increase in arterial pH (pHa) that would normally be expected to increase blood-O_2 affinity via the Bohr effect. At pHa 7.583, the higher predicted *in vivo* P_{50} (lower blood-O_2 affinity) for postprandial alligators relative to fasting alligators demonstrates how the allosteric regulation of Hb-O_2 affinity by bicarbonate ions counteracts the effects of increased pH during the alkaline tide. These results suggest that the bicarbonate-sensitivity of crocodilian Hb helps sustain tissue O_2 delivery in support of the increased rate of aerobic metabolism during digestion.

Data from Busk et al. (2000).

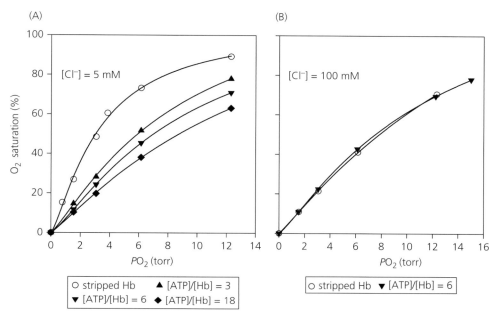

Fig. 7.12. Hb of American alligator (*Alligator mississippiensis*) is sensitive to the allosteric effects of organic phosphates only at low Cl⁻ concentrations, as would be the case in red blood cells during the alkaline tide. (A) At low Cl⁻ concentration (5 mM), Hb-O_2 affinity is progressively reduced (curve is right-shifted) with increasing [ATP]/[Hb] ratio. (B) At high Cl⁻ concentration (100 mM), Hb-O_2 affinity is unaffected by the presence of ATP. O_2 equilibrium curves were measured at pH 7.1, 25°C, in 0.1 M HEPES buffer, heme concentration [tetramer basis] = 0.17 mM). Curves generated from data reported in Weber and Wells (1994).

In large, aquatic crocodilians, bicarbonate-induced Hb-O_2 unloading may be physiologically advantageous because of its role in fueling aerobic metabolism during breath-hold submergence and/or during the postprandial alkaline tide. It is unclear whether similar advantages would accrue to smaller, semiterrestial crocodilians with different feeding habits.

7.3.1 Insights into the structural basis of bicarbonate sensitivity

On the basis of structural modeling, Perutz et al. (1981) hypothesized that the positively charged cleft between the β-chain subunits that binds organic phosphates in most vertebrate Hbs was co-opted for bicarbonate binding in crocodilian Hb, a shift that required a stereochemical reconfiguration of multiple residues. Perutz (1983) suggested that the preferential binding of bicarbonate ions may have involved as few as three amino acid substitutions, two of which simultaneously eliminated residues directly involved in phosphate binding. An add-

itional two amino acid substitutions (1 in the α-chain, 1 in the β-chain) were required to completely abolish the ancestral sensitivity to organic phosphates. These hypotheses were tested by engineering crocodile-specific mutations into recombinant human Hb (Komiyama et al., 1995). Contrary to Perutz's predictions, the study by Komiyama et al. revealed that the bicarbonate sensitivity of crocodilian Hb could be transplanted into human Hb by introducing a total of 12 amino acid substitutions (7 in the α-chain, 5 in the β-chain), most of which were concentrated in the symmetrical $\alpha_1\beta_2$ and $\alpha_2\beta_1$ intersubunit contact surfaces. Komiyama et al. (Komiyama et al., 1995) hypothesized that bicarbonate ions act as "molecular clamps" at the $\alpha_1\beta_2/\alpha_2\beta_1$ interfaces of deoxyHb that stabilize the low-affinity T-state. Results of these protein engineering experiments suggest the hypothesis that two croc-specific residues, Lysβ38 and Tyrβ41, bind a single bicarbonate ion in coordination with the highly conserved Tyrα42 (located on the α-chain subunit of the opposing αβ dimer) (Fig. 7.13). This would account for the

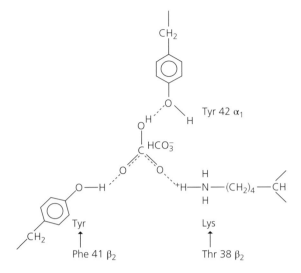

Fig. 7.13. Hypothesized binding site for bicarbonate ion (HCO$_3^-$) at the symmetrical $\alpha_1\beta_2$ and $\alpha_2\beta_1$ interdimeric interfaces of the Hb tetramer. Whereas the Tyr at α42 is conserved among contemporary archosaurs, the substitutions Fβ41Y and Tβ38K occurred in the stem lineage of crocodilians. Experiments by Komiyama et al. (1995) demonstrated that the Fβ41Y and Tβ41K replacements were necessary but not sufficient to confer a high level of bicarbonate sensitivity when engineered into recombinant human Hb.

binding of two bicarbonate ions per Hb tetramer, one at each of the two symmetrical interdimer interfaces ($\alpha_1\beta_2$ and $\alpha_2\beta_1$), consistent with experimental results for native crocodilian Hbs which documented that bicarbonate ions bind deoxyHb with a 2:1 stoichiometry (Bauer et al., 1981, Jensen et al., 1998b). The study by Komiyama et al. (1995) revealed that two amino acid substitutions, Tβ38K and Fβ41Y, were necessary but not sufficient to confer a high level of bicarbonate sensitivity in human Hb. If the structural hypothesis shown in Fig. 7.13 is correct, then it appears that effects of the Tβ38K and Fβ41Y substitutions on oxygenation-linked bicarbonate binding are modulated by one or more ancillary changes at other sites.

The results of Komiyama et al. (1995) indicated that the gain of bicarbonate sensitivity and the loss of phosphate sensitivity were attributable to substitutions at different sites. This raises the question of whether bicarbonate sensitivity evolved before or after the loss of phosphate sensitivity. The results of this study also suggest that ligand-binding affinities

can be altered by the indirect effects of amino acid substitutions at structurally remote sites.

7.4 Reduced or reversed thermal sensitivities of Hb-O$_2$ affinity in heterothermic endotherms

7.4.1 Hbs of heterothermic fishes

As we saw in Chapter 2, tissue O$_2$ delivery is typically augmented by the acidification of red blood cells during capillary transit, as the reduced pH triggers O$_2$ unloading by Hb (the Bohr effect). In addition to the reduced pH, the increased temperature of metabolically active tissue also generally enhances O$_2$ unloading because heme oxygenation is an exothermic reaction (the binding of O$_2$ absorbs heat and the release of O$_2$ releases heat). Hb-O$_2$ affinity therefore changes as an inverse function of temperature, with the result that more O$_2$ is released to the metabolically active, heat-producing tissues that need it most (Barcroft and King, 1909). The temperature-sensitivity of Hb-O$_2$ affinity is generally beneficial for tissue O$_2$ delivery, but it can be detrimental in some situations, most notably in heterothermic fishes that can maintain body core temperatures up to 20°C higher than ambient water temperatures (Carey and Teal, 1969, Carey et al., 1971, Dewar et al., 1994, Harter and Brauner, 2017).

Fast-swimming pelagic fishes like lamnid sharks (white sharks, porbeagle sharks, and makos) and scombroid teleosts like tunas retain metabolically produced heat in their lateral swimming muscles by means of countercurrent heat exchange. The swimming muscles are well-supplied with a dense capillary network (a *rete mirabile*) whereby afferent vessels carrying cold, O$_2$-rich arterial blood from the gills are closely juxtaposed with countercurrent, efferent vessels carrying warm, O$_2$-poor venous blood. Since blood is heated as it flows through the muscle capillary bed, heat from the warmed venous blood is transferred via conduction to the cooler arterial blood entering the muscle. This countercurrent heat exchange allows the swimming muscles to be maintained at temperatures well above the ambient water temperature, which raises rates of muscle force development, contraction, and relaxation (Bennett, 1985).

Billfish (marlin, swordfish, and spearfish) have evolved a more anatomically specific form of "cranial endothermy" whereby elevated temperatures in the brain and eyes are maintained via countercurrent heat exchange in noncontractile, extraocular muscle tissue with a specialized thermogenic function (Carey, 1982, Block, 1986, Block et al., 1993). In the swordfish, *Xiphais gladius*, this mechanism warms the brain and eyes 10–15°C above the ambient water temperature (Fritsches et al. 2005). Elevated cranial temperature may play a general role in buffering the central nervous system to rapid changes in temperature (van den Burg et al. 2005) and the elevated retinal temperature has been shown to enhance visual acuity and temporal resolution, which is likely important for capturing fast-moving prey in the cold, dark depths of the open ocean (Block, 1986, Fritsches et al., 2005).

In fishes that maintain regional heterothermy via countercurrent heat exchange, the possession of Hb with a normal temperature sensitivity could be maladaptive because it could short-circuit tissue O_2 delivery in the *rete* of the swimming muscle or cranial heater muscle. Due to the arteriovenous PO_2 gradient in the *rete mirabile*, some fraction of O_2 would be unloaded from cool arterial blood in the afferent arterioles of the *rete* to warmer venous blood in the efferent arterioles, thereby bypassing the perfused tissue. Moreover, Hb with normal temperature sensitivity would also impede the maintenance of regional heterothermy because the absorbance of heat associated with O_2 unloading would dissipate metabolically produced heat in swimming muscles or brain heaters. This may explain why lamnid sharks and the various lineages of regionally heterothermic teleosts have independently evolved Hbs with reduced or even reversed temperature sensitivities (Weber and Campbell, 2011, Morrison et al., 2016). These reductions of temperature sensitivity have typically been accomplished by evolving altered binding interactions with allosteric effectors such that the endothermic contributions of oxygenation-linked proton or anion dissociation offset the exothermic effect of heme-O_2 binding.

The temperature sensitivity of Hb-O_2 binding can be quantified by the overall enthalpy (or apparent heat) of oxygenation, $\Delta H'$, which can be calculated as a temperature-induced change in P_{50} at constant pH using the van't Hoff equation:

$$\Delta H' = 2.303R \left(\frac{\Delta \log P_{50}}{\Delta \frac{1}{T}} \right) \qquad (7.1)$$

where R is the gas constant and T is the absolute temperature (Wyman, 1964). Numerically negative $\Delta H'$ values denote an exothermic release of heat upon oxygenation. The overall enthalpy of Hb-O_2 binding includes the heat of heme oxygenation, the heat of solution of O_2 (–12.6 kJ mol^{-1}), the heat of the T→R transition in quaternary structure, and the heats of reaction of allosteric ligands (protons, organic phosphates, and Cl$^-$ ions). Since the prosthetic heme groups of all vertebrate Hbs are chemically identical, the exothermic contribution of the heat of heme oxygenation is invariant among Hbs of different species. It is clear that variation in the temperature sensitivity of vertebrate Hbs is largely attributable to variation in the heats of reactions with allosteric effectors (Weber and Campbell, 2011). In most cases, the exothermic heat of heme oxygenation is only partially offset by the endothermic release of allosteric effectors such that the calculated enthalpy value is negative ($\Delta H'<0$) at physiological pH. However, certain fish species have evolved Hbs in which the exothermic heat of heme oxygenation is more than counteracted by the endothermic release of allosteric effectors at physiological pH, resulting in a reversed thermal effect (Hb-O_2 affinity is increased by an increase in temperature), and the overall enthalpy value is positive ($\Delta H'>0$).

In contrast to the Hbs of most ectothermic teleosts, those of bluefin tuna, *Thunnus thynnus*, exhibit a low temperature sensitivity (a numerically low $\Delta H'$ value) in the physiological pH range. The low overall $\Delta H'$ value is attributable to a reversed temperature sensitivity that is only manifest at O_2 saturations above 50 percent (Rossi-Fanelli and Antonini, 1960, Carey and Gibson, 1977, Ikeda-Saito et al., 1983). This saturation-dependent temperature sensitivity suggests that the exothermic heat of heme oxygenation is offset by the endothermic release of protons that occurs late in the oxygenation process, after two or three O_2 molecules have bound (Weber and Campbell, 2011). Interestingly, a similarly low temperature-dependence of blood oxygenation has been documented in the chub mackerel, *Scomber japonicas*, an ectothermic species closely related to tunas in the same family (Scombridae) (Clark et al., 2010). This finding suggests that reduced or reversed Hb temperature effects preceded the origin of regional heterothermy in scombrid fishes, and may have originally evolved as a thermoregulatory mechanism for reducing oxygenation-linked heat loss.

Whereas the reduced temperature sensitivity of tuna Hb is attributable to the oxygenation-linked endothermic dissociation of protons, the reduced or reversed temperature sensitivities (positive $\Delta H'$ values) of lamnid shark and billfish Hbs are primarily attributable to the endothermic dissociation of organic phosphates (Larsen et al., 2003, Weber et al., 2010) (Fig. 7.14). In the case of billfish, $\Delta H'$ values for stripped hemolysates of three different species

are all quite similar (~62 kJ mol^{-1}) when measured in the absence of anionic effectors at pH greater than 8 where the Bohr effect is not operative (Fig. 7.15). Under these conditions, the measured $\Delta H'$ values approximate the intrinsic enthalpy of heme oxygenation. By contrast, $\Delta H'$ values are dramatically reduced in the presence of ATP at pH values in the physiological range. At pH 7.4, the exothermic oxygenation reactions in the absence of ATP ($\Delta H' = -39, -49,$ and -44 kJ mol^{-1} for Hbs of blue marlin, striped marlin, and short-billed spearfish, respectively) become endothermic or nearly so in the presence of ATP ($\Delta H'= +26, +4,$ and -7) (Weber et al., 2010) (Fig. 7.15). The structural mechanisms underlying variation in effector interactions among the Hbs of elasmobranchs and teleosts are known in some cases (Jensen et al., 1998a), although it has proven difficult to identify the specific amino acid residues involved in oxygenation-linked proton-binding (Yokoyama et al., 2004, Berenbrink, 2006).

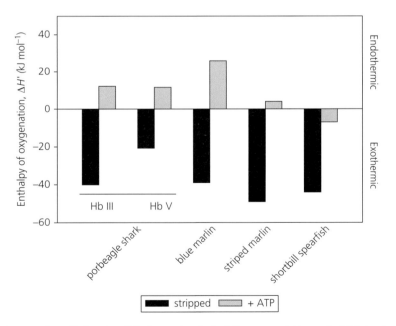

Fig. 7.14. Enthalpies of oxygenation, $\Delta H'$, for stripped Hbs from porbeagle shark (*Lamna nasus*), blue marlin (*Makaira nigricans*), striped marlin (*Tetrapturus audux*), and shortbill spearfish (*Tetrapturus angustirostris*), measured in the absence ("stripped") and presence of saturating concentrations of ATP. In the case of the porbeagle shark, O_2-binding measurements were conducted on purified isoHbs (Hb III and Hb V) at temperatures of 10°C and 26°C, pH 7.3, in 0.1 M HEPES buffer, [heme] = 0.29–0.30 mM). In the case of the three billfish species, measurements were conducted on purified of hemolysates at temperatures of 10°C and 25°C, pH 7.4, in 0.1 M HEPES buffer, [heme] = 0.22–0.30 mM).
Based on data reported in Larsen et al. (2003) and Weber et al. (2010).

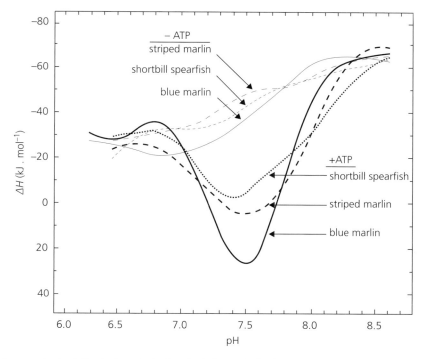

Fig. 7.15. Heats of oxygenation (excluding the heat of solvation of O_2) of stripped Hbs of blue marlin, striped marlin, and shortbill spearfish in the absence and presence of ATP.

Based on data reported in Weber et al. (2010).

7.4.2 Mammalian Hbs

Similar to the case with heterothermic fishes, the possession of Hbs with normal temperature sensitivities may also be detrimental in heterothermic endotherms that live in cold environments because O_2 unloading may be impaired in peripheral tissues and extremities that are maintained at much lower temperatures than the body core. Under such conditions, a number of authors have suggested that it may be adaptive for endotherms to have Hbs with reduced temperature sensitivities (Brix et al., 1990, Coletta et al., 1992, De Rosa et al., 2004, Weber and Campbell, 2011). Seemingly consistent with this idea, several studies have demonstrated that the Hbs of Arctic mammals tend to have low temperature sensitivities (less negative $\Delta H'$ values) in comparison with human Hb. Much of this work has focused on ruminant artiodactyls such as caribou (*Rangifer tarandus*) and muskox (*Ovibos moschatus*) (Brix et al., 1989, Giardina et al., 1989, Brix et al., 1990, Coletta et al., 1992). As discussed in Chapter 4, the adult Hbs of rumin-

ants differ from those of most other mammals in that they exhibit low intrinsic O_2 affinities and low sensitivities to DPG (additionally, their red cells contain very low concentrations of DPG). Accordingly, the relatively low temperature sensitivities of ruminant Hbs are attributable to endothermic contributions from the oxygenation-linked dissociation of Cl^- ions rather than DPG, and they also have a reduced heat of T→R transition (Weber et al., 2014). The problem with ascribing adaptive significance to the patterns observed for these Arctic species is that relatively low $\Delta H'$ values appear to be a general feature of artiodactyl Hbs and are not uniquely derived features of cold-adapted species.

A more informative comparison involves Hbs of the extinct woolly mammoth (*Mammuthus primigenius*) and its closest living relative, the Asian elephant (*Elephas maximus*) which, along with African elephant, is exclusively restricted to tropical latitudes. This comparison was made possible by the recovery of adult α- and β-globin DNA sequences from a 43,000-year-old permafrost-preserved mammoth specimen in Siberia. Campbell et al. (2010) cloned the mammoth globin sequences into a plasmid construct and resurrected mammoth Hb via recombinant expression in *E. coli* host cells. O_2-binding experiments revealed that woolly mammoth Hb evolved a derived reduction in temperature sensitivity due to the net effect of three mammoth-specific amino acid substitutions in the adult β-chain subunit. At physiological pH and in the presence of normal concentrations of Cl^- and DPG, the overall $\Delta H'$ of woolly mammoth Hb is reduced 8.8 kJ mol^{-1} relative to that of Asian elephant, which can be expected to facilitate O_2 unloading to cold peripheral tissues. This allows extremities to be maintained at lower temperatures, thereby contributing to the conservation of metabolic heat by minimizing conductive/convective heat loss. Unlike the case with the ruminant Hbs, the relatively low $\Delta H'$ value for mammoth Hb is attributable to the endothermic dissociation of both DPG and Cl^- ions upon deoxygenation (Campbell et al., 2010, Yuan et al., 2011).

In most hetereothermic mammals and birds, the adaptive significance of evolved changes in the thermal sensitivity of Hb-O_2 binding remains a matter of speculation. Phylogenetically informed comparisons involving more taxa with polar and nonpolar distributions could reveal whether there is a consistent trend.

This survey of novel Hb functions has focused on the evolution of qualitatively distinct allosteric properties— gains or losses of sensitivities to specific allosteric effectors, or changes in the thermal sensitivity of O_2 affinity. In Chapter 8 we will focus on quantitative changes in O_2-binding properties as adaptations to changes in environmental O_2 availability.

References

Barcroft, J. and King, W. O. R. (1909). The effect of temperature on the dissociation curve of blood. *Journal of Physiology-London*, **39**, 374–84.

Bauer, C., Forster, M., Gros, G., and Vogel, D. (1981). Analysis of bicarbonate binding to crocodilian hemoglobin. *Journal of Biological Chemistry*, **256**, 8429–35.

Bellelli, A. and Brunori, M. (2011). Hemoglobin allostery: variations on the theme. *Biochimica Et Biophysica Acta-Bioenergetics*, **1807**, 1262–72.

Bennett, A. F. (1985). Temperature and muscle. *Journal of Experimental Biology*, **115**, 333–44.

Berenbrink, M. (2006). Evolution of vertebrate haemoglobins: histidine side chains, specific buffer value and Bohr effect. *Respiratory Physiology and Neurobiology*, **154**, 165–84.

Berenbrink, M. (2007). Historical reconstructions of evolving physiological complexity: O_2 secretion in the eye and swimbladder of fishes. *Journal of Experimental Biology*, **210**, 1641–52.

Berenbrink, M. (2011). Root effect: molecular basis, evolution of the Root effect and rete systems. *In:* Farrell, A. P. (ed.) *Encyclopedia of Fish Physiology: From Genome to Environment*, pp. 929–34. Amsterdam, Elsevier.

Berenbrink, M. and Bridges, C. R. (1994). Catecholamine-activated sodium/proton exchange in the red blood cells of the marine teleost *Gadus morhua*. *Journal of Experimental Biology*, **192**, 253–67.

Berenbrink, M., Koldkjaer, P., Kepp, O., and Cossins, A. R. (2005). Evolution of oxygen secretion in fishes and the emergence of a complex physiological system. *Science*, **307**, 1752–7.

Betancur-R, R., Wiley, E. O., Arratia, G., et al. (2017). Phylogenetic classification of bony fishes. *BMC Evolutionary Biology*, **17**(1), 162.

Biot, M. (1807). Sur la nature de l'air continue dans la vessie natatoire des poisons. *Memoires de Physique et de Chimie de la Societe d'Arcuel*, **1**, 252–81.

Block, B. A. (1986). Structure of the brain and eye heater tissue in marlins, sailfish, and spearfishes. *Journal of Morphology*, **190**, 169–89.

Block, B. A., Finnerty, J. R., Stewart, A. F. R., and Kidd, J. (1993). Evolution of endothermy in fish—mapping physiological traits on a molecular phylogeny. *Science*, **260**, 210–14.

Brauner, C. J. and Berenbrink, M. (2007). Gas transport and exchange. *In:* McKenzie, D. J., Farrell, A. P., and Brauner, C. J. (eds.) *Fish Physiology, Vol. 26: Primitive Fishes*. New York, NY, Academic Press.

Brauner, C. J. and Randall, D. J. (1998). The linkage between oxygen and carbon dioxide transport. *In:* Perry, S. F. and Tufts, B. L. (eds.) *Fish Physiology, Vol. 17: Fish Respiration*, pp. 213–82. New York, NY, Academic Press.

Brittain, T. (1987). The Root effect. *Comparative Biochemistry and Physiology B-Biochemistry and Molecular Biology*, **86**, 473–81.

Brittain, T. (2005). Root effect hemoglobins. *Journal of Inorganic Biochemistry*, **99**, 120–9.

Brittain, T. (2008). Extreme pH sensitivity in the binding of oxygen to some fish hemoglobins: the Root effect. *In:* Ghosh, A. (ed.) *The Smallest Biomolecules: Diatomics and their Interactions with Heme Proteins*, p. 219. Amsterdam, Elsevier.

Brittain, T. and Wells, R. M. (1991). An investigation of the cooperative functioning of the hemoglobin of the crocodile, *Crocodylus porosus*. *Comparative Biochemistry and Physiology B-Biochemistry and Molecular Biology*, **98**, 641–6.

Brix, O., Bardgard, A., Mathisen, S., et al. (1989). Arctic life adaptation. 2. The function of musk ox (*Ovibos muschatos*) hemoglobin. *Comparative Biochemistry and Physiology B-Biochemistry and Molecular Biology*, **94**, 135–8.

Brix, O., Bardgard, A., Mathisen, S., et al. (1990). Oxygen-transport in the blood of Arctic mammals—adaptation to local heterothermia. *Journal of Comparative Physiology B-Biochemical Systemic and Environmental Physiology*, **159**, 655–60.

Brunori, M., Coletta, M., Giardina, B., and Wyman, J. (1978). Macromolecular transducer as illustrated by trout hemoglobin IV. *Proceedings of the National Academy of Sciences of the United States of America*, **75**, 4310–12.

Busk, M., Overgaard, J., Hicks, J. W., Bennet, A. F., and Wang, T. (2000). Effects of feeding on arterial blood gases in the American alligator *Alligator mississippiensis*. *Journal of Experimental Biology*, **203**, 3117–24.

Camardella, L., Caruso, C., Davino, R., et al. (1992). Hemoglobin of the Antarctic fish *Pagothenia bernacchii*. Amino acid sequence, oxygen equilibria and crystal structure of its carbonmonoxy derivative. *Journal of Molecular Biology*, **224**, 449–60.

Campbell, K. L., Roberts, J. E. E., Watson, L. N., et al. (2010). Substitutions in woolly mammoth hemoglobin confer biochemical properties adaptive for cold tolerance. *Nature Genetics*, **42**, 536–40.

Carey, F. G. (1982). A brain heater in the swordfish. *Science*, **216**, 1327–9.

Carey, F. G. and Gibson, Q. H. (1977). Reverse temperature-dependence of tuna hemoglobin oxygenation. *Biochemical and Biophysical Research Communications*, **78**, 1376–82.

Carey, F. G. and Teal, J. M. (1969). Mako and porbeagle: warm-bodied sharks. *Comparative Biochemistry and Physiology*, **28**, 199–204.

Carey, F. G., Teal, J. M., Kanwishe, J. W., Lawson, K. D., and Beckett, J. S. (1971). Warm-bodied fish. *American Zoologist*, **11**, 137–43.

Clark, T. D., Rummer, J. L., Sepulveda, C. A., Farrell, A. P., and Brauner, C. J. (2010). Reduced and reversed temperature dependence of blood oxygenation in an ectothermic scombrid fish: implications for the evolution of regional heterothermy? *Journal of Comparative Physiology B-Biochemical Systemic and Environmental Physiology*, **180**, 73–82.

Coletta, M., Clementi, M. E., Ascenzi, P., et al. (1992). A comparative study of the temperature dependence of the oxygen-binding properties of mammalian hemoglobins. *European Journal of Biochemistry*, **204**, 1155–7.

Coulson, R. A. and Hernandez, T. (1983). Alligator metabolism—studies on chemical reactions *in vivo*. *Comparative Biochemistry and Physiology B-Biochemistry and Molecular Biology*, **74**, R1–R182.

Coulson, R. A., Hernandez, T., and Dessauer, H. C. (1950). Alkaline tide of the alligator. *Proceedings of the Society for Experimental Biology and Medicine*, **74**, 866–9.

De Rosa, M. C., Castagnola, M., Bertonati, C., Galtieri, A., and Giardina, B. (2004). From the Arctic to fetal life: physiological importance and structural basis of an 'additional' chloride-binding site in haemoglobin. *Biochemical Journal*, **380**, 889–96.

Decker, H. and Nadja, H. (2007). Negative cooperativity in Root-effect hemoglobins: role of heterogeneity. *Integrative and Comparative Biology*, **47**, 656–61.

Dewar, H., Graham, J. B., and Brill, R. W. (1994). Studies of tropical tuna swimming performance in a large water tunnel. 2. Thermoregulation. *Journal of Experimental Biology*, **192**, 33–44.

Farmer, M., Fyhn, H. J., Fyhn, U. E. H., and Noble, R. W. (1979). Occurrence of Root effect hemoglobins in Amazonian fishes. *Comparative Biochemistry and Physiology A -Physiology*, **62**, 115–24.

Fritsches, K. A., Brill, R. W., and Warrant, E. J. (2005). Warm eyes provide superior vision in swordfishes. *Current Biology*, **15**, 55–8.

Giardina, B., Brix, O., Nuutinen, M., Elsherbini, S., et al. (1989). Arctic adaptation in reindeer—the energy saving of a hemoglobin. *FEBS Letters*, **247**, 135–8.

Harter, T. S. and Brauner, C. J. (2017). The O_2 and CO_2 transport system in teleosts and the specialized mechanisms that enhance Hb-O_2 unloading to tissues. In: Gamperl, A. K., Gillis, T. E., Farrell, A. P., and Brauner, C. J. (eds.) *Fish Physiology, Vol. 36B: The Cardiovascular System: Development, Plasticity, and Physiological Responses*. San Diego, CA, Elsevier Science.

Henry, R. P. and Swenson, E. R. (2000). The distribution and physiological significance of carbonic anhydrase in vertebrate gas exchange organs. *Respiration Physiology*, **121**, 1–12.

Ikeda-Saito, M., Yonetani, T., and Gibson, Q. H. (1983). Oxygen equilibrium studies on hemoglobin from the bluefin tuna (*Thunnus thynnus*). *Journal of Molecular Biology*, **168**, 673–86.

Ito, N., Komiyama, N. H., and Fermi, G. (1995). Structure of deoxyhemoglobin of the Antarctic fish *Pagothenia bernacchii* with an analysis of the structural basis of the Root effect by comparison of the liganded and unliganded hemoglobin structures. *Journal of Molecular Biology*, **250**, 648–58.

Jensen, F. B. (1989). Hydrogen ion equilibria in fish hemoglobins. *Journal of Experimental Biology*, **143**, 225–34.

Jensen, F. B., Fago, A., and Weber, R. E. (1998a). Hemoglobin structure and function. In: Perry, S. F. and Tufts, B. L. (eds.) *Fish Physiology, Vol. 17: Fish Respiration*, pp. 1–40. New York, NY, Academic Press.

Jensen, F. B., Wang, T., Jones, D. R., and Brahm, J. (1998b). Carbon dioxide transport in alligator blood and its erythrocyte permeability to anions and water. *American Journal of Physiology-Regulatory Integrative and Comparative Physiology*, **274**, R661–71.

Kobayashi, H., Pelster, B., and Scheid, P. (1990). CO_2 back-diffusion in the rete aids O_2 secretion in the swimbladder of the eel. *Respiration Physiology*, **79**, 231–42.

Komiyama, N. H., Miyazaki, G., Tame, J., and Nagai, K. (1995). Transplanting a unique allosteric effect from crocodile into human haemoglobin. *Nature*, **373**, 244–6.

Larsen, C., Malte, H., and Weber, R. E. (2003). ATP-induced reverse temperature effect in isohemoglobins from the endothermic porbeagle shark (*Lamna nasus*). *Journal of Biological Chemistry*, **278**, 30741–7.

Lauder, G. V. and Liem, K. F. (1983). The evolution and interrelationships of the actinopterygian fishes. *Bulletin of the Museum of Comparative Zoology*, **150**, 95–197.

Leclercq, F., Schnek, A. G., Braunitzer, G., Stangl, A., and Schrank, B. (1981). Direct reciprocal allosteric interaction of oxygen and hydrogen carbonate: sequence of the haemoglobins of the caiman (*Caiman crocodylus*), the Nile crocodile (*Crocodylus niloticus*) and the Mississippi crocodile (*Alligator mississippiensis*). *Hoppe Seyler Zeitchsrift fur Physiologische Chemie*, **362**, 1151–8.

Luisi, B. F. and Nagai, K. (1986). Crystallographic analysis of mutant human hemoglobins made in *Escherichia coli*. *Nature*, **320**, 555–6.

Mazzarella, L., Bonomi, G., Lubrano, M. C., et al. (2006a). Minimal structural requirements for root effect: crystal structure of the cathodic hemoglobin isolated from the Antarctic fish *Trematomus newnesi*. *Proteins-Structure Function and Bioinformatics*, **62**, 316–21.

Mazzarella, L., D'Avino, R., Di Prisco, G., et al. (1999). Crystal structure of Trematomus newnesi haemoglobin re-opens the Root effect question. *Journal of Molecular Biology*, **287**, 897–906.

Mazzarella, L., Vergara, A., Vitagliano, L., et al. (2006b). High resolution crystal structure of deoxy hemoglobin from *Trematomus bernacchii* at different pH values: the role of histidine residues in modulating the strength of the root effect. *Proteins-Structure Function and Bioinformatics*, **65**, 490–8.

McKenzie, D. J., Wong, S., Randall, D. J., et al. (2004). The effects of sustained exercise and hypoxia upon oxygen tensions in the red muscle of rainbow trout. *Journal of Experimental Biology*, **207**, 3629–37.

Merlino, A., Vitagliano, L., Balsamo, A., et al. (2010). Crystallization, preliminary X-ray diffraction studies and Raman microscopy of the major haemoglobin from the sub-Antarctic fish *Eleginops maclovinus* in the carbomonoxy form. *Acta Crystallographica Section F-Structural Biology Communications*, **66**, 1536–40.

Morrison, P. R., Gilmour, K. M., and Brauner, C. J. (2016). Oxygen and carbon dioxide transport in elasmobranchs. *In:* Shadwick, R., Farrell, A. P., and Brauner, C. J.

(eds.) *Fish Physiology, Vol. 34B: Physiology of Elasmobranch Fishes: Internal Processes.* New York, NY, Elsevier.

Motais, R., Borgese, F., Fievet, B., and Garciaromeu, F. (1992). Regulation of Na^+/H^+ exchange and pH in erythrocytes of fish. *Comparative Biochemistry and Physiology A-Physiology*, **102**, 597–602.

Nagai, K., Perutz, M. F., and Poyart, C. (1985). Oxygen binding properties of human mutant hemoglobins synthesized in *Escherichia coli*. *Proceedings of the National Academy of Sciences of the United States of America*, **82**, 7252–5.

Near, T. J., Eytan, R. I., Dornburg, A., et al. (2012). Resolution of ray-finned fish phylogeny and timing of diversification. *Proceedings of the National Academy of Sciences of the United States of America*, **109**, 13698–703.

Nikinmaa, M. (1992). Membrane transport and control of hemoglobin-oxygen affinity in nucleated erythrocytes. *Physiological Reviews*, **72**, 301–21.

Nikinmaa, M. (1997). Oxygen and carbon dioxide transport in vertebrate erythrocytes: an evolutionary change in the role of membrane transport. *Journal of Experimental Biology*, **200**, 369–80.

Pelster, B. and Randall, D. J. (1998). The physiology of the Root effect. *In:* Perry, S. F. and Tufts, B. L. (eds.) *Fish Physiology, Vol. 17: Fish Respiration.* New York, NY, Academic Press.

Pelster, B. and Scheid, P. (1992). Countercurrent concentration and gas secretion in the fish swim bladder. *Physiological Zoology*, **65**, 1–16.

Pelster, B. and Weber, R. E. (1991). The physiology of the Root effect. *Advances in Comparative and Environmental Physiology*, **8**, 51–77.

Perutz, M. F. (1983). Species adaptation in a protein molecule. *Molecular Biology and Evolution*, **1**, 1–28.

Perutz, M. F., Bauer, C., Gros, G., et al. (1981). Allosteric regulation of crocodilian haemoglobin. *Nature*, **291**, 682–4.

Perutz, M. F. and Brunori, M. (1982). Stereochemistry of cooperative effects in fish and amphibian hemoglobins. *Nature*, **299**, 421–6.

Randall, D. J., Rummer, J. L., Wilson, J. M., Wang, S., and Brauner, C. J. (2014). A unique mode of tissue oxygenation and the adaptive radiation of teleost fishes. *Journal of Experimental Biology*, **217**, 1205–14.

Regan, M. D. and Brauner, C. J. (2010a). The evolution of Root effect hemoglobins in the absence of intracellular pH protection of the red blood cell: insights from primitive fishes. *Journal of Comparative Physiology B-Biochemical Systemic and Environmental Physiology*, **180**, 695–706.

Regan, M. D. and Brauner, C. J. (2010b). The transition in hemoglobin proton-binding characteristics within the basal actinopterygian fishes. *Journal of Comparative Physiology B-Biochemical Systemic and Environmental Physiology*, **180**, 521–30.

Root, R. W. (1931). The respiratory function of the blood of marine fishes. *Biological Bulletin*, **61**, 427–56.

Rossi-Fanelli, A. and Antonini, E. (1960). Oxygen equilibrium of haemoglobin from *Thunnus thynnus*. *Nature*, **186**, 895–6.

Rummer, J. L. and Brauner, C. J. (2011). Plasma-accessible carbonic anhydrase at the tissue of a teleost fish may greatly enhance oxygen delivery: *in vitro* evidence in a rainbow trout, *Oncorynchus mykiss*. *Journal of Experimental Biology*, **214**, 2319–28.

Rummer, J. L. and Brauner, C. J. (2015). Root effect haemoglobins in fish may greatly enhance general oxygen delivery relative to other vertebrates. *PloS One*, **10**, e0139477.

Rummer, J. L., McKenzie, D. J., Innocenti, A., Supuran, C. T., and Brauner, C. J. (2013). Root effect hemoglobin may have evolved to enhance general tissue oxygen delivery. *Science*, **340**, 1327–9.

Schmidt-Nielsen, K. (1997). *Animal Physiology: Adaptation and Environment*, Cambridge, Cambridge University Press.

Scholander, P. F. (1954). Secretion of gases against high pressures in the swimbladder of deep sea fishes. 2. The rete mirabile. *Biological Bulletin*, **107**, 260–77.

Scholander, P. F. and Van Dam, L. (1954). Secretion of gases against high pressures in the swimbladder of deep sea fishes. 1. Oxygen dissociation in blood. *Biological Bulletin*, **107**, 247–59.

Steen, J. B. (1963). Physiology of swimbladder in the eel *Anguilla vulgaris*. 3. Mechanism of gas secretion. *Acta Physiologica Scandinavica*, **59**, 221–41.

Unzai, S., Imai, K., Park, S.-Y., et al. (2008). Mutagenic studies on the origins of the Root effect. *In*: Bolognesi, M., De Prisco, G., and Verde, C. (eds.) *Dioxygen Binding and Sensing Proteins: A Tribute to Beatrice and Jonathan Wittenberg*, pp. 67–78. Milan, Springer.

Van Den Burg, E. H., Peeters, R. R., Verhoye, M., et al. (2005). Brain responses to ambient temperature fluctuations in fish: reduction of blood volume and initiation of a whole-body stress response. *Journal of Neurophysiology*, **93**, 2849–55.

Verde, C., Berenbrink, M., and Di Prisco, G. (2008). Evolutionary physiology of oxygen secretion in the eye of fishes of the suborder Notothenioidei. In: Bolognesi, M., Di Prisco, G., and Verde, C. (eds.) *Dioxygen Binding and Sensing Proteins: A Tribute to Beatrice and Jonathan Wittenberg*, pp. 49–65. Milan, Springer.

Wang, T., Busk, H., and Overgaard, J. (2001). The respiratory consequences of feeding in amphibians and reptiles. *Comparative Biochemistry and Physiology A-Molecular and Integrative Physiology*, **128**, 535–49.

Wang, Y. X., Henry, R. P., Wright, P. M., Heigenhauser, G. J. F., and Wood, C. M. (1998). Respiratory and metabolic functions of carbonic anhydrase in exercised white muscle of trout. *American Journal of Physiology-Regulatory Integrative and Comparative Physiology*, **275**, R1766–79.

Waser, W. (2011). Root effect: Root effect definition, functional role in oxygen delivery to the eye and swimbladder. *In*: Farrell, A. P. (ed.) *Encyclopedia of Fish Physiology: From Genome to Environment*, pp. 929–34. Amsterdam, Elsevier.

Waser, W. and Heisler, N. (2005). Oxygen delivery to the fish eye: Root effect as crucial factor for elevated retinal O_2. *Journal of Experimental Biology*, **208**, 4035–47.

Weber, R. E. and Campbell, K. L. (2011). Temperature dependence of haemoglobin-oxygen affinity in heterothermic vertebrates: mechanisms and biological significance. *Acta Physiologica*, **202**, 549–62.

Weber, R. E., Campbell, K. L., Fago, A., Malte, H., and Jensen, F. B. (2010). ATP-induced temperature independence of hemoglobin-O_2 affinity in heterothermic billfish. *Journal of Experimental Biology*, **213**, 1579–85.

Weber, R. E., Fago, A., and Campbell, K. L. (2014). Enthalpic partitioning of the reduced temperature sensitivity of O_2-binding in bovine hemoglobin. *Comparative Biochemistry and Physiology A*, **176**, 20–5.

Weber, R. E., Fago, A., Malte, H., Storz, J. F., and Gorr, T. A. (2013). Lack of conventional oxygen-linked proton and anion binding sites does not impair allosteric regulation of oxygen binding in dwarf caiman hemoglobin. *American Journal of Physiology-Regulatory Integrative and Comparative Physiology*, **305**, R300–12.

Weber, R. E. and White, F. N. (1986). Oxygen binding in alligator blood related to temperature, diving, and alkaline tide. *American Journal of Physiology*, **251**, R901–8.

Weber, R. E. and White, F. N. (1994). Chloride-dependent organic phosphate sensitivity of the oxygenation reaction in crocodilian hemoglobins. *Journal of Experimental Biology*, **192**, 1–11.

Wittenberg, J. B. and Wittenberg, B. A. (1962). Active secretion of oxygen into eye of fish. *Nature*, **194**, 106–7.

Wittenberg, J. B. and Wittenberg, B. A. (1974). Choroid rete mirabile of fish eye. 1. Oxygen secretion and structure—comparison with swimbladder rete mirabile. *Biological Bulletin*, **146**, 116–36.

Wyman, J. (1964). Linked functions and reciprocal effects in hemoglobin: a second look. *Advances in Protein Chemistry*, **19**, 223–86.

Yokoyama, T., Chong, K. T., Miyazaki, G., et al. (2004). Novel mechanisms of pH sensitivity in tuna hemoglobin. A structural explanation of the Root effect. *Journal of Biological Chemistry*, **279**, 28632–40.

Yuan, Y., Shen, T. J., Gupta, P., et al. (2011). A biochemical-biophysical study of hemoglobins from woolly mammoth, Asian elephant, and humans. *Biochemistry*, **50**, 7350–60.

CHAPTER 8

Biochemical adaptation to environmental hypoxia

"Because Hb functions at the interface between organisms and their environment, it is ideally suited to study physiological adaptation."
—Powers (1980)

8.1 Challenges to respiratory gas transport under hypoxia

In order for air- or water-breathing vertebrates to cope with a reduction in ambient O_2 availability, blood O_2-transport capacity must be increased to sustain O_2 flux to the tissue mitochondria in support of aerobic ATP synthesis (Mairbäurl, 1994, Samaja et al., 2003, Storz et al., 2010b, Scott, 2011, Mairbäurl and Weber, 2012). Such changes complement physiological adjustments in other convective and diffusive steps in the O_2-transport pathway (Bouverot, 1985, Scott and Milsom, 2006). In combination with changes in the cardiorespiratory system and microcirculation, fine-tuned changes in Hb-O_2 affinity can enhance tissue O_2 delivery for a given difference in PO_2 at sites of O_2 loading and unloading.

Although an increased Hb-O_2 affinity helps to safeguard arterial O_2 saturation under environmental hypoxia, it can hinder O_2 unloading in the systemic circulation. For this reason, vertebrates living in hypoxic environments face the physiological challenge of optimizing the trade-off between O_2 loading at the respiratory surfaces and O_2 unloading in the tissue capillaries.

8.2 Changes in Hb-O_2 affinity

8.2.1 Evolutionary changes in Hb-O_2 affinity

Evolutionary changes in Hb-O_2 affinity can involve changes in the intrinsic O_2 affinity of Hb and/or changes in the responsiveness to allosteric effectors. The former mechanism can involve changes in the equilibrium constants of heme-O_2 binding in the R or T state or changes in the allosteric equilibrium constant for the R↔T transition. These genetically based modifications of Hb function are attributable to amino acid replacements in the α- and/or β-type subunits (Perutz, 1983, Weber, 1995, Weber, 2007, Bellelli et al., 2006, Storz and Moriyama, 2008).

8.2.2 Reversible changes in Hb-O_2 affinity

Reversible changes in Hb-O_2 affinity can be achieved by modulating red cell pH and/or the concentration of organic phosphates or other allosteric effectors (Nikinmaa, 2001, Jensen, 2004, Jensen, 2009). These changes in the chemical milieu of the red blood cell alter the operating conditions for Hb, but are not associated with structural changes in the Hb protein itself.

Hemoglobin: Insights into Protein Structure, Function, and Evolution. Jay F. Storz, Oxford University Press (2019).
© Jay F. Storz 2019. DOI: 10.1093/oso/9780198810681.001.0001

In principle, reversible adjustments can also be produced by cellular changes in isoHb composition. As discussed in Chapter 6, all vertebrates possess multiple α- and β-type globin genes, and therefore express multiple, structurally distinct isoHbs during different stages of prenatal development and post-natal life. Such Hb multiplicity suggests a potential mechanism for modulating blood-O_2 affinity via changes in the relative abundance of distinct isoHbs with different O_2-binding properties (Hiebl et al., 1988, Weber et al., 1988a, Weber and Jensen, 1988, Weber, 1996, Grispo et al., 2012, Opazo et al., 2015).

Thus, reversible modulation of Hb-O_2 affinity via metabolically induced changes in the red cell microenvironment and regulatory changes in isoHb expression represent potential mechanisms of phenotypic plasticity. These mechanisms could complement genetically based changes in Hb function, or they could obviate the need for such changes in the first place.

8.3 Is it physiologically advantageous to have an increased Hb-O$_2$ affinity under conditions of environmental hypoxia?

Let us now address the question of whether the optimal Hb-O_2 affinity varies as a function of ambient PO_2. This is central to the question of whether we should generally expect natural selection to favor different Hb-O_2 affinities in populations or species that inhabit environments with different ambient O_2 availabilities.

8.3.1 Theoretical results

A reduction in the PO_2 of inspired air or water generally results in a concomitant reduction in the PO_2 of arterial blood (P_aO_2). Under such conditions, adjustments in Hb-O_2 affinity provide an energetically efficient means of matching a reduced O_2 supply to an undiminished metabolic demand by shifting the steepest portion of the O_2-equilibrium curve within the range of the *in vivo* PO_2s at the sites of O_2 loading (P_aO_2) and unloading (P_vO_2, the PO_2 of mixed venous blood) (Bouverot, 1985, Samaja et al., 2003).

At a given P_aO_2, tissue O_2 delivery is enhanced by increasing cardiac output (Q) and/or the blood capacitance coefficient (βbO_2; Fig. 8.1A). As we saw in Chapter 2 (section 2.3), βbO_2 is defined as the

arterial-venous difference in O_2 concentration divided by the arterial-venous difference in PO_2: $([O_2]_a - [O_2]_v)/(P_aO_2 - P_vO_2)$. The capacitance coefficient therefore quantifies the amount of O_2 unloaded to the tissues for a given arterial-venous difference in PO_2 (Dejours, 1981). It is determined by both the quantity of Hb (its concentration in the blood) and the quality of Hb (its O_2-binding properties).

In mature red blood cells, the concentration of Hb is typically close to the solubility limit (Riggs, 1976), so whole-blood [Hb] is typically strongly correlated with hematocrit, the fraction of total blood volume occupied by red cells. Under environmental hypoxia, an increased blood Hb concentration increases $[O_2]_a$, thereby increasing blood O_2 conductance if P_aO_2, cardiac output (Q), and the rate of O_2 consumption (VO_2) remain constant. This is a common means of enhancing blood O_2 capacitance under conditions of environmental hypoxia as well as exercise-induced tissue hypoxia (Monge and Leon-Velarde, 1991, Mairbäurl, 1994, Samaja et al., 2003, Harter and Brauner, 2017). Although a moderate increase in [Hb] can improve aerobic performance (Kanstrup and Ekblom, 1984, Ekblom and Berglund, 1991), an excessively high [Hb] (polycythemia) can be counterproductive because the associated increase in blood viscosity compromises cardiac output, thereby offsetting the benefit of the elevated $[O_2]_a$ for systemic O_2 delivery. The nonlinear relationship between blood [Hb] and either blood O_2 transport capacity or aerobic exercise performance is well-documented in water-breathing fishes (Wells and Weber, 1991, Gallaugher et al., 1995) and air-breathing, terrestrial vertebrates (Schuler et al., 2010, Simonson et al., 2015, Wagner et al., 2015). From this point forward, we will focus on changes in blood O_2 capacitance that are mediated by changes in the oxygenation properties of Hb.

Let us now consider how changes in the shape and position of the O_2-equilibrium curve can be expected to affect systemic O_2 delivery under different degrees of hypoxia. We can see that under conditions of moderate hypoxia (P_aO_2 >45 torr), a reduced Hb-O_2 affinity (right-shifted O_2-equilibrium curve) maximizes βbO_2 (Fig. 8.1B). That is, it produces the largest increase in the slope of the line connecting the arterial and venous points on the curve, thereby maximizing tissue O_2 delivery ($C_aO_2 - C_vO_2$) for a given difference in PO_2 between the sites of O_2 loading and unloading ($P_aO_2 - P_vO_2$). By contrast, under

(A)

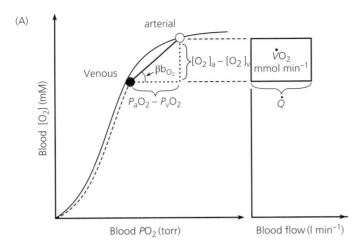

(B)

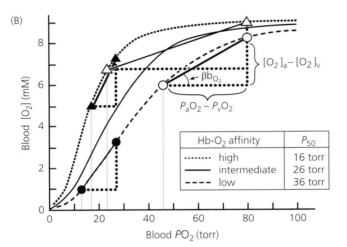

Fig. 8.1. Schematic illustration of blood O_2 transport. (A) An O_2-equilibrium curve under the physicochemical conditions prevailing in arterial blood (a, solid curve, open symbol) and venous blood (v, dashed curve, closed symbol). The curve is a plot of blood O_2 content (y-axis) versus PO_2 (x-axis), with paired values for arterial and venous blood connected by a solid line. $C_aO_2 - C_vO_2$ denotes the arterial–venous difference in blood O_2 content, $P_aO_2 - P_vO_2$ denotes the corresponding difference in PO_2, βbO_2 denotes the blood O_2 capacitance coefficient (see text for details), Q denotes cardiac output, and VO_2 denotes the rate of O_2 consumption. On the right-hand side of the graph, the area of the rectangle is proportional to total O_2 consumption, which can be enhanced by increasing Q and/or by increasing βbO_2. Increases in βbO_2 produce a corresponding increase in $C_aO_2 - C_vO_2$ through shifts in the shape or position of the O_2-equilibrium curve. (B) O_2-equilibrium curves showing the effect of changes in Hb-O_2 affinity on tissue O_2 delivery under conditions of moderate hypoxia (open symbols) and severe hypoxia (filled symbols). For each pair of arterial and venous points, the PO_2 for venous blood (P_vO_2) is marked by a vertical grey line that extends to the x-axis. The sigmoid O_2-equilibrium curves are shown for high, intermediate, and low Hb-O_2 affinities; P_{50}, the PO_2 at which Hb is 50% saturated. Each change in Hb-O_2 affinity produces a shift in P_vO_2, but the PO_2 of arterial blood (P_aO_2) is assumed to remain constant. Note that under conditions of moderate hypoxia the right-shifted curve maximizes βbO_2 and preserves a higher P_vO_2 (an overall index of tissue oxygenation). Under severe hypoxia, by contrast, the left-shifted curve maximizes βbO_2 and preserves a higher P_vO_2 relative to the right-shifted curve. When the kinetics of O_2 transfer across the alveolar gas–blood barrier is a limiting step (diffusion limitation), a left-shifted O_2-equilibrium curve may also be advantageous under less severe hypoxia (Bencowitz et al., 1982).

severe hypoxia, an increased Hb-O_2 affinity (left-shifted curve) produces the largest increase in βbO_2 because the arterial-venous difference in PO_2 spans a steeper portion of the curve (Fig. 8.1B). Figure 8.1B also shows that a left-shifted curve preserves a higher P_vO_2 under such conditions (Woodson, 1988). The venous PO_2, which reflects the pressure gradient for O_2 diffusion to the cells of perfused tissue, can be expressed as

$$P_vO_2 = P_aO_2 - \frac{1}{\left(\beta bO_2 \times \dfrac{Q}{VO_2} \right)} \quad (8.1)$$

where VO_2 is the rate of O_2 consumption and the product $\beta bO_2 \times (Q/VO_2)$ is the specific blood O_2 conductance (Dejours, 1981, Bouverot, 1985). Under

hypoxia, an increased circulatory O_2 conductance can also be achieved via increases in cardiac output, but increasing βbO_2 via adjustments in Hb-O_2 affinity is far more energetically efficient (Mairbäurl, 1994, Samaja et al., 2003).

In summary, theory predicts that a reduced Hb-O_2 affinity is generally beneficial under moderate hypoxia, whereas an increased Hb-O_2 affinity is beneficial under severe hypoxia. This is broadly consistent with theoretical investigations of tissue O_2 delivery at rest and during exercise in air-breathing vertebrates (Turek et al., 1973, West and Wagner, 1980, Bencowitz et al., 1982, Willford et al., 1982, Samaja et al., 1986, Samaja et al., 2003, Scott and Milsom, 2006) and in water-breathing fish (Brauner and Wang, 1997, Wang and Malte, 2011).

Several theoretical treatments have calculated optimal values of P_{50} that maximize the arterial-venous difference in O_2 content or O_2 saturation for a given arterial-venous difference in PO_2 (Bencowitz et al., 1982, Willford et al., 1982, Brauner and Wang, 1997, Wang and Malte, 2011). The arterial-venous difference in O_2 saturation can be expressed as

$$S_aO_2 - S_vO_2 = \left[\frac{P_aO_2^n}{P_aO_2^n - P_{50}^n}\right] - \left[\frac{P_vO_2^n}{P_vO_2^n - P_{50}^n}\right], \quad (8.2)$$

where n is the Hill coefficient.

Taking the first derivative of the maximum arterial-venous saturation difference with respect to P_{50} indicates that the optimal P_{50} ($P_{50}*$) can be expressed as

$$P_{50}* = \left(P_aO_2 \times P_vO_2\right)^{0.5}. \quad (8.3)$$

Equation 2 can also be solved for P_vO_2, but the same conditions that maximize $S_aO_2 - S_vO_2$ also maximize P_vO_2 (Willford et al., 1982).

Under normoxia ($P_aO_2 = 90$ torr, $P_vO_2 = 30$ torr), and assuming that pH, partial pressure of CO_2 (PCO_2), cardiac output, and Hb concentration remain constant, Eqn.8.2 predicts that tissue O_2 delivery increases as n increases and as P_{50} increases to its optimum (Fig. 8.2A). According to Eqn. 8.3, the optimal P_{50} under these conditions is $(90 \times 30)^{0.5}$ = 52.0 torr. This is the point on the plot where the slope with respect to P_{50} is zero. By contrast, under severe hypoxia ($P_aO_2 = 40$ torr, $P_vO_2 = 20$ torr), with other assumptions as just described, the optimal P_{50} is predicted to be $(40 \times 20)^{0.5}$ = 28.3 torr, indicating

that tissue O_2 delivery is maximized at a far higher Hb-O_2 affinity (Fig. 8.2B). In both cases, tissue O_2 delivery generally increases as a positive function of n, demonstrating the physiological significance of cooperative O_2 binding. The only exception occurs under extremely severe hypoxia, when the venous point on the O_2-equilibrium curve drops down to the lower curvilinear asymptote.

Figure 8.3 illustrates predicted relationships among P_{50}, P_aO_2, and P_vO_2, while keeping P_{50} constant. Assuming 50 percent tissue O_2 extraction (as might occur during exercise), the figure shows that a higher P_{50} maintains a higher P_vO_2 under normoxia ($P_aO_2 = $ ~90 torr), resulting in improved tissue oxygenation. By contrast, a lower P_{50} maintains a higher P_vO_2 under severe hypoxia ($P_aO_2 = $ ~40 torr). If the "critical P_vO_2" for tissue oxygenation is, for example, 10 torr, then a blood P_{50} of 20 torr would allow P_aO_2 to fall as low as 25 torr, whereas a blood P_{50} of 50 torr would not allow P_aO_2 to fall below 52 torr. Results are qualitatively similar under the assumption of 25 percent tissue O_2 extraction, corresponding to the situation at rest (Willford et al., 1982).

Variation in Hb-O_2 affinity is relevant to fitness only to the extent that it affects whole-animal physiological performance. Several modeling analyses have investigated how changes in Hb-O_2 affinity influence VO_2max, a measure of maximal aerobic performance capacity, over a broad range of inspired PO_2 values (Wagner, 1996, Scott and Milsom, 2006, Wang and Malte, 2011). Results demonstrate that the optimal P_{50} declines with decreasing

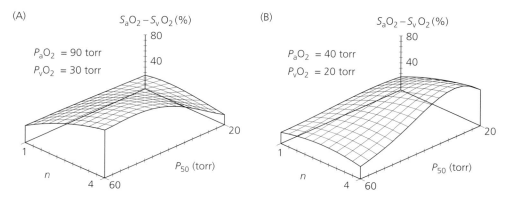

Fig. 8.2. Blood P_{50} and Hill's cooperativity coefficient, n, influence blood O_2 transport (indexed by the difference in arterial and venous O_2 saturation) under normoxia (A) and hypoxia (B). In these three-dimensional plots, the difference in arterial and venous O_2 saturation ($S_aO_2 - S_vO_2$) is indicated by the height of the projection above the reference plane. The higher the projection, the greater the difference in O_2 saturation.

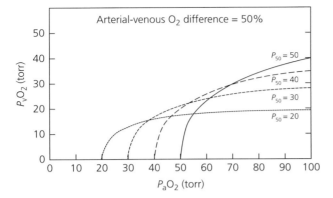

Fig. 8.3. Isobars showing predicted values of venous PO_2 (P_vO_2, an index of tissue oxygenation) as a function of arterial PO_2 (P_aO_2) at different values of blood P_{50}, assuming 50 percent tissue O_2 extraction and constant cardiac output. At normal or moderately reduced P_aO_2, a higher P_{50} results in a higher P_vO_2 (and, hence, improved tissue oxygenation). Under more severe hypoxemia, by contrast, a lower P_{50} results in a higher P_vO_2 while still maintaining constant O_2 extraction.

inspired PO_2, but effects on VO_2max are generally modest over a broad range of values. Wang and Malte (2011) used physiological data from rainbow trout (*Oncorhynchus mykiss*) to parameterize a quantitative model of the O_2-transport pathway in water-breathing fishes. In addition to confirming that the optimal P_{50} declines with decreasing inspired PO_2, the modeling results also revealed that the performance curve relating VO_2max to blood P_{50} is asymmetrical, as VO_2max declines more drastically at low P_{50} values below the optimum than at high P_{50} values above the optimum (Fig. 8.4A). The benefit of reducing blood P_{50} (increasing Hb-O_2 affinity) is most pronounced under conditions of

severe hypoxia (PO_2 <20 torr, which approximates the estimated blood P_{50} used in the model) (Fig. 8.4B).

8.3.2 Experimental results

The most direct means of testing theoretical predictions about how the optimal Hb-O_2 affinity varies in relation to ambient PO_2 is to experimentally evaluate how titrated changes in blood P_{50} affect relevant measures of physiological performance under normoxia and varying degrees of hypoxia. Experiments on rats with pharmacologically manipulated Hb-O_2 affinities have confirmed theoretical

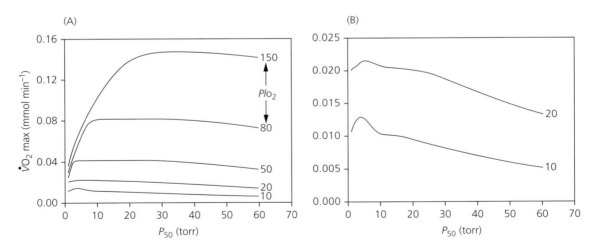

Fig. 8.4. The predicted effect of blood P_{50} (torr) on the maximal rate of O_2 consumption (VO_2max) in water-breathing fishes, calculated using the model of Wang and Malte (2011). VO_2max was calculated over a broad range of ambient PO_2 values. (A) The optimal P_{50} declines with decreasing inspired PO_2 (PIO_2), but the relationship between P_{50} and VO_2max is highly asymmetric. (B) The benefit of increasing Hb-O_2 affinity (decreasing P_{50}) is most pronounced under conditions of severe hypoxia (PO_2 <20 torr).

predictions that an increased P_{50} improves tissue O_2 delivery under normoxia and moderate hypoxia, and that a reduced P_{50} is beneficial under severe hypoxia (Turek et al., 1978a, Turek et al., 1978b). Similarly, reciprocal-transplant experiments involving wild-derived strains of deer mice (*Peromyscus maniculatus*) revealed that high-altitude natives with high Hb-O_2 affinities have higher capacities for

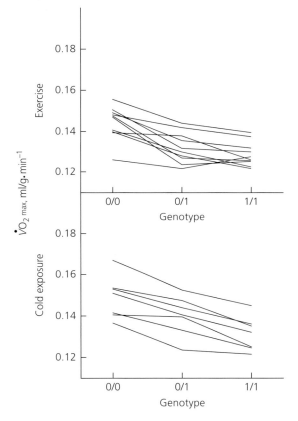

thermogenesis and aerobic exercise under severe hypoxia, whereas lowland natives with lower Hb-O_2 affinities exhibit superior performance under normoxia (Chappell and Snyder, 1984) (Fig. 8.5). The adaptive significance of this variation in whole-animal physiological performance was summarized by Chappell and Snyder (1984) as follows: "A mouse capable of attaining a higher VO_2max can exercise more vigorously without incurring debilitating oxygen debt and/or it can maintain body temperature by means of aerobic thermogenesis at lower ambient temperatures." Experiments involving other mammals have reported qualitatively similar findings, suggesting that a reduced Hb-O_2 affinity improves tissue O_2 delivery under normoxia or moderate hypoxia, whereas an increased O_2 affinity provides the greatest improvement under more severe hypoxia (Dawson and Evans, 1966, Banchero and Grover, 1972).

Experiments involving rats with pharmacologically manipulated Hb-O_2 affinities have also demonstrated that reductions in blood P_{50} significantly increase the survival of animals subjected to acute, severe hypoxia (Eaton et al., 1974, Penney and Thomas, 1975). Similar results were reported in studies of physiological performance under hypoxia in other mammals with naturally occurring variation in Hb-O_2 affinity (Dawson and Evans, 1966, Hall, 1966, Hebbel et al., 1978). In addition to studies of survival and whole-animal physiological performance, *ex vivo* studies of microvascular O_2 transport and tissue perfusion have demonstrated that an increased Hb-O_2 affinity enhances O_2 delivery under severe hypoxia (Bakker et al., 1976, Stein and Ellsworth, 1993, Yalcin and Cabrales, 2012).

Fig. 8.5. Variation in the maximal rate of O_2 consumption (VO_2max) among wild-derived strains of deer mice (*Peromyscus maniculatus*) with α-globin genotypes associated with different Hb-O_2 affinities. The experiments revealed that genetic variation in Hb-O_2 affinity was positively associated with aerobic exercise performance (A) and thermogenesis under hypoxia (B). Mice were tested at 3,800 m elevation, and VO_2max was elicited by treadmill exercise (top) or cold exposure (bottom). Mice used in the experiments carried alternative α-globin alleles (denoted "0" and "1") that were associated with different Hb-O_2 affinities (rank order of blood P_{50}'s among the three tested genotypes: 0/0 <0/1 <1/1). Breeding experiments ensured that the alternative allelic variants were tested on randomized genetic backgrounds.

Data from Chappell and Snyder (1984).

8.3.3 Threshold PO_2

The theoretical and experimental results reviewed here indicate that the optimal Hb-O_2 affinity varies according to ambient PO_2. The relationship is nonlinear and depends critically on the magnitude of diffusion limitation (the extent to which O_2 equilibration at the blood-gas interface is limited by the kinetics of O_2 exchange; Bencowitz et al., 1982). For any given species, theory predicts that there must be some threshold level of ambient PO_2 at which it becomes beneficial to have an increased Hb-O_2 affinity. This threshold would be largely dictated by P_aO_2 (which is determined by ventilation and O_2 equilibration at the blood-gas interface) and other diffusive and convective steps in the O_2-transport pathway that can be expected to vary from species to species (Bencowitz et al., 1982, Scott and Milsom, 2006).

For humans living at high altitude, modeling results suggest that an increased Hb-O_2 affinity only confers a benefit to tissue O_2 delivery at elevations greater than 5,000–5,400 m (Samaja et al., 1986, Samaja et al., 2003). At 5,400 m above sea level, which is roughly the elevation of the South Everest base camp in Nepal, the standard barometric pressure is 53 kPa (399 torr), meaning that the ambient PO_2 is 52 percent of that at sea level. The highest human settlements in the Himalayas and the Andes are generally at elevations of less than 4,900 m (the Peruvian mining town, La Rinconada, is situated at 5,100 m above sea level, and most mine workers have homes at lower elevation). The highest permanent settlements in the Ethiopian highlands are less than 3,500 m above sea level. The fact that humans do not live at elevations above the theoretically predicted 5,000–5,400 m threshold provides a possible explanation for why increased Hb-O_2 affinities have not evolved in indigenous mountain dwellers.

8.4 Insights from comparative studies

8.4.1 Life ascending: Hb adaptation to hypoxia in high-altitude mammals and birds

Since vertebrates at high altitude face the physiological challenge of optimizing the trade-off between arterial O_2 loading and peripheral O_2 unloading, it is not always clear whether an increased or decreased Hb-O_2 affinity should be expected to improve tissue O_2 delivery, so the adaptive significance of elevational patterns has been the subject of considerable debate (Barcroft et al., 1923, Aste-Salazar and Hurtado, 1944, Lenfant et al., 1968, Eaton et al., 1969, Lenfant et al., 1969, Torrance et al., 1970/71, Lenfant et al., 1971, Bullard, 1972, Turek et al., 1973, Eaton et al., 1974, Dempsey et al., 1975, Frisancho, 1975, West and Wagner, 1980, Bencowitz et al., 1982, Willford et al., 1982, Samaja et al., 1986, Mairbäurl, 1994, Samaja et al., 2003, Storz et al., 2010b, Storz, 2016c).

Since natural selection is "the ultimate arbiter of what constitutes an adaptation" (Snyder, 1982), a systematic survey of altitude-related changes in Hb-O_2 affinity—as revealed by comparisons among extant species—can provide insights into the possible adaptive significance of such changes. If high-altitude taxa have generally evolved increased Hb-O_2 affinities relative to lowland sister taxa—and if the elevational pattern is too consistent to be ascribed to chance (i.e., genetic drift)—this would be consistent with the hypothesis that the elevational differences reflect a history of natural selection. Ideally, comparative studies that exploit the outcomes of natural experiments (e.g., the independent colonization of high-altitude environments by multiple species) complement insights derived from controlled laboratory experiments. For example, comparative studies have documented that high-altitude rodents often have higher Hb- and/or blood-O_2 affinities than their lowland relatives (Hall et al., 1936, Bullard et al., 1966, Ostojic et al., 2002, Storz, 2007, Storz et al., 2009, Storz et al., 2010a, Natarajan et al., 2013, Natarajan et al., 2015a, Jensen et al., 2016), and these observations complement the results of experiments demonstrating that increases in blood-O_2 affinity enhance tissue O_2 delivery and measures of physiological performance in rodents subjected to environmental hypoxia (Eaton et al., 1974, Turek et al., 1978a, Turek et al., 1978b, Chappell and Snyder, 1984). Such consilience of evidence from comparative and experimental studies can greatly strengthen conclusions about the adaptive significance of evolutionary changes in Hb-O_2 affinity.

8.4.1.1 The importance of accounting for phylogenetic history

In comparative analyses of phenotypic variation, it is important to account for the fact that trait values from different species are not statistically independent because the sampled species did not evolve independently of one another; the phylogenetic history of any set of species is represented by a hierarchically nested pattern of relationships (Garland et al., 2005). If a phylogeny is available for a given set of species, then phylogenetically independent contrasts (PIC) (Felsenstein, 1985) can be used to test for a relationship between native elevation and

Hb-O_2 affinity. The PIC approach uses phylogenetic information and a model of trait evolution (typically a stochastic, Brownian motion-like model) to transform the data for the set of surveyed species into values that are statistically independent and identically distributed. This approach was used to document a strong positive correlation between Hb-O_2 affinity and native elevation in Andean hummingbirds (Projecto-Garcia et al., 2013).

An alternative to using PIC is the paired-lineage test, which restricts comparisons to phylogenetically replicated pairs of taxa that are chosen so that there is no overlap in evolutionary paths of descent (Fig. 8.6). A nonrandom association between Hb-O_2 affinity and native elevation can then be assessed using a sign test (a nonparametric test that contrasts matched pairs of samples with respect to a continuous outcome).

If the comparative analysis includes a phylogenetically diverse range of taxa, an advantage of the paired-lineage test is that comparisons can be restricted to closely related species by excluding pairs with long paths between them. To determine whether there is a relationship between Hb-O_2 affinity and native elevation, we want to make

comparisons between close relatives so that we can minimize the number of potentially confounding differences in other aspects of their physiology.

There are two additional issues to consider in comparative studies of Hb function in relation to native elevation. The first relates to the effect of environmentally induced variation, which can obscure phylogenetic signal in trait values (Garland et al., 2005, Storz et al., 2015). Measurements of an environmentally labile trait (like the O_2 affinity of whole blood, which is influenced by red cell metabolism and acid-base status) may prevent an accurate assessment of the extent to which phenotypic similarity between a given pair of species is attributable to shared phylogenetic heritage versus a shared, plastic response to similar environmental conditions (i.e., exposure to hypoxia). This problem can be avoided if the trait is measured under common-garden conditions to control for environmentally induced variation, or if measurements are restricted to genetically based trait variation (e.g., O_2 affinity of purified Hb rather than O_2 affinity of whole blood or red cell suspensions).

Another issue concerns genealogical discordance between the phylogeny of the examined species and the phylogenies of the genes that underlie the measured trait (Hahn and Nakhleh, 2016, Storz, 2016a). In comparative studies of Hb evolution involving orthologous genes from a diversity of species, it may often be the case that phylogenies of the α- and β-globin genes are not congruent with one another or with the assumed species tree (Storz et al., 2007, Hoffmann et al., 2008a, Hoffmann et al., 2008b, Opazo et al., 2008, Opazo et al., 2009, Runck et al., 2009, Runck et al., 2010, Natarajan et al., 2015a, Zhu et al., 2018). This genealogical discordance can have multiple biological causes, including interparalog gene conversion (a form of nonreciprocal recombinational exchange between duplicated genes; Chapter 5, section 5.9.1), introgressive hybridization (incorporation of allelic variants from one species into the gene pool of another species by means of hybridization and repeated back-crossing), and incomplete lineage sorting (the retention of ancestral polymorphism from one split between populations to the next, followed by stochastic sorting of allelic lineages among the descendant species). A given amino acid substitution may have occurred

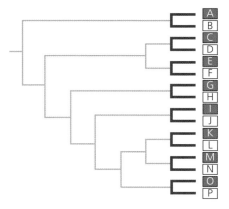

Fig. 8.6. Sampling design for paired-lineage test (Felsenstein, 1985a) where members of each species pair are grouped according to some binary cistinction (e.g., they are native to high- vs. low-altitude), denoted by black and white branch-tip labels. Pairs of contrasting species are chosen that have nonintersecting paths of descent, indicated by branches shown in bold. Since the species pairs are phylogenetically independent of one another, a sign test can be used to determine whether evolved changes in a continuously distributed character (e.g., Hb-O_2 affinity) changes in a nonrandom, directional fashion with respect to the categorical variable (e.g., native altitude).

a single time on an internal branch of the gene tree, but it can present the appearance of having occurred twice independently when mapped onto a discordant species tree, resulting in spurious inferences. Hahn and Nakhleh (2016) discuss possible solutions to the problem of species tree/gene tree discordance in comparative studies of trait evolution.

8.4.1.2 Considerations of zoogeographic history

In addition to making comparisons between high- and low-altitude taxa that are as closely related as possible, it is also important to consider the evolutionary histories of study species with regard to their current elevational distributions. Many alpine and subalpine natives may have predominantly lowland ancestries, possibly reflecting postglacial range shifts. Alternatively, residence at high altitude may represent the ancestral condition for members of groups that diversified in mountainous regions. For example, Andean hummingbirds in the Brilliants/Coquettes clade diversified during a period of rapid uplift of the Andean massif in the period between ~10 and ~6 million years ago (McGuire et al., 2014). Within this group, many species with lowland distributions may have descended from highland ancestors. In such cases, it is important to consider that any altitude-related species differences in Hb-O_2 affinity could be attributable to derived increases in O_2 affinity in highland species and/or secondarily derived reductions in O_2 affinity in lowland species (Projecto-Garcia et al., 2013).

8.4.1.3 Evaluating evidence for an empirical generalization regarding the relationship between native altitude and Hb-O_2 affinity

The theoretical and experimental results reviewed earlier suggest that it is generally beneficial to have an increased Hb-O_2 affinity under conditions of severe hypoxia. An obvious prediction is that derived increases in Hb-O_2 affinity will have evolved repeatedly in disparate vertebrate taxa that have independently colonized extreme altitudes, provided that their range limits exceed the elevational threshold at which an increased Hb-O_2 affinity becomes beneficial. Let us now test this prediction using available comparative data for mammals and birds.

Here we will restrict the analysis to data based on standardized measurements of purified Hbs, so the variation in P_{50} values is purely genetic, reflecting evolved changes in the amino acid sequences of the α- and/or β-chain subunits. This focus on purified Hbs avoids problems associated with the confounding effects of environmentally induced variation. However, an analysis based on *in vitro* measurements of purified proteins involves its own interpretative challenges because evolved changes in the inherent properties of Hb are physiologically relevant to circulatory O_2 transport only to the extent that such changes affect the oxygenation properties of blood (Berenbrink, 2006). In the following, we will focus on data for purified Hbs while recognizing that species differences in Hb-O_2 affinity may not perfectly reflect *in vivo* differences in blood-O_2 affinity.

Fig. 8.7. Phylogenetic relationships of fourteen mammalian taxa used in comparative analyses of Hb function. Rows corresponding to high-altitude taxa are shaded. Branches in bold connect pairs of high- and low-altitude taxa that were used to test for a relationship between Hb-O_2 affinity and native elevation. Since there are no overlaps in the paths of descent connecting each designated pair of high- and low-altitude taxa, the seven pairwise comparisons are statistically independent. For information regarding elevational ranges, see Storz et al. (2009, 2010a), Revsbech et al. (2013), Janecka et al. (2015), Natarajan et al. (2015a), and Tufts et al. (2015).

In the case of mammals, we can summarize data from fourteen taxa representing seven high- versus low-altitude pairwise comparisons (Fig. 8.7). These comparisons include rodents (marmotine ground squirrels and *Peromyscus* mice), lagomorphs (pikas), and carnivores (Storz et al., 2009, Storz et al., 2010a, Revsbech et al., 2013, Janecka et al., 2015, Natarajan et al., 2015a, Tufts et al., 2015). Six of the comparisons involve closely related species with contrasting elevational ranges, and one comparison involves high- and low-altitude populations of the broadly distributed deer mouse, *Peromyscus maniculatus*. Each of these pairwise comparisons involves a pronounced elevational contrast between an alpine or subalpine taxon and a closely related lowland taxon. For example, the high-altitude ground squirrels, deer mice, and pikas occur at elevations more than 4,300 m (the highest elevations that occur within the limits of their geographic distributions in North America). Since an increased Hb-O_2 affinity is only expected to be physiologically beneficial above a given threshold elevation, potentially adaptive differences in Hb-O_2 affinity can only be detected if the high- and low-altitude members of each taxon pair have range limits on opposite sides of that threshold.

In each taxon, O_2 affinities of purified Hbs were measured in the presence and absence of Cl^- ions (added as KCl) and DPG (for experimental details, see legend for Fig. 8.8). The "KCl+DPG" treatment is most relevant to *in vivo* conditions in mammalian red blood cells, and we can therefore focus primarily on measures of $P_{50(KCl+DPG)}$. However, measurements under each of the experimental treatments are valuable because they can provide insights into the functional mechanisms responsible for observed differences in Hb-O_2 affinity.

In the presence of anionic effectors, high-altitude taxa have higher Hb-O_2 affinities than their lowland counterparts in some cases (e.g., deer mice and pikas; Fig. 8.8A,B), but in other cases there are no appreciable differences (e.g., marmots and big cats; Fig. 8.8C,D). In comparisons involving deer mice, pikas, and some ground squirrels, integrated analyses of Hb function and sequence divergence revealed that the high-altitude member of each pair evolved a derived increase in Hb-O_2 affinity (i.e., the phenotype of the lowland taxon represents the ancestral condition).

In the comparisons between conspecific populations of deer mice and between the golden-mantled ground squirrel (*Callospermophilus lateralis*) and thirteen-lined ground squirrel (*Ictidomys tridecemlineatus*), the evolved changes in Hb function involved an increase in intrinsic O_2 affinity in combination with a suppressed sensitivity to anionic effectors (Storz et al., 2009, Storz et al., 2010a, Natarajan et al., 2013, Revsbech et al., 2013, Natarajan et al., 2015a). In the case of the deer mice, this is indicated by the fact that the high-altitude Hb variant exhibits a slightly lower P_{50} in the absence on anions ("stripped") and the P_{50} difference is further augmented in the presence of Cl^- and DPG (Fig. 8.8A). By contrast, in the comparison between the two pika species (*Ochotona princeps* and *O. collaris*), the difference in Hb function was exclusively attributable to an evolved change in intrinsic O_2 affinity (Tufts et al., 2015) (Fig. 8.8B).

Phylogenetically independent comparisons involving the full set of mammalian taxa revealed no significant association between Hb-O_2 affinity and elevation (Wilcoxon's signed-rank test, $W = 6.5$ $P > 0.05$, $N = 7$; Fig. 8.9). Previous studies involving experimental measurements on whole blood indicate that high-altitude mammals have lower P_{50} values than their lowland relatives in some cases (Chiodi, 1971, Leon-Velarde et al., 1996, Ostojic et al., 2002) but not in others (Lechner, 1976). Unless measures of blood-O_2 affinities are integrated with measurements on purified Hbs (Petschow et al., 1977, Campbell et al., 2010), components of environmental and genetic variation are confounded and it is unclear to what extent the measured differences represent evolved changes in Hb function and/or red cell metabolism. Overall, evidence for an altitudinal trend in Hb-O_2 affinity in mammals is equivocal; data from additional taxa may eventually reveal a clearer relationship.

In the case of birds, we can summarize data from 70 taxa representing thirty-five matched pairs of high- versus low-altitude taxa (Fig. 8.10). These taxa include ground-doves, nightjars, hummingbirds, passerines, and waterfowl (Projecto-Garcia et al., 2013, Cheviron et al., 2014, Galen et al., 2015, Natarajan et al., 2015b, Natarajan et al., 2016, Kumar et al., 2017, Jendroszek et al., 2018, Zhu et al., 2018). In the case of the passerines and waterfowl, we can make comparisons between closely related species

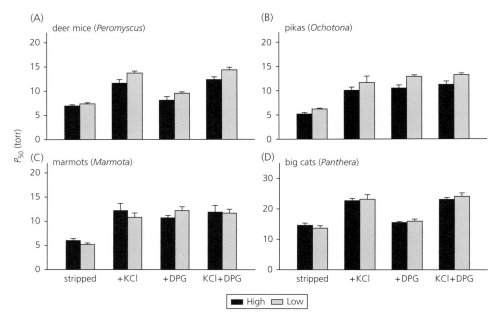

Fig. 8.8. O_2 affinities of purified Hbs from representative pairs of high- and low-altitude mammals. O_2 equilibria were measured at pH 7.40, 37°C, in the presence and absence of allosteric effectors ([Cl⁻], 0.10 mol l⁻1; [HEPES], 0.1 mol l⁻1; DPG:tetrameric Hb ratio, 2.0: [heme], 0.2–0.3 mmol l⁻1). For each taxon, P_{50} values (± SE) are reported for stripped Hbs in the absence of added anions, in the presence of Cl⁻ alone (added as KCl), in the presence of DPG alone and in the presence of both anions combined. This latter "KCl+DPG" treatment is most relevant to *in vivo* conditions in mammalian red blood cells, but measurements of O_2 affinity under each of the four standardized treatments can provide insights into the functional mechanism responsible for observed differences in $P_{50(KCl+DPG)}$. (A) Comparison between Hb variants of high- and low-altitude deer mice, *Peromyscus maniculatus*, from the Rocky Mountains and Great Plains, respectively [data from Natarajan et al., 2015a; for additional details, see Storz et al. (2009, 2010); Jensen et al. (2016)]. (B) Comparison between Hbs of the high-altitude American pika (*Ochotona princeps*) and the low-altitude collared pika (*O. collaris*) (data from Tufts et al., 2015). (C) Comparison between Hbs of the high-altitude yellow-bellied marmot (*Marmota flaviventris*) and the low-altitude hoary marmot (*M. caligata*) (data from Revsbech et al. 2013). (D) Comparison between Hbs of the high-altitude snow leopard (*Panthera uncia*) and the low-altitude African lion (*P. leo*) (data from Janecka et al., 2015). For both cat species, P_{50} is shown as the mean value for two coexpressed isoforms, HbA and HbB, that are present at roughly equimolar concentrations (Janecka et al., 2015).

as well as conspecific populations. All pairwise comparisons involved dramatic elevational contrasts; high-altitude taxa native to very high elevations (3,500–5,000 m above sea level) were paired with close relatives that typically occur at or near sea level. As with the analysis of mammalian Hbs, O_2 affinities of purified avian Hbs were measured under standardized conditions (see legend for Fig. 8.11) in the presence and absence of anionic effectors. However, to obtain measurements that are physiologically relevant to *in vivo* conditions in avian red cells, IHP (a chemical analog of IPP) was used instead of DPG. As mentioned in Chapter 6 (section 6.6), most bird species express two main isoHbs, the major HbA and the minor HbD. In species that expressed both

isoHbs, the O_2-binding properties of isolated, purified HbA and HbD were measured separately.

In the overwhelming majority of pairwise comparisons, the high-altitude taxon exhibited a higher Hb-O_2 affinity across all treatments, as illustrated by representative examples involving the major HbA isoform (Fig. 8.11). Phylogenetically independent comparisons revealed that highland natives generally have an increased Hb-O_2 affinity relative to their lowland counterparts, a pattern consistent for both HbA (Wilcoxon's signed-rank test, $Z = -4.6844$, $P < 0.0001$, $N = 35$; Fig. 8.12A) and HbD ($Z = -3.3144$, $P = 0.0009$, $N = 26$; Fig. 8.12B). In all pairwise comparisons in which the high-altitude taxa exhibited significantly higher Hb-O_2 affinities relative to the

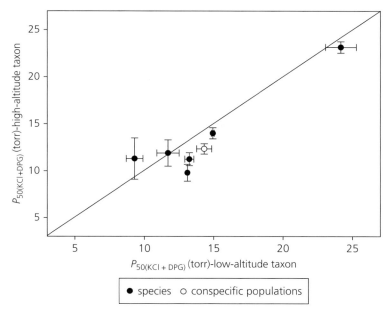

Fig. 8.9. No evidence for a significant elevational trend in the Hb-O_2 affinities of mammals. The plot shows measures of Hb-O_2 affinity in the presence of anionic effectors [$P_{50(KCl+IHP)}$ (±SE)] for seven matched pairs of high- and low-altitude taxa. Data points that fall below the diagonal (x=y) denote cases in which the high-altitude member of a given taxon pair possesses a higher Hb-O_2 affinity (lower P_{50}). The paired-lineage design ensures that all data points are statistically independent (see text for details). Black symbols denote comparisons between species, whereas the open symbol denotes a comparison between high- versus low-altitude populations of the same species (*Peromyscus maniculatus*).

lowland taxa ($N = 35$ taxon pairs for HbA, $N = 26$ for HbD), the measured differences were entirely attributable to differences in intrinsic O_2 affinity rather than differences in sensitivity to Cl⁻ ions or IHP (Natarajan et al., 2015b, Natarajan et al., 2016, Zhu et al., 2018). Comparisons between the high-flying bar-headed goose (*Anser indicus*) and lowland congeners based on O_2 affinity measurements of whole blood, purified hemolysates, and recombinantly expressed Hbs (Petschow et al., 1977, Black and Tenney, 1980, Jessen et al., 1991, Natarajan et al., 2018) are also consistent with the relationships shown in Fig. 8.12.

8.4.1.4 The role of isoHb switching in hypoxia adaptation

As discussed in Chapter 6, regulatory changes in the expression of isoHbs with different oxygenation properties could—in principle—provide an effective means of reversibly modulating blood-O_2 affinity in response to changes in O_2 availability or metabolic demand. This regulatory mechanism could poten-

tially complement genetically based changes in the O_2 affinity of individual isoHbs.

In birds, consistent differences in O_2 affinity between HbA and HbD suggest that an increased blood-O_2 affinity could be achieved by upregulating the expression of HbD (Hiebl et al., 1988, Weber et al., 1988a, Grispo et al., 2012, Opazo et al., 2015). A comparison involving sixty-four taxa representing thirty-two matched high/low-altitude pairs revealed no evidence for an altitude-related difference in relative isoHb abundance (Wilcoxon's signed-rank test, $Z = -0.5245$, $P = 0.6031$, $N = 32$; Fig. 8.13). It therefore appears that regulatory changes in red cell isoHb composition do not represent an important general mechanism of hypoxia adaptation in birds. It remains to be seen whether isoHb switching plays a role in the seasonal acclimatization to acute hypoxia, for example, in species that undergo trans-Himalayan migratory flights like bar-headed geese, ruddy shelducks (*Tadoma ferruginea*) and demoiselle cranes (*Anthropoides virgo*).

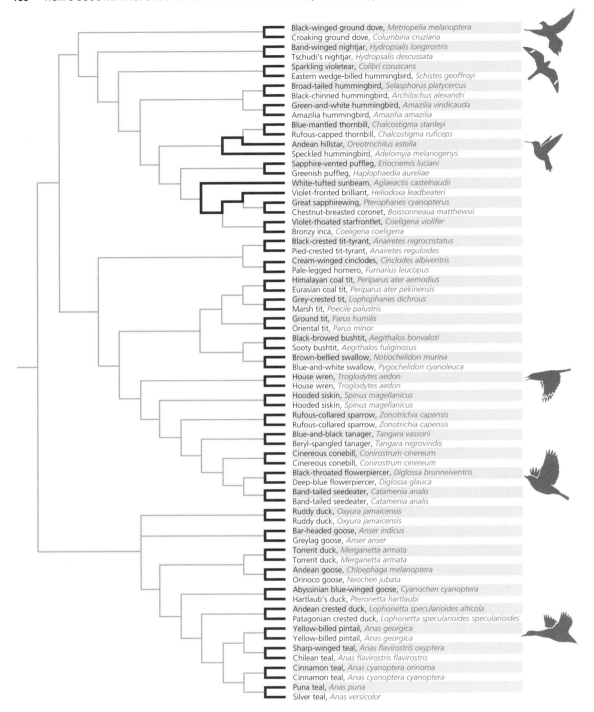

Fig. 8.10. Phylogenetic relationships of 70 avian taxa used in comparative analyses of Hb function. Rows corresponding to high-altitude taxa are shaded. Branches in bold connect pairs of high- and low-altitude taxa that were used to test for a relationship between Hb-O$_2$ affinity and native elevation. As in the case with the mammals, the 35 pairwise comparisons are phylogenetically independent. For information regarding elevational ranges, see Projecto-Garcia et al. (2013), Cheviron et al. (2014), Galen et al. (2015), Natarajan et al. (2015b, 2016), and Zhu et al. (2018).

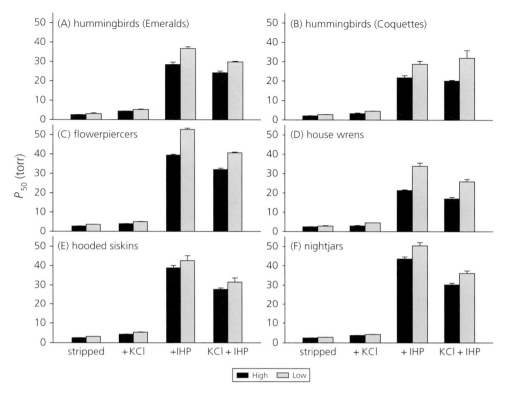

Fig. 8.11. Comparison of oxygenation properties of the major Hb isoform (HbA) between pairs of high- and low-altitude birds in the Andes. O_2 equilibria were measured at pH 7.40, 37°C in the presence and absence of allosteric effectors ([Cl⁻], 0.10 M; [HEPES], 0.1 M; IHP:tetrameric Hb ratio, 2.0: [heme], 0.3 mM). For each taxon, P_{50} values (± SE) are reported for stripped Hbs in the absence of added anions, in the presence of Cl⁻ alone (added as KCl), in the presence of IHP alone and in the presence of both anions combined. As explained in the main text, the "KCl+IHP" treatment is most relevant to *in vivo* conditions in avian red blood cells, but measurements of O_2 affinity under each of the four standardized treatments provide insights into the functional mechanism responsible for observed differences in $P_{50(KCl+IHP)}$. In each pairwise comparison shown here, slight differences in intrinsic Hb-O_2 affinity (reflected by values $P_{50(stripped)}$) become more pronounced in the presence of IHP. (A) Comparison of HbA O_2 affinities between high- and low-altitude hummingbirds in the Emeralds clade (Trochilidae: Apodiformes): the green-and-white humming-bird, *Amazilia viridicauda*, and the amazilia hummingbird, *A. amazilia*, respectively. (B) Comparison of HbA O_2 affinities between high- and low-altitude hummingbirds in the Coquettes clade (Trochilidae: Apodiformes): the Andean hillstar, *Oreotrochilus estella*, and the speckled hummingbird, *Adelomyia melanogenys*, respectively. (C) Comparison of HbA O_2 affinities between high- and low-altitude flowerpiercers (Thraupidae: Passeriformes): the black-throated flowerpiercer, *Diglossa brunneiventris*, and deep-blue flowerpiercer, *D. glauca*, respectively. (D) Comparison of HbA O_2 affinities between high- and low-altitude populations of the house wren, *Troglodytes aedon* (Troglodytidae: Passeriformes). (E) Comparison of HbA O_2 affinities between high- and low-altitude populations of the hooded siskin, *Spinus magellanica* (Fringillidae: Passeriformes). (F) Comparison of HbA O_2 affinities between high- and low-altitude nightjars (Caprimulgidae: Caprimulgiformes): the band-winged nightjar, *Hydropsalis longirostris*, and Tschudi's nightjar, *H. decussata*, respectively.

Data from Projecto-Garcia et al. (2013), Galen et al. (2015), Natarajan et al. (2016), and Kumar et al. (2017).

In contrast to birds and other sauropsid taxa, most mammals do not coexpress functionally distinct isoHbs during postnatal life. Since fetally expressed isoHbs typically have higher O_2 affinities than normal adult Hbs, it is possible that the continued expression of such isoHbs into adulthood could contribute to an enhanced blood-O_2 affinity in response to environmental hypoxia (a form of biochemical paedomorphosis). As discussed in Chapter 6 (section 6.2), stage-specific expression of fetal isoHbs evolved independently in simian primates (New World monkeys, Old World monkeys, apes, and humans) and in bovid artiodactyls (cattle, antelope, sheep, and goats). In humans, retention of

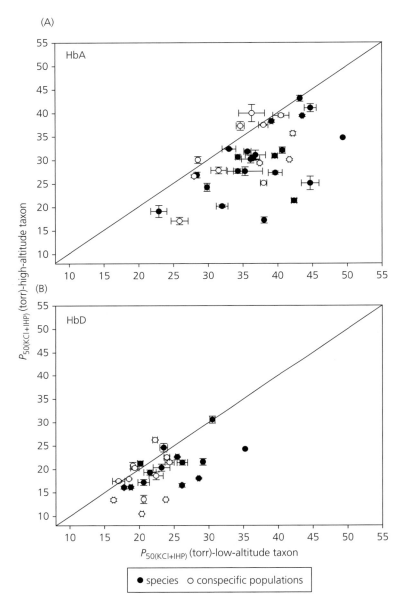

Fig. 8.12. Convergent increases in Hb-O$_2$ affinity in high-altitude birds. (A) Plot of $P_{50(KCl+IHP)}$ (± SE) for HbA in thirty-five matched pairs of high- and low-altitude taxa. Data points that fall below the diagonal (x = y) denote cases in which the high-altitude member of a given taxon pair possesses a higher Hb-O$_2$ affinity (lower P_{50}). Comparisons involve phylogenetically replicated pairs of taxa, so all data points are statistically independent. (B) Plot of $P_{50(KCl+IHP)}$ (± SE) for the minor HbD isoform in a subset of the same taxon pairs in which both members of the pair express HbD. Sample sizes are larger for HbA than for HbD because the two ground dove species (*Metriopelia melanoptera* and *Columbina cruziana*) expressed no trace of HbD, and several hummingbird species expressed HbD at exceedingly low levels (Projecto-Garcia et al., 2013, Natarajan et al., 2016). In such cases, sufficient quantities of HbD could not be purified for measures of O$_2$ equilibria. Black symbols denote comparisons between species, whereas open symbols denote comparisons between high- versus low-altitude populations of the same species. Data from Natarajan et al. (2015b, 2016).

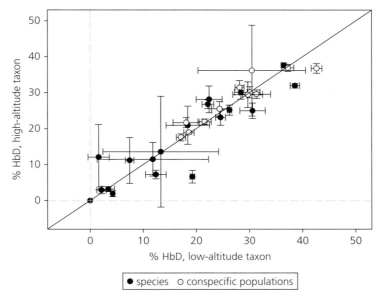

Fig. 8.13. No evidence for altitude-related differences in the relative abundance of HbA and HbD isoforms in pairs of high- and low-altitude bird species. Phylogenetically independent comparisons involving thirty-two pairs of high- and low-altitude taxa revealed no systematic difference in the relative expression level of the minor HbD isoform. The diagonal represents the line of equality ($x = y$). Black symbols denote comparisons between species, whereas open symbols denote comparisons between high- versus low-altitude populations of the same species.

Data from Natarajan et al. (2015, 2016) and Zhu et al. (2018).

fetal Hb expression into adulthood is known to ameliorate the pathological effects of sickle-cell anemia (Akinsheye et al., 2011, Pack-Mabien and Imran, 2013), but there is no evidence for the existence of a similar expression pattern in indigenous highlanders. In yaks, high-affinity fetal isoHbs are expressed at high levels in one-month-old calves (Weber et al., 1988b), but it is not known whether high expression is retained into adulthood.

In summary, there is currently no compelling evidence to suggest that isoHb switching represents an important mechanism of physiological acclimatization to hypoxia in mammals or birds (Storz, 2016b). However, aside from recent studies of Andean and Sino-Himalayan birds (Natarajan et al., 2015b, Natarajan et al., 2016, Zhu et al., 2018), it is also true that hypoxia-induced isoHb switching has not been systematically investigated as a mechanism of phenotypic plasticity in terrestrial vertebrates.

8.5 Regulation of blood O_2 capacitance *in vivo*

8.5.1 Maladaptive plasticity and hypoxia acclimatization

On balance, theoretical and experimental results suggest that an increased Hb-O_2 affinity is generally beneficial under severe hypoxia, and a synthesis of available comparative data suggests that natural selection has favored increased Hb-O_2 affinities in numerous high-altitude mammals and birds. It therefore seems paradoxical that the acclimatization response to environmental hypoxia in humans and other lowland mammals typically involves a reduction in blood-O_2 affinity (Samaja et al., 1986, Mairbäurl et al., 1993, Mairbäurl, 1994, Samaja et al., 2003, Storz et al., 2010b, Storz, 2016c). This is largely attributable to an increase in red cell DPG concentration (Torrance et al., 1970/71, Lenfant et al., 1971,

Mairbäurl et al., 1986, Mairbäurl, 1994, Mairbäurl et al., 1993), or—more specifically—an increase in the relative concentration of Hb liganded with DPG. The increased concentration of DPG-liganded Hb at high altitude is largely attributable to a hypoxia-induced increase in ventilation; the resultant respiratory alkalosis stimulates red cell glycolytic activity which, in turn, increases DPG synthesis (Rapoport et al., 1977). At the whole-blood level, the hypoxia-induced increase in [DPG] is also attributable to the stimulation of erythropoiesis because this produces a downward shift in the mean age of circulating red blood cells, and newly produced red cells have higher [DPG] than older cells (Mairbäurl, 1994, Samaja et al., 2003). Several previous authors interpreted the hypoxia-induced increase in red cell [DPG] and the associated reduction in blood-O_2 affinity as an adaptive response (Aste-Salazar and Hurtado, 1944, Lenfant et al., 1968, Eaton et al., 1969, Lenfant et al., 1969, Lenfant et al., 1971, Frisancho, 1975). The theoretical and empirical results reviewed in sections 8.3.1–8.3.2 suggest that such a response may be beneficial under moderate hypoxia. However, in humans and other mammals, the hypoxia-induced increase in red cell [DPG] continues at elevations well above the threshold at which further reductions in Hb-O_2 affinity become counterproductive due to arterial desaturation (Winslow et al., 1984, Samaja et al., 2003). Even when all Hb is fully liganded with DPG (at a DPG:Hb ratio ≥2–3), further increases in [DPG] continue to reduce Hb-O_2 affinity indirectly because the increased erythrocytic concentration of nondiffusible anions reduces cellular pH due to a shift in the Donnan equilibrium (the ionic equilibrium between diffusible and nondiffusible ions across the red cell membrane), thereby reducing Hb-O_2 affinity via the Bohr effect (Duhm, 1971, Samaja and Winslow, 1979, Mairbäurl, 1994). Consequently, the increase in plasma pH caused by respiratory alkalosis has offsetting effects in mammalian red cells: the Bohr effect promotes an increased Hb-O_2 affinity, but this is counterbalanced by the increase in intracellular [DPG] (Winslow et al., 1984, Samaja et al., 1997).

In mammals that have acclimatized to chronic hypoxia, the seemingly maladaptive increase in red cell [DPG] may represent a miscued response to environmental hypoxia in species whose ancestors evolved in a lowland environment (Storz et al., 2010b, Tufts et al., 2013). In such species, hematological responses to environmental hypoxia may have originally evolved as a response to anemia (Hebbel et al., 1978, Storz, 2010). Environmental hypoxia and anemia both result in reduced levels of tissue oxygenation, but they have different root causes. In the case of anemia, reduced tissue oxygenation is caused by a diminution of blood O_2 transport capacity, and can therefore be rectified by increasing [Hb] and by decreasing Hb-O_2 affinity, which enhances O_2 unloading without compromising pulmonary O_2 loading (at least under normoxia). In the case of environmental hypoxia, by contrast, reduced tissue oxygenation has an external cause: the reduced PO_2 of inspired air, and the concomitant reduction in arterial PO_2. In this situation, safeguarding arterial O_2 saturation is at a premium so a reduction in Hb-O_2 affinity can be counterproductive and may further compromise tissue O_2 delivery. In humans and other lowland mammals whose ancestors were never forced to contend with the physiological challenges of low-O_2 environments, the hypoxia-induced increase in red cell [DPG] may represent a misdirected acclimatization response that originally evolved for a different purpose.

It remains to be seen whether birds and other terrestrial vertebrates with nucleated red cells acclimatize to environmental hypoxia via changes in red cell organic phosphate concentrations, but available evidence indicates that [IPP] in avian red cells is highly constant and is unresponsive to changes in temperature or PO_2 (Jaeger and McGrath, 1974, Lutz, 1980).

8.5.2 The life aquatic: regulation of blood O_2 capacitance in teleost fishes

In general, teleost fishes inhabiting waters that are chronically or episodically hypoxic tend to have higher Hb-O_2 affinities relative to species that inhabit well-aerated waters (Powers, 1980, Jensen, 2004, Wells, 2009, Harter and Brauner, 2017). Moreover, acclimatization to environmental hypoxia typically involves erythrocytic changes that increase Hb-O_2 affinity, thereby enhancing O_2 uptake at the gills (Wells, 2009, Harter and Brauner, 2017).

Teleosts typically respond to environmental hypoxia by increasing ventilation as a first line of defense. Increasing water flow over the gills increases the mean PO_2 gradient across the branchial epithelium and prevents P_aO_2 from falling more than it would otherwise. Another effect of hyperventilation is respiratory alkalosis, which results in an increased red cell pH and, thus, an increased Hb-O_2 affinity due to the Bohr effect (Randall and Daxboeck, 1984, Malte and Weber, 1985, Perry and Wood, 1989, Piiper, 1998, Perry and Gilmour, 2002, Perry et al., 2009). In many teleost species, red cell pH may be further increased via catecholamine-induced activation of the β-NHE in the red cell membrane (Randall and Perry 1992, Thomas and Perry 1992).

The activation of β-NHE increases red cell pH and therefore directly increases Hb-O_2 affinity via the Bohr effect. The associated increase in red cell volume also indirectly contributes to an increase in Hb-O_2 affinity by diluting the effective concentration of the nucleotide triphosphates (NTPs), ATP and GTP, that are important allosteric effectors (Nikinmaa 1992).

In addition to the hypoxia-induced increase in red cell pH caused by hyperventilation and β-NHE activation, the other important mechanism for reversibly increasing Hb-O_2 affinity involves a

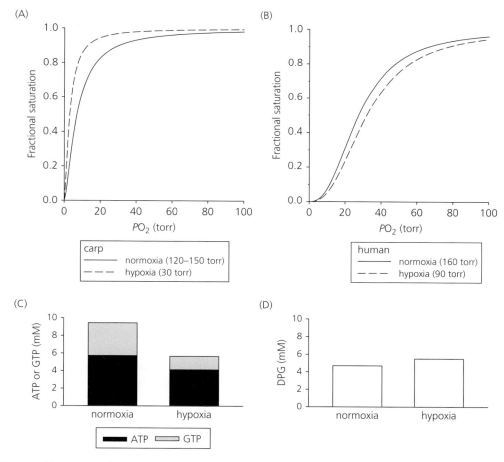

Fig. 8.14. O_2-equilibrium curves of whole blood (A and B) and red cell concentrations of organic phosphates (C and D) in carp (*Cyprinus carpio*) and human. Curves for carp and human blood were measured in normoxia (solid lines) and hypoxia (dashed lines). Panels A and C: Carp were acclimated to normoxia (PO_2 ~120–150 torr) and hypoxia (PO_2 ~30 torr) for >1 month, and blood O_2-equilibrium curves were measured at pH 7.9, 20°C (Weber and Lykkeboe, 1978). Panels B and D: Human subjects were exposed to acute hypoxia upon ascent to 4,559 m altitude and blood P_{50} was estimated at a plasma pH 7.4, PCO_2 40 torr, 37°C (Mairbäurl et al., 1993).

Modified from Fago (2017) with permission from the American Physiological Society.

reduction in the red cell concentration of NTPs (Weber, 1982, Weber and Jensen, 1988, Boutilier and Ferguson, 1989, Jensen et al., 1993, Weber, 1996, Gallaugher and Farrell, 1998, Jensen et al., 1998, Nikinmaa and Salama, 1998, Val, 2000, Nikinmaa, 2001, Nikinmaa, 2002, Jensen, 2004, Wells, 2009, Harter and Brauner, 2017). Reducing the red cell concentration of NTPs directly reduces blood P_{50} by reducing NTP-binding to deoxyHb, but it also has an indirect effect by increasing red cell pH due to a shift in the Donnan equilibrium. When both ATP and GTP are present, the latter typically exerts a more potent allosteric effect due to its higher binding affinity for deoxyHb (Chapter 4, section 4.6.4). Whereas ATP is the exclusive or predominant organic phosphate in the red cells of many teleost species, some especially hypoxia-tolerant taxa, like carp and eels, have similar concentrations of ATP and GTP (Weber et al., 1976, Weber and Lykkeboe, 1978). In such species, the higher responsiveness of Hb to GTP relative to ATP means that a given reduction in total [NTP] can produce a disproportionately large increase in Hb-O_2 affinity (Nikinmaa and Salama, 1998, Soivio et al., 1980).

The contrasting hematological responses to environmental hypoxia in representative teleost fishes and mammals are illustrated in Fig 8.14. Hypoxia-tolerant species like carp, *Cyprinus carpio*, respond to a drop in water PO_2 by reducing red cell [NTP], thereby increasing Hb-O_2 affinity (Fig. 8.14A). By contrast, humans and most other mammals with lowland ancestries respond to hypoxia by increasing red cell [DPG], which results in a reduced Hb-O_2 affinity (a right-shifted curve; Fig. 8.14B), a potentially maladaptive response.

As discussed in Chapter 6 (section 6.6.2), Hb multiplicity confers a clear advantage to many teleost species by permitting a division of labor between isoHbs with quantitatively different O_2 affinities and/or qualitatively different allosteric properties. However, there is currently little evidence to suggest that hypoxia-induced changes in red cell isoHb composition contribute to reversible changes in blood-O_2 affinity in teleost fishes (Ingermann, 1997, Wells, 2009). More experiments like those of Rutjes et al. (2007) are needed to settle the question. Overall, the modulation of red cell pH and [NTP] appears to play a far more important role in regulat-

ing blood O_2 capacitance *in vivo* (Weber and Jensen, 1988, Weber, 1996, Val, 2000, Weber, 2000, Nikinmaa, 2001, Brauner and Val, 2006, Wells, 2009).

8.6 Conclusion

Theoretical results indicate that—for air-breathers and water-breathers alike—it is generally beneficial to have an increased Hb-O_2 affinity under conditions of severe hypoxia. This prediction is supported by experimental studies of survival and whole-animal physiological performance and by *ex vivo* studies of microvascular O_2 transport and tissue perfusion. Case studies involving high-altitude mammals and birds have provided compelling evidence that evolved increases in Hb-O_2 affinity are adaptive. Evolutionary changes in Hb-O_2 affinity involve a variety of functional mechanisms. In high-altitude mammals, evolved increases in Hb-O_2 affinity are attributable to changes in the intrinsic O_2 affinity and, in some cases, suppressed sensitivities to anionic effectors. In birds, evolved increases in Hb-O_2 affinity are consistently attributable to changes in intrinsic affinity that do not compromise allosteric regulatory capacity. Available evidence suggests that regulatory changes in isoHb composition do not play a general role in adaptation to high-altitude hypoxia in birds or mammals.

In mammals, the evidence for a positive relationship between Hb-O_2 affinity and native elevation is equivocal. In birds, by contrast, there is a remarkably strong positive relationship between Hb-O_2 affinity and native elevation. In fact, the case studies of high-altitude birds provide some of the most compelling examples of convergent biochemical adaptation in vertebrates. An important question for future work concerns the reason for the apparent difference between mammals and birds, two amniote lineages that independently evolved endothermy and which diversified in parallel during periods of increasing atmospheric O_2 in the late Mesozoic. In teleost fishes, species that exploit chronically or transiently hypoxic habitats tend to have elevated Hb-O_2 affinities, and at the level of red cell function, the typical acclimatization response to environmental hypoxia is directed toward increasing Hb-O_2 affinity.

An important goal for future research is to elucidate the specific physiological mechanisms by which changes in the oxygenation properties of Hb translate into enhancements of whole-animal aerobic performance.

In air-breathing and water-breathing vertebrates alike, fine-tuned adjustments in Hb-O₂ affinity provide an energetically efficient means of mitigating the effects of arterial hypoxemia. The adaptive significance of such changes is indicated by evolved changes in Hb function in high-altitude mammals and birds, and erythrocytic acclimatization responses to environmental hypoxia in teleost fishes. In Chapter 9 we will further explore mechanisms of biochemical adaptation and protein evolution with an overview of lessons learned from research on Hb function.

References

Akinsheye, I., Alsultan, A., Solovieff, N., et al. (2011). Fetal hemoglobin in sickle cell anemia. *Blood*, **118**, 19–27.

Aste-Salazar, H. and Hurtado, A. (1944). The affinity of hemoglobin for oxygen at sea level and at high altitudes. *American Journal of Physiology*, **142**, 733–43.

Bakker, J. C., Gortmaker, G. C., Vrolijk, A. C. M., and Offerijns, F. G. J. (1976). Influence of position of oxygen dissociation curve on oxygen-dependent functions of isolated perfused rat liver. 1. Studies at different levels of hypoxic hypoxia. *Pflügers Archiv-European Journal of Physiology*, **362**, 21–31.

Banchero, N. and Grover, R. F. (1972). Effect of different levels of simulated altitude on O₂ transport in llama and sheep. *American Journal of Physiology*, **222**, 1239–45.

Barcroft, J., Binger, C. A., Bock, A. V., et al. (1923). Observations upon the effect of high altitude on the physiological process of the human body, carried out in the Peruvian Andes, chiefly at Cerro de Pasco. *Philosophical Transactions of the Royal Society of London B*, **211**, 351–480.

Bellelli, A., Brunori, M., Miele, A. E., Panetta, G., and Vallone, B. (2006). The allosteric properties of hemoglobin: insights from natural and site directed mutants. *Current Protein and Peptide Science*, **7**, 17–45.

Bencowitz, H. Z., Wagner, P. D., and West, J. B. (1982). Effect of change in P50 on exercise tolerance at high-altitude—a theoretical study. *Journal of Applied Physiology*, **53**, 1487–95.

Berenbrink, M. (2006). Evolution of vertebrate haemoglobins: histidine side chains, specific buffer value and Bohr effect. *Respiratory Physiology and Neurobiology*, **154**, 165–84.

Black, C. P. and Tenney, S. M. (1980). Oxygen-transport during progressive hypoxia in high-altitude and sea-level waterfowl. *Respiration Physiology*, **39**, 217–39.

Boutilier, R. G. and Ferguson, R. A. (1989). Nucleated red cell function—metabolism and pH regulation. *Canadian Journal of Zoology*, **67**, 2986–93.

Bouverot, P. (1985). *Adaptation to Altitude-Hypoxia in Vertebrates*, Berlin, Springer-Verlag.

Brauner, C. J. and Val, A. L. (2006). Oxygen transfer. *In:* Val, A. L. and Almeida-Val, V. M. F. (eds.) *Fish Physiology, Vol. 21: The Physiology of Tropical Fishes*, pp. 277–306. Amsterdam, Elsevier.

Brauner, C. J. and Wang, T. (1997). The optimal oxygen equilibrium curve: a comparison between environmental hypoxia and anemia. *American Zoologist*, **37**, 101–8.

Bullard, R. W. (1972). Vertebrates at altitude. *In:* Yousef, M. K., Horvath, S. M., and Bullard, R. W. (eds.) *Physiological Adaptations*, pp. 209–25. New York, NY, Academic Press.

Bullard, R. W., Broumand, C., and Meyer, F. R. (1966). Blood characteristics and volume in two rodents native to high altitude. *Journal of Applied Physiology*, **21**, 994–8.

Campbell, K. L., Storz, J. F., Signore, A. V., et al. (2010). Molecular basis of a novel adaptation to hypoxic-hypercapnia in a strictly fossorial mole. *BMC Evolutionary Biology*, **10**, 214.

Chappell, M. A. and Snyder, L. R. G. (1984). Biochemical and physiological correlates of deer mouse α-chain hemoglobin polymorphisms. *Proceedings of the National Academy of Sciences of the United States of America*, **81**, 5484–8.

Cheviron, Z. A., Natarajan, C., Projecto-Garcia, J., et al. (2014). Integrating evolutionary and functional tests of adaptive hypotheses: a case study of altitudinal differentiation in hemoglobin function in an Andean sparrow, *Zonotrichia capensis. Molecular Biology and Evolution*, **31**, 2948–62.

Chiodi, H. (1971). Comparative study of the blood gas transport in high altitude and sea level camelidae and goats. *Respiration Physiology*, **11**, 84–93.

Dawson, T. J. and Evans, J. V. (1966). Effect of hypoxia on oxygen transport in sheep with different hemoglobin types. *American Journal of Physiology*, **210**, 1021–5.

Dejours, P. 1981. *Principles of Comparative Respiratory Physiology*, Amsterdam, Elsevier/North-Holland Biomedical Press.

Dempsey, J. A., Thomson, J. M., Forster, H. V., Cerny, F. C., and Chosy, L. W. (1975). Hb-O₂ dissociation in man during prolonged work in chronic hypoxia. *Journal of Applied Physiology*, **38**, 1022–9.

Duhm, J. (1971). Effects of 2,3-diphosphoglycerate and other organic phosphate compounds on oxygen affinity and intracellular pH of human erythrocytes. *Pflügers Archiv-European Journal of Physiology*, **326**, 341–56.

Eaton, J. W., Brewer, G. J., and Groover, R. F. (1969). Role of red cell 2,3-diphosphoglycerate in the adaptation of man to altitude. *Journal of Laboratory and Clinical Medicine*, **73**, 603–9.

Eaton, J. W., Skelton, T. D., and Berger, E. (1974). Survival at extreme altitude—protective effect of increased hemoglobin-oxygen affinity. *Science*, **183**, 743–4.

Ekblom, B. and Berglund, B. (1991). Effect of erythropoietin administration on mammal aerobic power. *Scandanavian Journal of Medicine and Science in Sports*, **1**, 88–93.

Fago, A. (2017). Functional roles of globin proteins in hypoxia-tolerant ectothermic vertebrates. *Journal of Applied Physiology*, **123**, 926–34.

Felsenstein, J. (1985). Phylogenies and the comparative method. *American Naturalist*, **125**, 1–15.

Frisancho, A. R. (1975). Functional adaptation to high altitude hypoxia. *Science*, 187, 313–19.

Galen, S. C., Natarajan, C., Moriyama, H., et al. (2015). Contribution of a mutational hotspot to adaptive changes in hemoglobin function in high-altitude Andean house wrens. *Proceedings of the National Academy of Sciences of the United States of America*, **112**, 13958–63.

Gallaugher, P. and Farrell, A. P. (1998). Hematocrit and blood-oxygen carrying capacity. *In*: Perry, S. F. and Tufts, B. L. (eds.) *Fish Physiology, Vol. 17: Fish Respiration*, pp. 185–227. New York, NY, Academic Press.

Gallaugher, P., Thorarensen, H., and Farrell, A. P. (1995). Hematocrit in oxygen transport and swimming in rainbow trout (*Oncorhynchus mykiss*). *Respiration Physiology*, **102**, 279–92.

Garland, T., JR., Bennett, A. F., and Rezende, E. L. (2005). Phylogenetic approaches in comparative physiology. *Journal of Experimental Biology*, **208**, 3015–35.

Grispo, M. T., Natarajan, C., Projecto-Garcia, J., et al. (2012). Gene duplication and the evolution of hemoglobin isoform differentiation in birds. *Journal of Biological Chemistry*, **287**, 37647–58.

Hahn, M. W. and Nakhleh, L. (2016). Irrational exuberance for resolved species trees. *Evolution*, **70**, 7–17.

Hall, F. G. (1966). Minimal utilizable oxygen and oxygen dissociation curve of blood of rodents. *Journal of Applied Physiology*, **21**, 375–8.

Hall, F. G., Dill, D. B., and Barron, E. S. G. (1936). Comparative physiology in high altitudes. *Journal of Cellular and Comparative Physiology*, **8**, 301–13.

Harter, T. S. and Brauner, C. J. (2017). The O_2 and CO_2 transport system in teleosts and the specialized mechanisms that enhance Hb-O_2 unloading to tissues. *In*: Gamperl, A. K., Gillis, T. E., Farrell, A. P., and Brauner, C. J. (eds.) *Fish Physiology, Vol. 36B: The Cardiovascular System: Development, Plasticity, and Physiological Responses*, p. 1–106. San Diego, CA, Elsevier Science.

Hebbel, R. P., Eaton, J. W., Kronenberg, R. S., et al. (1978). Human llamas—adaptation to altitude in subjects with high hemoglobin oxygen affinity. *Journal of Clinical Investigation*, **62**, 593–600.

Hiebl, I., Weber, R. E., Schneeganss, D., Kosters, J., and Braunitzer, G. (1988). High-altitude respiration of birds—structural adaptations in the major and minor hemoglobin components of adult Ruppells griffon (*Gryps rueppellii*)—a new molecular pattern for hypoxia tolerance. *Biological Chemistry Hoppe-Seyler*, **369**, 217–32.

Hoffmann, F. G., Opazo, J. C., and Storz, J. F. (2008a). New genes originated via multiple recombinational pathways in the beta-globin gene family of rodents. *Molecular Biology and Evolution*, **25**, 2589–600.

Hoffmann, F. G., Opazo, J. C., and Storz, J. F. (2008b). Rapid rates of lineage-specific gene duplication and deletion in the alpha-globin gene family. *Molecular Biology and Evolution*, **25**, 591–602.

Ingermann, R. L. (1997). Vertebrate hemoglobins. *Handbook of Physiology, Comparative Physiology*, **30**, 357–408.

Jaeger, J. J. and McGrath, J. J. (1974). Hematologic and biochemical effects of simulated high-altitude on Japanese quail. *Journal of Applied Physiology*, **37**, 357–61.

Jendroszek, A., Malte, H., Overgaard, C. B., et al. (2018) Allosteric mechanisms underlying the adaptive increase in hemoglobin-oxygen affinity of the bar-headed goose. *Journal of Experimental Biology*, doi: 10.1242/jeb.185470.

Janecka, J. E., Nielsen, S. S. E., Andersen, S. D., et al. (2015). Genetically based low oxygen affinities of felid hemoglobins: lack of biochemical adaptation to high-altitude hypoxia in the snow leopard. *Journal of Experimental Biology*, **218**, 2402–9.

Jensen, B., Storz, J. F., and Fago, A. (2016). Bohr effect and temperature sensitivity of hemoglobins from highland and lowland deer mice. *Comparative Biochemistry and Physiology A-Molecular and Integrative Physiology*, **195**, 10–14.

Jensen, F. B. (2004). Red blood cell pH, the Bohr effect, and other oxygenation-linked phenomena in blood O_2 and CO_2 transport. *Acta Physiologica Scandinavica*, **182**, 215–27.

Jensen, F. B. (2009). The dual roles of red blood cells in tissue oxygen delivery: oxygen carriers and regulators of local blood flow. *Journal of Experimental Biology*, **212**, 3387–93.

Jensen, F. B., Fago, A., and Weber, R. E. (1998). Hemoglobin structure and function. *In*: Perry, S. F. and Tufts, B. L. (eds.) *Fish Physiology, Vol. 17: Fish Respiration*. New York, NY, Academic Press.

Jensen, F. B., Nikinmaa, M., and Weber, R. E. (1993). Environmental perturbations of oxygen transport in teleost fishes: causes, consequences and compensations. *In*: Rankin, J. C. and Jensen, F. B. (eds.) *Fish Ecophysiology*, pp. 161–79. London, Chapman and Hall.

Jessen, T. H., Weber, R. E., Fermi, G., Tame, J., and Braunitzer, G. (1991). Adaptation of bird hemoglobins to high-altitudes—demonstration of molecular mechanism by protein engineering. *Proceedings of the National Academy of Sciences of the United States of America*, **88**, 6519–22.

Kanstrup, I. L. and Ekblom, B. (1984). Blood volume and hemoglobin concentration as determinants of maximal aerobic power. *Medicine and Science in Sports and Exercise,* **16**, 256–62.

Kumar, A., Natarajan, C., Moriyama, H., et al. (2017). Stability-mediated epistasis restricts accessible mutational pathways in the functional evolution of avian hemoglobin. *Molecular Biology and Evolution,* **34**, 1240–51.

Lechner, A. J. (1976). Respiratory adaptations in burrowing pocket gophers from sea-level and high-altitude. *Journal of Applied Physiology,* **41**, 168–73.

Lenfant, C., Torrance, J., English, E., et al. (1968). Effect of altitude on oxygen binding by hemoglobin and on organic phosphate levels. *Journal of Clinical Investigation,* **47**, 2652–6.

Lenfant, C., Torrance, J. D., and Reynafar, C. (1971). Shift of the O_2-Hb dissociation curve at altitude: mechanism and effect. *Journal of Applied Physiology,* **30**, 625–31.

Lenfant, C., Ways, P., Aucutt, C., and Cruz, J. (1969). Effect of chronic hypoxic hypoxia on the O_2-Hb dissociation curve and respiratory gas transport in man. *Respiration Physiology,* **7**, 7–29.

Leon-Velarde, F., Demuizon, C., Palacios, J. A., Clark, D., and Mongec, C. (1996). Hemoglobin affinity and structure in high-altitude and sea-level carnivores from Peru. *Comparative Biochemistry and Physiology A-Molecular and Integrative Physiology,* **113**, 407–11.

Lutz, P. L. (1980). On the oxygen-affinity of bird blood. *American Zoologist,* **20**, 187–98.

Mairbäurl, H. (1994). Red blood cell function in hypoxia at altitude and exercise. *International Journal of Sports Medicine,* **15**, 51–63.

Mairbäurl, H., Oelz, O., and Bartsch, P. (1993). Interactions between Hb, Mg, DPG, ATP, and Cl determine the change in Hb-O_2 affinity at high-altitude. *Journal of Applied Physiology,* **74**, 40–8.

Mairbäurl, H., Schobersberger, W., Hasibeder, W., et al. (1986). Regulation of red-cell 2,3-DPG and Hb-O_2 affinity during acute exercise. *European Journal of Applied Physiology and Occupational Physiology,* **55**, 174–80.

Mairbäurl, H. and Weber, R. E. (2012). Oxygen transport by hemoglobin. *Comprehensive Physiology,* **2**, 1463–89.

Malte, H. and Weber, R. E. (1985). A mathematical model for gas exchange in the fish gill based on nonlinear blood-gas equilibrium curves. *Respiration Physiology,* **62**, 359–74.

McGuire, J. A., Witt, C. C., Remsen, J. V. Jr, et al. (2014). Molecular phylogenetics and the diversification of hummingbirds. *Current Biology,* **24**, 910–16.

Monge, C. and Leon-Velarde, F. (1991). Physiological adaptation to high-altitude—oxygen-transport in mammals and birds. *Physiological Reviews,* **71**, 1135–72.

Natarajan, C., Hoffman, F. G., Lanier, H. C., et al. 2015a. Intraspecific polymorphism, interspecific divergence,

and the origins of function-altering mutations in deer mouse hemoglobin. *Molecular Biology and Evolution,* **32**, 978–97.

Natarajan, C., Hoffmann, F. G., Weber, R. E., et al. (2016). Predictable convergence in hemoglobin function has unpredictable molecular underpinnings. *Science,* **354**, 336–40.

Natarajan, C., Inoguchi, N., Weber, R. E., et al. (2013). Epistasis among adaptive mutations in deer mouse hemoglobin. *Science,* **340**, 1324–7.

Natarajan, C., Jendroszek, A., Kumar, A., et al. (2018). Molecular basis of hemoglobin adaptation in the high-flying bar-headed goose. *PLoS Genetics,* e1007331.

Natarajan, C., Projecto-Garcia, J., Moriyama, H., et al. (2015b). Convergent evolution of hemoglobin function in high-altitude Andean waterfowl involves limited parallelism at the molecular sequence level. *PLoS Genetics,* **11**, e1005681.

Nikinmaa, M. (1992). Membrane transport and control of hemoglobin-oxygen affinity in nucleated erythrocytes. *Physiological Reviews,* **72**, 301–21.

Nikinmaa, M. (2001). Haemoglobin function in vertebrates: evolutionary changes in cellular regulation in hypoxia. *Respiration Physiology,* **128**, 317–29.

Nikinmaa, M. (2002). Oxygen-dependent cellular functions—why fishes and their aquatic environment are a prime choice of study. *Comparative Biochemistry and Physiology A-Molecular and Integrative Physiology,* **133**, 1–16.

Nikinmaa, M. and Salama, A. (1998). Oxygen transport in fish. *In:* Perry, S. F. and Tufts, B. L. (eds.) *Fish Physiology, Vol. 17: Fish Respiration,* pp. 141–84. New York, NY, Academic Press.

Opazo, J. C., Hoffman, F. G., Natarajan, C., et al. (2015). Gene turnover in the avian globin gene family and evolutionary changes in hemoglobin isoform expression. *Molecular Biology and Evolution,* **32**, 871–87.

Opazo, J. C., Hoffmann, F. G., and Storz, J. F. (2008). Differential loss of embryonic globin genes during the radiation of placental mammals. *Proceedings of the National Academy of Sciences of the United States of America,* **105**, 12950–5.

Opazo, J. C., Sloan, A. M., Campbell, K. L., and Storz, J. F. (2009). Origin and ascendancy of a chimeric fusion gene: the β/δ-globin gene of paenungulate mammals. *Molecular Biology and Evolution,* **26**, 1469–78.

Ostojic, H., Cifuentes, V., and Monge, C. (2002). Hemoglobin affinity in Andean rodents. *Biological Research,* **35**, 27–30.

Pack-Mabien, A. V. and Imran, H. (2013). Benefits of delayed fetal hemoglobin (HbF) switching in sickle cell disease (SCD): a case report and review of the literature. *Journal of Pediatric Hematology Oncology,* **35**, E347–9.

Penney, D. and Thomas, M. (1975). Hematological alterations and response to acute hypobaric stress. *Journal of Applied Physiology*, **39**, 1034–7.

Perry, S. F. and Gilmour, K. M. (2002). Sensing and transfer of respiratory gases at the fish gill. *Journal of Experimental Zoology*, **293**, 249–63.

Perry, S. F., Jonz, M., and Gilmour, K. M. (2009). Oxygen sensing and the hypoxic ventilatory response. In: Farrell, A. P. and Brauner, C. J. (eds.) *Fish Physiology, Vol. 27: Hypoxia*, pp. 193–253. New York, NY, Academic Press.

Perry, S. F. and Wood, C. M. (1989). Control and coordination of gas transfer in fishes. *Canadian Journal of Zoology*, **67**, 2961–70.

Perutz, M. F. (1983). Species adaptation in a protein molecule. *Molecular Biology and Evolution*, **1**, 1–28.

Petschow, D., Wurdinger, I., Baumann, R., et al. (1977). Causes of high blood O_2 affinity of animals living at high-altitude. *Journal of Applied Physiology*, **42**, 139–43.

Piiper, J. (1998). Brachial gas transfer models. *Comparative Biochemistry and Physiology A -Molecular and Integrative Physiology*, **119**, 125–30.

Powers, D. A. (1980). Molecular ecology of teleost fish hemoglobins—Strategies for adapting to changing environments. *American Zoologist*, **20**, 139–62.

Projecto-Garcia, J., Natarajan, C., Moriyama, H., et al. (2013). Repeated elevational transitions in hemoglobin function during the evolution of Andean hummingbirds. *Proceedings of the National Academy of Sciences of the United States of America*, **110**, 20669–74.

Randall, D. J. and Daxboeck, C. (1984). Oxygen and carbon dioxide transfer across fish gills. In: Hoar, W. S. and Randall, D. J. (eds.) *Fish Physiology, Vol. 10A: Gills*, pp. 263–314. New York, NY, Academic Press.

Randall, D. J. and Perry, S. F. (1992). Catecholamines. In: Randall, D. J. and Hoar, W. S. (eds.) *Fish Physiology, Vol. 12: The Cardiovascular System*, pp. 255–300. New York, NY, Academic Press.

Rapoport, I., Berger, H., Elsner, R., and Rapoport, S. (1977). pH-dependent changes of 2,3-biphosphoglycerate in human red-cells during transitional and steady states in vitro. *European Journal of Biochemistry*, **73**, 421–7.

Revsbech, I. G., Tufts, D. M., Projecto-Garcia, J., et al. (2013). Hemoglobin function and allosteric regulation in semi-fossorial rodents (family Sciuridae) with different altitudinal ranges. *Journal of Experimental Biology*, **216**, 4264–71.

Riggs, A. (1976). Factors in evolution of hemoglobin function. *Federation Proceedings*, **35**, 2115–18.

Runck, A. M., Moriyama, H., and Storz, J. F. (2009). Evolution of duplicated β-globin genes and the structural basis of hemoglobin isoform differentiation in *Mus*. *Molecular Biology and Evolution*, **26**, 2521–32.

Runck, A. M., Weber, R. E., Fago, A., and Storz, J. F. (2010). Evolutionary and functional properties of a two-locus β-globin polymorphism in Indian house mice. *Genetics*, **184**, 1121–31.

Rutjes, H. A., Nieveen, M. C., Weber, R. E., Witte, F., and Van Den Thillart, G. E. E. J. M. (2007). Multiple strategies of Lake Victoria cichlids to cope with lifelong hypoxia include hemoglobin switching. *American Journal of Physiology-Regulatory Integrative and Comparative Physiology*, **293**, R1376–83.

Samaja, M., Crespi, T., Guazzi, M., and Vandegriff, K. D. (2003). Oxygen transport in blood at high altitude: role of the hemoglobin-oxygen affinity and impact of the phenomena related to hemoglobin allosterism and red cell function. *European Journal of Applied Physiology*, **90**, 351–9.

Samaja, M., Diprampero, P. E., and Cerretelli, P. (1986). The role of 2,3-DPG in the oxygen transport at altitude. *Respiration Physiology*, **64**, 191–202.

Samaja, M., Mariani, C., Prestini, A., and Cerretelli, P. (1997). Acid-base balance and O_2 transport at high altitude. *Acta Physiologica Scandanavica*, **159**, 249–56.

Samaja, M. and Winslow, R. M. (1979). Separate effects of H^+ and 2,3-DPG on the oxygen equilibrium curve of human blood. *British Journal of Haematology*, **41**, 373–81.

Schuler, B., Arras, M., Keller, S., et al. (2010). Optimal hematocrit for maximal exercise performance in acute and chronic erythropoietin-treated mice. *Proceedings of the National Academy of Sciences of the United States of America*, **107**, 419–23.

Scott, G. R. (2011). Elevated performance: the unique physiology of birds that fly at high altitudes. *Journal of Experimental Biology*, **214**, 2455–62.

Scott, G. R. and Milsom, W. K. (2006). Flying high: a theoretical analysis of the factors limiting exercise performance in birds at altitude. *Respiratory Physiology and Neurobiology*, **154**, 284–301.

Simonson, T. S., Wei, G., Wagner, H. E., et al. (2015). Low haemoglobin concentration in Tibetan males is associated with greater high-altitude exercise capacity. *Journal of Physiology-London*, **593**, 3207–18.

Snyder, L. R. G. (1982). 2,3-diphosphoglycerate in high-altitude and low-altitude populations of the deer mouse. *Respiration Physiology*, **48**, 107–23.

Soivio, A., Nikinmaa, M., and Westman, K. (1980). The blood-oxygen binding properties of hypoxic *Salmo gairdneri*. *Journal of Comparative Physiology*, **136**, 83–7.

Stein, J. C. and Ellsworth, M. L. (1993). Capillary oxygen-transport during severe hypoxia—Role of hemoglobin-oxygen affinity. *Journal of Applied Physiology*, **75**, 1601–7.

Storz, J. F. (2007). Hemoglobin function and physiological adaptation to hypoxia in high-altitude mammals. *Journal of Mammalogy*, **88**, 24–31.

Storz, J. F. (2010). Genes for high altitudes. *Science*, **329**, 40–1.

Storz, J. F. (2016a). Causes of molecular convergence and parallelism in protein evolution. *Nature Reviews Genetics*, **17**, 239–50.

Storz, J. F. (2016b). Gene duplication and evolutionary innovations in hemoglobin-oxygen transport. *Physiology*, **31**, 223–32.

Storz, J. F. (2016c). Hemoglobin-oxygen affinity in high-altitude vertebrates: is there evidence for an adaptive trend? *Journal of Experimental Biology*, **219**, 3190–203.

Storz, J. F., Baze, M., Waite, J. L., et al. (2007). Complex signatures of selection and gene conversion in the duplicated globin genes of house mice. *Genetics*, **177**, 481–500.

Storz, J. F., Bridgham, J. T., Kelly, S. A., and Garland, T., JR. (2015). Genetic approaches in comparative and evolutionary physiology. *American Journal of Physiology-Regulatory Integrative and Comparative Physiology*, **309**, R197–R214.

Storz, J. F. and Moriyama, H. (2008). Mechanisms of hemoglobin adaptation to high altitude hypoxia. *High Altitude Medicine and Biology*, **9**, 148–57.

Storz, J. F., Runck, A. M., Moriyama, H., Weber, R. E., and Fago, A. (2010a). Genetic differences in hemoglobin function between highland and lowland deer mice. *Journal of Experimental Biology*, **213**, 2565–74.

Storz, J. F., Runck, A. M., Sabatino, S. J., et al. (2009). Evolutionary and functional insights into the mechanism underlying high-altitude adaptation of deer mouse hemoglobin. *Proceedings of the National Academy of Sciences of the United States of America*, **106**, 14450–5.

Storz, J. F., Scott, G. R., and Cheviron, Z. A. (2010b). Phenotypic plasticity and genetic adaptation to high-altitude hypoxia in vertebrates. *Journal of Experimental Biology*, **213**, 4125–36.

Thomas, S. and Perry, S. F. (1992). Control and consequences of adrenergic activation of red blood cell Na+/H+ exchange on blood-oxygen and carbon-dioxide transport in fish. *Journal of Experimental Zoology*, **263**, 160–75.

Torrance, J. D., Lenfant, C., Cruz, J., and Marticorena, E. (1970/71). Oxygen transport mechanism in residents at high altitude. *Respiration Physiology*, **11**, 1–15.

Tufts, D. M., Natarajan, C., Revsbech, I. G., et al. (2015). Epistasis constrains mutational pathways of hemoglobin adaptation in high-altitude pikas. *Molecular Biology and Evolution*, **32**, 287–98.

Tufts, D. M., Revsbech, I. G., Cheviron, Z. A., et al. (2013). Phenotypic plasticity in blood-oxygen transport in highland and lowland deer mice. *Journal of Experimental Biology*, **216**, 1167–73.

Turek, Z., Kreuzer, F., and Hoofd, L. J. C. (1973). Advantage or disadvantage of a decrease in blood-oxygen affinity for tissue oxygen supply at hypoxia—theoretical study comparing man and rat. *Pflügers Archiv-European Journal of Physiology*, **342**, 185–97.

Turek, Z., Kreuzer, F., and Ringnalda, B. E. M. (1978a). Blood-gases at several levels of oxygenation in rats with a left-shifted blood-oxygen dissociation curve. *Pflügers Archiv-European Journal of Physiology*, **376**, 7–13.

Turek, Z., Kreuzer, F., Turekmaischeider, M., and Ringnalda, B. E. M. (1978b). Blood O_2 content, cardiac output, and flow to organs at several levels of oxygenation in rats with a left-shifted blood-oxygen dissociation curve. *Pflügers Archiv-European Journal of Physiology*, **376**, 201–7.

Val, A. L. (2000). Organic phosphates in the red blood cells of fish. *Comparative Biochemistry and Physiology A-Molecular and Integrative Physiology*, **125**, 417–35.

Wagner, P. D. (1996). A theoretical analysis of factors determining VO_2max at sea level and altitude. *Respiration Physiology*, **106**, 329–43.

Wagner, P. D., Simonson, T. S., Wei, G., et al. (2015). Sea-level haemoglobin concentration is associated with greater exercise capacity in Tibetan males at 4200m. *Experimental Physiology*, **100**, 1256–62.

Wang, T. and Malte, H. (2011). O_2 uptake and transport: the optimal P50. *In:* Farrell, A. P. (ed.) *Encyclopedia of Fish Physiology: From Genome to Environment*, pp. 1845–55. Amsterdam, Elsevier.

Weber, R. E. (1982). Intraspecific adaptation of hemoglobin function in fish to oxygen availability. *In:* Addink, A. D. F. and Spronk, N. (eds.) *Exogenous and Endogenous Influences on Metabolic and Neural Control*, pp. 87–102. Oxford, Pergamon Press.

Weber, R. E. (1995). Hemoglobin adaptations to hypoxia and altitude—the phylogenetic perspective. *In:* Sutton, J. R., Houston, C. S., and Coates, G. (eds.) *Hypoxia and the Brain*, pp. 31–44. Burlington, VT, Queen City Printers.

Weber, R. E. (1996). Hemoglobin adaptations in Amazonian and temperate fish with special reference to hypoxia, allosteric effectors and functional heterogeneity. *In:* Val, A. L., Almeida-Val, V. M. F., and Randall, D. J. (eds.) *Physiology and Biochemistry of the Fishes of the Amazon*, pp. 75–90. Manaus, INPA.

Weber, R. E. (2000). Adaptations for oxygen transport: lessons from fish hemoglobins. *In:* Di Prisco, G., Giardina, B., and Weber, R. E. (eds.) *Hemoglobin Function in Vertebrates. Molecular Adaptation to Extreme and Temperate Environments*, pp. 23–37. Berlin, Springer-Verlag.

Weber, R. E. (2007). High-altitude adaptations in vertebrate hemoglobins. *Respiratory Physiology and Neurobiology*, **158**, 132–42.

Weber, R. E., Hiebl, I., and Braunitzer, G. (1988a). High-altitude and hemoglobin function in the vultures *Gyps rueppellii* and *Aegypius monachus*. *Biological Chemistry Hoppe-Seyler*, **369**, 233–40.

Weber, R. E. and Jensen, F. B. (1988). Functional adaptations in hemoglobins from ectothermic vertebrates. *Annual Review of Physiology*, **50**, 161–79.

Weber, R. E., Lalthantluanga, R., and Braunitzer, G. (1988b). Functional characterization of fetal and adult yak hemoglobins—an oxygen binding cascade and its molecular basis. *Archives of Biochemistry and Biophysics*, **263**, 199–203.

Weber, R. E. and Lykkeboe, G. (1978). Respiratory adaptations in carp blood. Influences of hypoxia, red cell organic phosphates, divalent cations and CO_2 on hemoglobin-oxygen affinity. *Journal of Comparative Physiology B*, **128**, 127–37.

Weber, R. E., Lykkeboe, G., and Johansen, K. (1976). Physiological properties of eel hemoglobin—hypoxic acclimation, phosphate effects and multiplicity. *Journal of Experimental Biology*, **64**, 75–88.

Wells, R. M. G. (2009). Blood-gas transport and hemoglobin function: adaptations for functional and environmental hypoxia. *In:* Richards, J. G., Farrell, A. P., and Brauner, C. J. (eds.) *Fish Physiology, Vol. 27: Hypoxia*, pp. 25–99. Amsterdam, Elsevier.

Wells, R. M. G. and Weber, R. E. (1991). Is there an optimal haematocrit for rainbow trout, *Oncorhynchus mykiss* (Walbaum)? An interpretation of recent data based on blood viscosity measurements. *Journal of Fish Biology*, **38**, 53–65.

West, J. B. and Wagner, P. D. (1980). Predicted gas exchange on the summit of Mt. Everest. *Respiration Physiology*, **42**, 1–16.

Willford, D. C., Hill, E. P., and Moores, W. Y. (1982). Theoretical analysis of optimal P_{50}. *Journal of Applied Physiology*, **52**, 1043–8.

Winslow, R. M., Samaja, M., and West, J. B. (1984). Red cell function at extreme altitudes on Mount Everest. *Journal of Applied Physiology*, **56**, 109–16.

Woodson, R. D. (1988). Evidence that changes in blood oxygen affinity modulate oxygen delivery: implications for control of tissue PO_2 gradients. *Advances in Experimental Medicine and Biology*, **222**, 309–13.

Yalcin, O. and Cabrales, P. (2012). Increased hemoglobin O_2 affinity protects during acute hypoxia. *American Journal of Physiology-Heart and Circulatory Physiology*, **303**, H271–81.

Zhu, X., Guan, Y., Signore, A. V., et al. (2018). Divergent and parallel routes of biochemical adaptation in high-altitude passerine birds from the Qinghai-Tibet Plateau. *Proceedings of National Academy of Sciences USA*, **115**, 1865–70.

CHAPTER 9

Darwin's molecule

Evolutionary insights into mechanisms of biochemical adaptation and protein evolution

9.1 Hb paradigms in molecular medicine and biochemical adaptation

The discovery of the causative mutation responsible for sickle-cell anemia (Pauling et al., 1949) and subsequent insights into the biophysical mechanism underlying the sickling phenomenon (Ingram, 1956) marked the beginning of the modern era of molecular medicine (Schechter, 2008, Steinberg, 2009). In the ensuing decades, clinically oriented research on Hb pathologies has generated a wealth of information about the phenotypic effects of amino acid mutations, and has established important paradigms in the disciplines of structural biology, biochemistry, molecular genetics, and medicine (Dickerson and Geis, 1983, Bunn and Forget, 1986, Steinberg et al., 2009, Thom et al., 2013).

Research on Hb function has an equally transformative role to play in our understanding of biochemical adaptation and protein evolution. Answers to fundamental questions about the genetics of adaptation require experimental data on the functional effects of individual mutations (Orr, 2005a) and an integrated understanding of the biology of adaptation requires insight into the causal mechanism by which genotype determines phenotype (Golding and Dean, 1998, Dean and Thornton, 2007, Storz and Wheat, 2010, Storz and Zera, 2011). Using Hb as a model molecule, we can exploit an unparalleled base of knowledge about structure-function relationships and we can characterize biophysical mechanisms of molecular adaptation at atomic resolution. It is therefore possible to document causal connections between genotype and biochemical phenotype at an unsurpassed level of rigor and detail. Moreover, since the oxygenation properties of Hb provide a direct link between ambient O_2 availability and aerobic metabolism, genetically based changes in protein function can be related to ecologically relevant aspects of organismal physiology. We therefore have a solid theoretical framework for making predictions and for interpreting observed associations between biochemical phenotype and fitness-related measures of whole-animal physiological performance (Turek et al., 1973, Turek et al., 1978a, Turek et al., 1978b, Bencowitz et al., 1982, Willford et al., 1982, Brauner and Wang, 1997, Scott and Milsom, 2006, Wang and Malte, 2011). This is exemplified by studies of deer mice (*Peromyscus maniculatus*) which documented that allelic variation in Hb-O_2 affinity contributes to adaptive variation in aerobic exercise capacity and thermogenic capacity under hypoxia (Chappell and Snyder, 1984, Chappell et al., 1988).

9.2 Key questions and conceptual issues in protein evolution

A long-standing question in molecular evolution concerns the relative fractions of new mutations that are deleterious, neutral, or beneficial with respect to fitness (Kimura, 1983, Gillespie, 1991). Clinically

Hemoglobin: Insights into Protein Structure, Function, and Evolution. Jay F. Storz, Oxford University Press (2019).
© Jay F. Storz 2019. DOI: 10.1093/oso/ 9780198810681.001.0001

oriented research on human Hb mutants has generated a wealth of information about the spectrum of phenotypic effects of spontaneous mutations in each class (Dickerson and Geis, 1983, Bunn and Forget, 1986, Nagel, 2001, Steinberg and Nagel, 2001, Steinberg and Nagel, 2009, Thom et al., 2013). Hb mutations that come to the attention of clinical researchers will tend to be enriched for those with deleterious effects—for example, amino acid mutations that destabilize the protein or affect ligand-binding behavior, thereby giving rise to various forms of cyanosis, erythrocytosis, or hemolytic anemia. However, most spontaneous mutations and low-frequency amino acid variants in human Hb have no clinically relevant effects and appear to be physiologically inconsequential. Such mutations are presumably invisible to natural selection. Finally, the various sickle-cell and thalassemia mutations provide rare and valuable examples of beneficial mutations (although the effects are best characterized as *conditionally* beneficial, given the well-documented trade-offs between malaria resistance and anemia).

Many important questions in evolutionary genetics can be addressed with information about the spectrum of phenotypic effects of spontaneous mutations in combination with information about the spectrum of effects of *substitutions*—that is, mutations that become "fixed" (attaining a frequency of 1.0 in the population, thereby supplanting the ancestral allele at the same site). Although only a minute fraction of spontaneous mutations are expected to have beneficial fitness effects, research on the molecular basis of Hb adaptations in a variety of vertebrate taxa is yielding valuable information about the nature of such mutations.

With the data that have been amassed, we are now in a position to address a long-standing question in evolutionary biology: Are particular mutations—or particular types of mutations—preferentially fixed during the course of adaptive protein evolution? This question is relevant to understanding the inherent repeatability (and, hence, predictability) of molecular adaptation. To address this and several other important corollary questions, we will first explore the relevant conceptual issues, and we will then attempt a synthesis of evolutionary lessons learned from experimental studies of vertebrate Hbs.

9.2.1 Predictions based on first principles

In a highly influential paper on biophysical mechanisms of protein evolution, Perutz (1983) predicted that adaptive changes in the functional properties of vertebrate Hb are typically attributable to "a few replacements at key positions." Perutz's argument was primarily based on a principled consideration of crystallographic models rather than experimental results, as relevant empirical data were quite sparse at the time. According to Perutz (1983), amino acid substitutions that could be expected to make especially important contributions to evolutionary changes in Hb-O_2 affinity involve heme-protein contacts (affecting intrinsic heme reactivity), intersubunit contacts (affecting the oxygenation-linked, allosteric transition in quaternary structure), and binding sites for allosteric effectors, which are mainly located at the N- and C-termini of the globin chains and in the central cavity of the Hb tetramer. A similar view was expressed by Riggs (1976): "A major part of the functional adaptations depends on only a small part of the hemoglobin. The NH_2- and COOH-terminal segments of the chains dominate in the control of functional properties." If Perutz and Riggs are correct that adaptive modifications of Hb function are typically attributable to a small number of substitutions at key positions, then the clear prediction is that the same restricted subset of causative mutations will be preferentially fixed in different species that independently evolved similar changes in Hb function. In other words, the hypothesis predicts that convergent and parallel amino acid substitutions should be pervasive. Before proceeding further, let us define these terms.

9.2.2 Molecular convergence and parallelism

In studies of organismal phenotypes, "convergence" generally suffices as a term to describe the independent acquisition of similar traits in different species (Arendt and Reznick, 2008). By contrast, the digital nature of molecular sequence data generally permits more refined inferences about the polarity of changes in character state, and it can be useful to make distinctions between different modes of replicated change. In comparisons between orthologous proteins from a given pair of species, convergent substitutions at a particular site refer to independent changes from

(A) Convergent substitutions   (B) Parallel substitutions

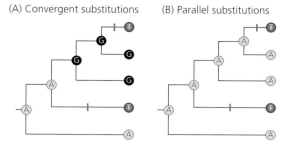

Fig. 9.1. Convergent and parallel substitutions: phylogenetically replicated changes that involve different mutational paths. (A) In comparisons among orthologous proteins from a given set of species, convergent substitutions at a particular site refer to independent changes from different ancestral amino acids to the same derived amino acid. In this case, there was a change from G (the ancestral state) to T (the derived state) in one species, and a change from A to T in another species. The convergent substitutions are denoted by hash-marks. (B) Parallel substitutions at a site refer to independent changes from the same ancestral amino acid to the same derived amino acid. In this case, changes from A to T occurred in two different species. The parallel substitutions are denoted by hash-marks. In sets of closely related species, parallelism is generally more common than convergence simply because—at any given site—close relatives will be more likely to share the same ancestral state prior to the occurrence of independent substitutions.

different ancestral amino acids to the same derived amino acid in both species (Fig. 9.1A), whereas parallel substitutions refer to independent changes from the same ancestral amino acid to the same derived amino acid (Fig. 9.1B) (Zhang and Kumar, 1997).

Strictly defined, convergent, and parallel substitutions involve the fixation of identical-by-state alleles in different lineages—the alleles have independent mutational origins. Comparative studies of naturally evolved proteins have documented several striking cases of convergence and parallelism at the amino acid level (Storz, 2016). In addition to the replicated fixation of identical-by-state alleles that have independent mutational origins, the sharing of identical-by-descent alleles between species may also be attributable to incomplete lineage sorting or introgressive hybridization (mentioned briefly in Chapter 8, section 8.4.1.1). In such cases, the shared alleles do not have independent mutational origins.

9.2.3 Forced options and causes of substitution bias

During the adaptive evolution of protein function, there are two main reasons we might expect evolu-

tion to repeatedly hit upon the same solutions to similar problems, resulting in convergent or parallel substitutions among species. First, it may be that there is simply a limited number of mutations capable of producing a given change in protein function, reflecting intrinsic constraints imposed by the nature of structure-function relationships. This could be considered the "forced option" hypothesis. Alternatively, there may be many possible mutations that can produce the requisite change in function (a many-to-one mapping of genotype to phenotype), but certain mutations or certain types of mutations are preferentially fixed. This substitution bias may be attributable to variation in rates of origin—that is, sites vary in rates of mutation to alleles that produce the beneficial change in phenotype (mutation bias)—or mutations with similar main effects on the selected phenotype may vary in their probability of fixation once they arise (fixation bias) due to variation in the magnitude of deleterious side effects. As a result of such side effects (or "mutational pleiotropy," whereby the same mutation affects multiple aspects of protein function), mutations with equivalent main effects on a positively selected phenotype may still be unequal in the eyes of natural selection (Streisfeld and Rausher, 2011). In the parlance of population genetics, such mutations will have different selection coefficients and, hence, different probabilities of fixation.

To formalize this reasoning, the probability of parallel substitution between two species, which both choose the next mutational step from the same distribution of n possible options (each with probability $p(i)$), is:

$$\Pr(//) = \sum_{i=1}^{n} p_i^2. \qquad (9.1)$$

It follows that

$$\Pr(//) = \frac{1}{n} + Vn, \qquad (9.2)$$

where n is the number of possible options and V is the variance in the probabilities of those options. Equivalently,

$$\Pr(//) = \frac{(C^2 + 1)}{n}, \qquad (9.3)$$

where C is the coefficient of variation in the probabilities. In biological terms, n is the number of possible beneficial mutations and C subsumes the effects of all mutational and selective factors that increase variability in the distribution of fixation probabilities of the n possible mutations. For a given n, the probability of parallel substitution is minimized if all n mutations occur at equal rates and if they all have identical selection coefficients; that is, $\Pr(//) = 1/n$ when $C = 0$. The probability of parallel substitution increases linearly with decreasing n, and it increases monotonically with increasing C.

How might we go about estimating these key parameters that influence the probability of parallel substitution? Experimental insights into the effective number of mutations that are capable of producing a given change in protein function can be obtained via reverse-genetic screens of engineered mutational libraries, and insights into the causes of substitution bias can be obtained via directed evolution experiments and/or comparative studies of naturally evolved proteins. As described next, comparative studies of orthologous proteins from multiple species can be used to assess evidence for the preferential fixation of particular types of mutation by testing whether average substitution rates are the same across sites (Streisfeld and Rausher, 2011).

9.2.3.1 Are some mutations preferentially fixed during the adaptive evolution of protein function?

To assess whether mutations that contribute to adaptive modifications of protein function represent a biased subset of all possible mutations that are capable of producing the same functional effect, we can test whether average substitution rates are the same for different mutation classes (e.g., mutations in the active site vs. mutations affecting protein allostery). In the simplest possible "origin-fixation" model of molecular evolution, the substitution rate is given by:

$$K = 2N\mu\lambda, \qquad (9.4)$$

where N is the size of a diploid population, μ is the per-copy rate of mutation, and λ is the probability that a new mutation becomes fixed once it has arisen. In this framework, we can specify the substitution rate as the product of the rate at which new alleles originate via mutation and the probability that they become fixed in the population once they arise (Orr, 1998, McCandlish and Stoltzfus, 2014). The origin-fixation formalism therefore describes a regime of mutation-limited evolution where the rate of evolution is directly proportional to the mutation rate.

An important implication is that a bias in mutation rates can produce a bias in substitution rates, even when the substitutions are driven by positive selection (Yampolsky and Stoltzfus, 2001, Stoltzfus, 2006, Stoltzfus and Yampolsky, 2009). In the absence of contributions from standing genetic variation, a bias in rates of origin affects the joint probability of origin and fixation. Since μ and λ can vary among different classes of mutations, site-specific substitution rates will vary accordingly, so the expected rate of substitution for mutations in mutation class i is $K_i = 2N\mu_i\lambda_i$. The class-specific mutation rate is $\mu_i = \mu\theta_i$, where μ is the overall rate of origin for mutations that produce a given change in phenotype and θ_i is the proportion of these mutations in class i, yielding $K_i = 2N\mu\theta_i\lambda_i$. Thus, the proportion of fixed mutations in class i is

$$r_i = \frac{\theta_i\lambda_i}{\sum_i \theta_i\lambda_i}. \qquad (9.5)$$

We can assess evidence for substitution bias by testing whether each mutation class contributes equally to evolutionary changes in a particular protein function; that is, we can test the null hypothesis that $r_i = 1/n$ for all i. In a comparison between two discrete mutation classes, i and j, substitution bias is indicated if the different types of mutation have unequal rates of origin ($\theta_i \neq \theta_j$) and/or unequal probabilities of fixation once they arise ($\lambda_i \neq \lambda_j$). The role of mutation bias can be assessed by testing the null hypothesis that $\theta_i = 1/n$ for all i. Using data from mutagenesis screens (or compilations of phenotypic data for human Hb mutants), the role of fixation bias can be assessed by testing the null hypothesis that the spectrum of spontaneous mutations that produce a given change in protein function is equal to the spectrum of substitutions that are responsible for evolutionary changes in protein function between species (Streisfeld and Rausher, 2011).

For comparisons between different classes of site (or different classes of mutational change or mutational effect), this framework provides a means of assessing the extent to which an observed substitution bias is attributable to biased mutation rates and/or biased fixation probabilities (Yampolsky and Stoltzfus, 2001, Stoltzfus and Yampolsky, 2009, Streisfeld and Rausher, 2011) and can therefore provide insights into why particular types of mutations are preferentially fixed.

9.2.3.2 Mutation bias as a cause of substitution bias

An important but underexplored question in evolutionary genetics concerns the extent to which mutation bias influences pathways of adaptive molecular evolution. In vertebrate genomes, transition:transversion bias results in especially high rates of change from one pyrimidine to another (C↔T) or from one purine to another (G↔A). In mammals and birds (and possibly other vertebrate groups), among-site variation in mutation rate is further augmented by CpG bias: If the DNA nucleotide cytosine (C) is immediately 5′ to guanine (G) on the same coding strand, forming a "CpG" dinucleotide, and if the C is methylated to form 5′-methylcytosine, then point mutations at both sites occur at an elevated rate relative to mutations at non-CpG sites. The genetic code determines how these biased rates of nucleotide change translate into propensities of amino acid change (Yampolsky and Stoltzfus, 2005).

Mutation bias can be an important orienting factor in both neutral and adaptive molecular evolution because an asymmetry in rates of origin can affect the joint probability of origin and fixation (Yampolsky and Stoltzfus, 2005, Stoltzfus, 2006, Stoltzfus and Yampolsky, 2009). Results of several experimental evolution studies have provided evidence that mutation bias can influence trajectories of adaptive protein evolution (Rokyta et al., 2005, Lozovsky et al., 2009, Weigand and Sundin, 2012, Wong et al., 2012, Couce et al., 2015, Stoltzfus and McCandlish, 2017), and in some cases the fixation probabilities of mutant alleles are more accurately predicted by site-specific mutation rates than by the magnitude of fitness effects (Rokyta et al., 2005).

Transition mutations at CpG dinucleotides also account for a disproportionate number of Hb pathologies involving both adult Hb (due to point mutations in the α- and/or β-globin genes) and fetal Hb (due to point mutations in the α- and/or γ-globin genes) (Perutz, 1990b). For example, Hb Köln (β98Val→Met), a highly pathogenic mutant associated with hemolytic anemia, is caused by a C→T transition mutation in a methylated CpG dinucleo-

tide in the adult β-globin gene. Relative to Hb pathologies caused by non-CpG mutations, Hb Köln has an unusually high rate of incidence.

9.2.3.3 Fixation bias as a cause of substitution bias: the role of mutational pleiotropy

Amino acid mutations commonly have pleiotropic effects on protein biochemistry as they can simultaneously affect multiple aspects of molecular function, structural stability, folding, solubility, and propensity for aggregation (Wang et al., 2002, DePristo et al., 2005, Bloom et al., 2006, Tokuriki et al., 2008, Harms and Thornton, 2013). Consequently, mutations that improve one aspect of protein function may simultaneously compromise other structural or functional properties. Within the set of mutations that have functionally equivalent effects on a selected biochemical phenotype, those that incur a lesser magnitude of deleterious pleiotropy should have a higher fixation probability (Otto, 2004, Chevin et al., 2010), and may therefore make a disproportionate contribution to phenotypic evolution.

Consider the increased Hb-O_2 affinities that have evolved independently in different high-altitude mammals and birds (Chapter 8). If similar increases in Hb-O_2 affinity can be produced by numerous possible mutations—involving different structural or functional mechanisms, but achieving equally serviceable results—then it would be surprising if replicated increases in Hb-O_2 affinity in different species were consistently caused by substitutions at a restricted subset of all possible sites. Mutagenesis experiments on recombinant human Hb have successfully adjusted O_2 affinity over a 100-fold range and functional studies of naturally occurring mutant Hbs have demonstrated that there are numerous possible mutations that can produce identical increases in Hb-O_2 affinity (Dickerson and Geis, 1983, Bellelli et al., 2006, Olson and Maillett, 2005, Maillett et al., 2008, Varnado et al., 2013). However, many affinity-altering mutations have deleterious pleiotropic effects on other aspects of protein structure or function. For example, "active site" mutations that alter the polarity or hydrophobicity of the distal heme pocket can produce significant changes in ligand-affinity, but such mutations can increase susceptibility to spontaneous heme oxidation ("autoxidation"). As stated by Riggs (1976), "any change in the protein that would enhance the rate of autoxidation would presumably be selected against." This is because oxidation of the ferrous (Fe^{2+}) heme iron to the ferric state (Fe^{3+}) releases superoxide (O_2^-) or perhydroxy ($HO_2 \bullet$) radical, and prevents reversible Fe-O_2 binding, rendering Hb functionally inert as an O_2-transport molecule. Moreover, heme oxidation promotes denaturation of the globin chains and propagates oxidative reactions that are damaging to the cell (Alayash et al., 2001, Olson and Maillett, 2005, Reeder and Wilson, 2005, Reeder, 2010, Bonaventura et al., 2013, Varnado et al., 2013).

A similar trade-off is associated with mutations that produce an increased Hb-O_2 affinity by suppressing sensitivity to allosteric effectors. Such mutations sacrifice an important mechanism of phenotypic plasticity because changes in red cell pH or concentrations of organic phosphates are rendered less effective as a means of modulating blood-O_2 affinity in response to changes in environmental O_2 availability and/or internal metabolic demands (Storz et al., 2010a, Storz et al., 2010b). The point is that numerous amino acid mutations could have identical main effects on a positively selected phenotype, such as Hb-O_2 affinity, but they may still have unequal selection coefficients due to variation in the magnitude of deleterious pleiotropy.

9.3 Intramolecular epistasis

9.3.1 Non-uniformity of mutational effects at orthologous sites

Protein engineering studies have documented that the functional effects of amino acid mutations often depend on the sequence context in which they occur, a phenomenon known as epistasis (Phillips, 2008). In extreme cases, this context dependence involves the sign of a mutation's effect ("sign epistasis") such that the same mutation can have opposite phenotypic effects on different genetic backgrounds (Weinreich et al., 2005). An important implication of sign epistasis is that evolved changes in sequence context can reduce the number of site-specific amino acid states that are unconditionally acceptable in the divergent backgrounds of orthologous proteins (Fig. 9.2). Thus, in comparisons between orthologous proteins in different species, sign epistasis for fitness should typically *decrease* the probability of convergence and parallelism because a mutation that has a beneficial effect on the genetic

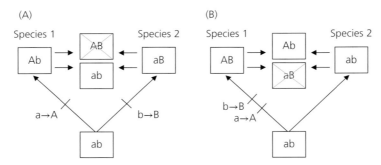

Fig. 9.2. Amino acid states that are allowable in one genetic background may be deleterious in other backgrounds. (A) Two species diverge from a common ancestor with the two-site genotype, *ab*. The substitution *a*→*A* occurs at the first site in species 1 (yielding *Ab*) and the substitution *b*→*B* occurs at the second site in species 2 (yielding *aB*). A negative epistatic interaction ("Dobzhansky–Müller incompatibility") is revealed by moving mutation *A* from species 1 into the orthologous background of species 2, or by moving mutation *B* from species 2 into the orthologous background of species 1. Mutations *A* and *B* are individually neutral on the genetic backgrounds in which they occurred during evolution, but they are deleterious in combination. Swapping mutations *a* or *b* to form genotype *ab* results in a reversion to the ancestral state. (B) The same type of incompatibility can arise if both substitutions occur in one lineage, while the other species retains the ancestral states at both sites. In this case, swapping mutations *a* or *B* yields a low-fitness genotype (*aB*). Swapping mutations *A* or *b* yields genotype *Ab*, one of two possible mutational intermediates in the ancestry of species 1. Note that since mutations *a* and *B* are deleterious in combination, substitution *a*→*A* must have preceded substitution *b*→*B* in the ancestry of species 1 because *Ab* is the only viable single-mutant intermediate connecting the ancestral (*ab*) and descendant (*AB*) genotypes. Modified from Storz (2018).

background of one species may have a neutral or deleterious effect on the divergent genetic backgrounds of other species (Storz, 2016). Adaptive convergence and parallelism would only be expected for (non-epistatic) mutations that retain their beneficial effects across all backgrounds.

Consider a pair of parallel substitutions that occur at the same site in orthologous proteins of two closely related species. If orthologs of the two species are identical or nearly identical in sequence, then the parallel substitution will likely have similar phenotypic effects. However, if the same substitution occurs in a more distantly related species, then it will be more likely to have a phenotypic effect that is different in magnitude or sign, simply because there would be more opportunity for divergence in sequence context (or, more specifically, there would be more opportunity for divergence at sites that epistatically interact with the focal residue). Convergent or parallel substitutions at orthologous sites in different species may have different fitness effects due to lineage-specific changes in the fitness landscapes of individual residue positions (Fig. 9.3A). Since a given site can be occupied by up to twenty different amino acids, the set of fitness values conferred by each possible variant defines a vector of site-specific amino acid propensities for a

given genetic background (Bazykin 2015, Shah et al., 2015). This single position fitness landscape can be considered a cross-section of the complete fitness landscape (Fig. 9.3B), and can change through time due to changes in the environment and/or changes in genetic background (Fig. 9.3C). In addition to reducing the probability of convergent and parallel substitutions at orthologous sites, context-dependent fitness effects of mutations can also make substitutions conditionally irreversible (Rogozin et al., 2008, Bridgham et al., 2009, Povolotskaya and Kondrashov, 2010, Naumenko et al., 2012, Pollock et al., 2012, Soylemez and Kondrashov, 2012, Kaltenbach et al., 2015, Shah et al., 2015). Even if a given substitution was neutral or nearly neutral when it initially occurred, mutational reversions to the ancestral state may eventually become maladaptive due to subsequent changes in sequence context—a phenomenon termed "entrenchment" (Shah et al., 2015).

These considerations suggest that epistasis may often reduce probabilities of molecular convergence, parallelism, and reversal because it reduces the number of possible mutations that have unconditionally acceptable effects in divergent genetic backgrounds (Storz, 2016). As explained next, research on compensatory mutations has provided

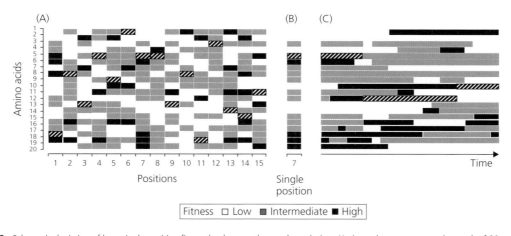

Fig. 9.3. Schematic depiction of how single-position fitness landscapes change through time. Horizontal rows correspond to each of 20 possible amino acids at each site in a protein. (A) At each site, the currently predominant amino acid (shown in cross-hatching) confers highest fitness. (B) The single-position fitness landscape of site 7, shown in isolation. (C) Temporal change in the single-position fitness landscape. The relative fitness of different amino acids at the position changes through time due to changes in the genetic background (i.e., substitutions at other sites) and/or changes in the environment. The depicted changes in fitness are modelled as a Poisson process.

Modified from Bazykin (2015) with permission from the Royal Society.

strong evidence for the pervasiveness of such context-dependent fitness effects.

9.3.2 Compensatory substitutions

Pleiotropic trade-offs can give rise to a context-dependence of mutational effects, as evidenced by cases where the fitness impact of a given mutation is determined by compensatory (conditionally beneficial) mutations at other sites in the same protein. For example, the selective fixation of function-altering mutations that confer a net fitness benefit may select for compensatory mutations to mitigate deleterious pleiotropic effects of the functional change (DePristo et al., 2005, Poon and Chao, 2005, Poon and Chao, 2006). Alternatively, function-altering mutations may be selectively permissible only on a background in which the requisite compensatory (or "permissive") mutations have already occurred (Bloom et al., 2005, Bloom et al., 2006, Bloom et al., 2007, Ortlund et al., 2007, Bloom et al., 2010, Gong et al., 2013). These compensatory mutations are neutral or deleterious on their own—they are only beneficial in combination with the function-altering mutation. An important implication is that the evolution of adaptive changes in protein function may be facilitated by neutral mutations that confer no benefit when they first arise but which lay the groundwork for subsequent function-altering mutations.

9.4 Testing for evidence of adaptive molecular convergence and parallelism

Convergent or parallel substitutions can occur by chance since each amino acid site in a protein has only twenty possible character states. In practice, a far more restricted number of amino acids can generally be tolerated at any given residue position, so the effective number of possible character states will typically be far less than 20 (Bazykin et al., 2007, Rokas and Carroll, 2008, Pollock et al., 2012, Goldstein et al., 2015, Usmanova et al., 2015, Zou and Zhang, 2015). Many parallel sequence changes in protein evolution involve neutral or nearly neutral back-and-forth exchanges between physico-chemically similar amino acids that are specified by mutationally adjacent codons (Bazykin et al., 2007,

Rokas and Carroll, 2008). Since nonrandom patterns of molecular convergence and parallelism can be produced by mutation bias and/or purifying selection that discriminates among a restricted number of exchangeable amino acids, clear evidence is required to invoke positive selection as a cause of such patterns.

Zhang (2003) listed four requirements for establishing that convergent or parallel substitutions are responsible for adaptive convergence in protein function: (1) replicated substitutions are observed in independent lineages; (2) the proteins under investigation have independently evolved derived changes in function; (3) replicated substitutions are responsible for the convergent changes in protein function; and (4) the number of replicated substitutions is greater than expected by chance alone. Most claims of adaptive molecular convergence and parallelism satisfy one or two of these requirements, and the third requirement on the list (establishing a causal connection between change in sequence and change in phenotype) is the one that most often remains unfulfilled because it requires experimental data on the functional effects of individual substitutions.

The comparative analyses of mammalian and avian Hbs highlighted in Chapter 8 provides a rich trove of experimental data for addressing questions about the predictability of molecular adaptation and the causes of substitution bias. This is because the original studies integrated functional analyses of native and recombinantly expressed Hbs with evolutionary analyses of sequence data, and site-directed mutagenesis experiments were used to measure the additive and nonadditive effects of specific mutations. Let us now synthesize these results and summarize the evolutionary lessons they teach.

9.5 Evolutionary lessons

9.5.1 Predictable convergence in Hb function has unpredictable molecular underpinnings

"We do not ask often enough why natural selection had homed in upon this *particular* optimum – and not another among a set of unrealized alternatives. In other words, we are dazzled by good design and therefore stop our inquiry too soon when we have answered, 'How does this feature work so well?' – when we should also be asking the historian's questions: 'Why *this* and not *that*?' or 'Why *this* over here, and *that* in a related creature living elsewhere?'"   —**Gould (1995)**

A long-standing question in evolutionary genetics concerns the extent to which adaptive phenotypic convergence is attributable to repeatable mutational changes at the molecular sequence level. One especially powerful means of addressing this question is to examine phylogenetically replicated changes in protein function that can be traced to specific amino acid substitutions. Using the experimental data summarized in Chapter 8, we can use this approach to examine replicated increases in Hb-O_2 affinity in bird species that independently colonized high-altitude environments. Having documented that high-altitude taxa have convergently evolved derived increases in Hb-O_2 affinity, we can assess the extent to which such changes are attributable to parallel amino acid substitutions. As we saw in Chapter 6, most bird species coexpress two functionally distinct Hb isoforms during adult life, the major HbA ($\alpha^A_2\beta^A_2$) and the minor HbD ($\alpha^D_2\beta^A_2$). Comparative sequence data for the set of seventy taxa shown in Fig. 8.10 revealed phylogenetically replicated replacements at numerous sites in the α^A- and α^D-globin genes (affecting HbA and HbD, respectively) and in the β^A-globin gene (affecting both HbA and HbD) (Projecto-Garcia et al., 2013, Natarajan et al., 2015b, Natarajan et al., 2016, Zhu et al., 2018). Although numerous parallel and convergent substitutions were observed, functional data from native Hb variants and engineered, recombinant Hb mutants revealed that only a subset of replicated replacements actually contributed to convergent increases in Hb-O_2 affinity in the different highland taxa. These included two replicated replacements in α^A-globin (α^AA34T and α^AP119A) and four in β^A-globin (β^AN/G83S, β^AA86S, β^AD94E, and β^AA116S) (Projecto-Garcia et al., 2013, Natarajan et al., 2015b, Natarajan et al., 2016, Zhu et al., 2018) (Fig. 9.4). Of these replacements, α^A34, α^A119, and β^A116 are located in the $\alpha_1\beta_1/\alpha_2\beta_2$ intradimer interface (which, in contrast to the $\alpha_1\beta_2/\alpha_2\beta_1$ interdimer interface, does not play a direct role in mediating the allosteric T↔R conformational switch) and β^A94 plays a key role in allosteric proton binding in human Hb. With the exception of β^A94, none of the

other replicated replacements—nor any of the non-replicated affinity-enhancing replacements—involved heme contacts, intersubunit contacts, or canonical binding sites for allosteric effectors (i.e., "sites that dominate in the control of functional properties" (Riggs, 1976)). In general, convergent increases in Hb-O_2 affinity were attributable to non-replicated replacements and/or replacements at sites that are not considered "key residues" according to the criteria of Riggs and Perutz. Clearly, evolutionary increases in Hb-O_2 affinity can be produced by amino acid replacements at numerous sites, and the effects are often subtle and indirect.

If fixation probabilities of affinity-enhancing mutations are influenced by the magnitude of deleterious pleiotropy, the expectation is that the spectrum of substitutions that produce evolutionary increases in Hb-O_2 affinity will be distinct from the spectrum of affinity-enhancing spontaneous mutations. We therefore want to know what properties distinguish actualized solutions from the much larger universe of non-actualized possibilities. Although available data do not permit a full quantitative assessment within the framework outlined in section 9.2.3.1, the data reveal a clear pattern related to trade-offs between Hb-O_2 affinity and allosteric regulatory capacity (i.e., the extent to which Hb-O_2 affinity can be reversibly modulated by adjustments in the concentration of allosteric effectors). Numerous spontaneous mutations in human Hb produce an increased O_2 affinity via suppression of phosphate sensitivity (Wajcman and Galacteros, 1996, Nagel, 2001, Wajcman and Galacteros, 2005, Steinberg and Nagel, 2009, Thom et al., 2013) (see Table 4.6). In most cases, phosphate binding is impaired by adding a negative charge in the central cavity (Hbs Shepherds Bush, Ohio) or by eliminating a positive charge (Hbs Rahere, Helsinki, Little Rock, Syracuse). As mentioned in Chapter 8, in all cases where high-altitude avian taxa evolved increased Hb-O_2 affinities, the changes were exclusively attributable to differences in intrinsic O_2 affinity rather than differences in sensitivity to Cl^- ions or IHP. Thus, in contrast to the deleterious pleiotropic effects documented for many affinity-enhancing mutations in human Hb, the mutations that have contributed to evolutionary changes in Hb-O_2 affinity in high-altitude birds do not compromise cooperativity or allosteric

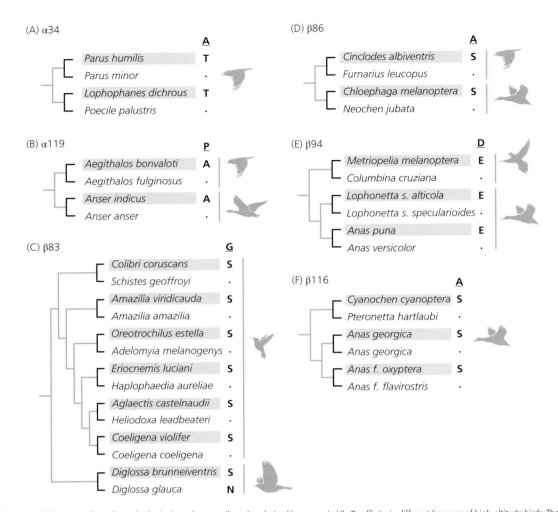

Fig. 9.4. Phylogenetically replicated substitutions that contributed to derived increases in Hb-O$_2$ affinity in different lineages of high-altitude birds. The select pairs of high- and low-altitude taxa are taken from the set of 70 taxa shown in Fig. 8.10. In each case, the high-altitude taxon is denoted by shading and the inferred ancestral amino acid is shown as the reference state. (A) Parallel substitutions at α34 in two high-altitude passerines from the Sino-Himalayan region. Paired comparisons between high- and low-altitude taxa involved ground tit, *Parus humilis* (high) vs. oriental tit, *P. minor* (low), and gray-crested tit, *Lophophanes dichrous* (high) vs. *Poecile palustris* (low) (Zhu et al., 2018). (B) Parallel substitutions at α119 in high-altitude passerine and waterfowl species from the Sino-Himalayan region. Comparisons involved black-browed bushtit, *Aegithalos bonvaloti* (high) vs. sooty bushtit, *A. fulginosus* (low), and bar-headed goose, *Anser indicus* (high) vs. greylag goose, *Anser anser* (low) (Natarajan et al., 2018, Zhu et al., 2018). (C) Parallel and convergent substitutions at β83 in high-altitude hummingbird and passerine species from the Andes. Comparisons involved sparkling violetear, *Colibri coruscans* (high) vs. Eastern wedge-billed hummingbird, *Schistes geoffroyi* (low), green-and-white hummingbird, *Amazilia viridicauda* (high) vs. amazilia hummingbird, *A. amazilia* (low), Andean hillstar, *Oreotrochilus estella* (high) vs. speckled hummingbird, *Adelomyia melanogenys* (low), sapphire-vented puffleg, *Eriocnemis luciani* (high) vs. greenish puffleg, *Haplophaedia aurelieae* (low), white-tufted sunbeam, *Aglaeactis castelnaudii* (high) vs. violet-fronted brilliant, *Heliodoxa leadbeateri* (low), violet-throated starfrontlet, *Coeligena violifer* (high) vs. bronzy inca, *C. coeligena* (low), and black-throated flowerpiercer, *Diglossa brunneiventris* (high) vs. deep-blue flowerpiercer, *Diglossa glauca* (low) (Projecto-Garcia et al., 2013, Natarajan et al., 2016). Note that the inferred ancestral state for β83 of hummingbirds is glycine (G), but the inferred ancestral state for this same site in *Diglossa* is asparagine (N). Thus, the βN83S substitution in *Diglossa brunneiventris* species represents a convergent substitution relative to the hummingbird clade. (D) Parallel βA86S substitutions in high-altitude suboscine passerines and waterfowl species from the Andes. Comparisons involved cream-winged cinclodes, *Cinclodes albiventris* (high) vs. pale-legged hornero, *Furnarius leucopus* (low), and Andean goose, *Chloephaga melanoptera* (high) vs. Orinoco goose, *Neochen jubata* (low) (Natarajan et al., 2015b, Natarajan et al., 2016). (E) Parallel βD94E substitutions in high-altitude doves and waterfowl species from the Andes. Comparisons involved black-winged ground-dove, *Metriopelia melanoptera* vs. croaking ground-dove, *Columbina cruziana*, Andean crested duck, *Lophonetta s. alticola* (high) vs. Patagonian crested duck, *L. s. specularioides* (low), and Puna teal, *Anas puna* (high) vs. silver teal, *A. versicolor* (low) (Natarajan et al., 2015b, Natarajan et al., 2016). (F) Replicated βA116S substitutions in high-altitude waterfowl. Comparisons involve Abyssinian blue-winged goose, *Cyanochen cyanoptera* (high) vs. Hartlaub's duck, *Pteronetta hartlaubi* (low), and high- vs. low-altitude populations of yellow-billed pintail, *Anas georgica*, and sharp-winged teal, *Anas flavirostris*. In the latter two taxon pairs, the derived Ser β116 variants do not have independent mutational origins in the highland taxa; the allele-sharing is attributable to a history of introgressive hybridization (Natarajan et al., 2015b).

regulatory capacity. The deleterious pleiotropy hypothesis predicts that such mutations should make disproportionate contributions to evolutionary changes in Hb-O_2 affinity (Otto, 2004, Streisfeld and Rausher, 2011), and when such changes are driven by positive directional selection, theory predicts that they are especially likely to evolve in parallel (Orr, 2005b, Unckless and Orr, 2009, Chevin et al., 2010).

What do the results of these comparative studies tell us about the repeatability and predictability of molecular adaptation? Where do we place the boundary between "predictability under invariant law and the multifarious possibilities of historical contingencies" (Gould, 1989)? In other words, at what scale or level of focus can evolution be characterized as "predictable"? Results of research on Hb evolution in high-altitude birds have exposed a clear demarcation between the realms of chance and necessity at different hierarchical levels. Experimental data for high-altitude birds from the Andes, Ethiopian Plateau, and the Tibetan Plateau document a striking pattern of convergence in biochemical phenotype, and in the underlying functional mechanism: high-altitude taxa generally have evolved derived increases in Hb-O_2 affinity, and these changes are invariably attributable to changes in intrinsic O_2 affinity (either via changes in the equilibrium constants of O_2 binding to ^{T}Hb or ^{R}Hb or via changes in the allosteric equilibrium constant for the oxygenation-linked transition between the T and R quaternary structures). At the level of biochemical phenotype and functional mechanism, the pattern of evolutionary change is highly predictable. However, convergent changes in protein function were generally not attributable to convergent or parallel changes at the amino acid level. With a few notable exceptions (Fig. 9.4), convergent increases in Hb-O_2 affinity were mainly attributable to divergent substitutions. So at the molecular sequence level, predictability breaks down—evolutionary outcomes are far more idiosyncratic than predicted by Perutz's hypothesis.

Parallel amino acid substitutions were not uncommon in the set of examined bird species, but only a small fraction of them contributed to convergent increases in Hb-O_2 affinity in high-altitude taxa and most were inconsequential with respect to oxygenation properties (Natarajan et al., 2015b, Natarajan

et al., 2016, Zhu et al., 2018). Thus, another important object lesson is that molecular parallelism should not be interpreted as *prima facie* evidence for a history of positive selection. Insights into the adaptive significance of observed sequence changes require experimental tests of functional effects.

9.5.2 Mutation bias may influence pathways of adaptive change

In studies of Hb evolution in high-altitude birds, site-directed mutagenesis experiments involving recombinant Hb mutants have documented multiple cases in which non-synonymous mutations at CpG dinucleotides contributed to presumably adaptive increases in Hb-O_2 affinity: β^AI55V in Andean house wrens (*Troglodytes aedon*) (Galen et al., 2015), and parallel α^AA34T substitutions in two different passerine species that are native to the Tibetan Plateau, the ground tit (*Parus humilis*) and the grey crested tit (*Lophophanes dichrous*) (Zhu et al., 2018). In each of these three highland taxa, there seems little reason to suppose that the causative amino acid mutations would have had larger selection coefficients (and, hence, higher fixation probabilities) than any number of other possible mutations that could have produced the same increase in Hb-O_2 affinity. However, if the rate of CpG mutation occurs at a higher rate than non-CpG mutations, then—in the absence of contributions from standing variation—the bias in mutation rate is expected to influence evolutionary outcomes in the same way as a commensurate bias in fixation probability (Yampolsky and Stoltzfus, 2001, Stoltzfus and Yampolsky 2009, McCandlish and Stoltzfus 2014). These case studies of avian Hb suggest that mutation bias may influence which mutations are most likely to contribute to molecular adaptation.

9.5.3 Compensatory substitutions may play an important role in adaptive protein evolution

During adaptive protein evolution, some fraction of selectively fixed mutations will be directly causative (contributing to adaptive improvement in the selected property) and some may be purely compensatory (alleviating problems that were created by initial attempts at solution). The relative contributions of these two classes of substitution depend on the prevalence and magnitude of antagonistic pleiotropy. If mutations that produce an adaptive improvement in one trait have adverse effects on other traits, then the fixation of such mutations will select for compensatory mutations to mitigate the deleterious side effects, and evolution will proceed as a "two steps forward, one step back" process (Otto, 2004).

To investigate the nature of adaptive substitutions and their pleiotropic effects, Natarajan et al. (2018) used a protein engineering approach to characterize the molecular basis of Hb adaptation in the high-flying bar-headed goose (*Anser indicus*). This hypoxia-tolerant species is renowned for its trans-Himalayan migratory flights (Hawkes et al., 2011, Hawkes et al., 2013, Bishop et al., 2015), and its elevated Hb-O_2 affinity is thought to make a key contribution to its capacity for powered flight at extreme elevations of 6,000–9,000 m above sea level (Petschow et al., 1977, Black and Tenney, 1980, Scott and Milsom, 2006, Scott, 2011, Scott et al., 2015). The major HbA isoform of the bar-headed goose has a significantly higher O_2 affinity than that of the closely related greylag goose (*Anser anser*), a strictly lowland species (Petschow et al., 1977, Rollema and Bauer, 1979, Jendroszek et al., 2018). The HbA isoforms of the two species differ at five amino acid sites: three in the α^A subunit and two in the β^A subunit. Of these five amino acid differences, Perutz (1983) predicted that the Pro→Ala replacement at $\alpha^A 119$ is primarily responsible for the adaptive increase in Hb-O_2 affinity in bar-headed goose. This site is located at an intradimer ($\alpha_1\beta_1/\alpha_2\beta_2$) interface where the ancestral Pro $\alpha 119$ forms a van der Waals contact with Met $\beta 55$ on the opposing subunit of the same $\alpha\beta$ dimer. Perutz predicted that the single $\alpha P119A$ substitution eliminated this intradimer contact, thereby destabilizing the T-state and shifting the conformational equilibrium in favor of the high-affinity R state. Jessen et al. (1991) and Weber et al. (1993) tested Perutz's hypothesis using site-directed mutagenesis of recombinant human Hb, and their experiments confirmed the predicted mechanism.

As a result of these experiments, bar-headed goose Hb is often held up as an example of a biochemical adaptation that is attributable to a single, large-effect substitution. However, several key questions were left unanswered: Do the other substitutions also contribute to the change in Hb-O_2 affinity? If not, do they compensate for deleterious pleiotropic effects of the affinity-enhancing $\alpha P119A$ substitution? To address these questions, Natarajan et al. (2018) used a protein engineering approach involving ancestral protein resurrection. Specifically, they statistically reconstructed the α^A and β^A sequences of the most recent common ancestor of bar-headed goose and its closest living relatives, all of which are lowland species in the genus *Anser*. This reconstruction step involves estimation of ancestral sequences using an alignment of orthologous sequences from contemporary species, an estimated phylogeny of those sequences, and a model of codon substitution (Hochberg and Thornton, 2017). The resurrection step involves the synthesis of plasmid constructs containing the estimated ancestral sequences and the subsequent expression and purification of the recombinant ancestral protein, which is then available for *in vitro* experiments. The value of ancestral protein resurrection is that it permits an assessment of the effects of historical substitutions on the genetic background in which they actually occurred during evolution (Harms and Thornton, 2010, Hochberg and Thornton, 2017).

Of the five substitutions that distinguish bar-headed goose and greylag goose Hbs, the ancestral sequence reconstructions revealed that each of the three α^A substitutions occurred in the bar-headed goose lineage and the two β^A substitutions occurred in the greylag goose lineage (Fig. 9.5). O_2-equilibrium experiments confirmed that the wildtype Hb of the bar-headed goose has a higher intrinsic O_2 affinity than that of the greylag goose, a difference that persisted in the presence of anions (Fig. 9.6A). The triangulated comparison involving recombinant Hbs from the two contemporary species (bar-headed goose and greylag goose) and their reconstructed ancestor ("AncAnser") revealed that the observed difference in Hb-O_2 affinity between the bar-headed goose and greylag goose is mainly attributable to a derived increase in affinity in the bar-headed goose lineage. Kinetic measurements demonstrated that the increased O_2 affinity of bar-headed goose rHb is associated with a lower apparent rate of O_2 dissociation, k_{off}, relative to the rHbs of both the greylag goose and AncAnser (Fig. 9.6B).

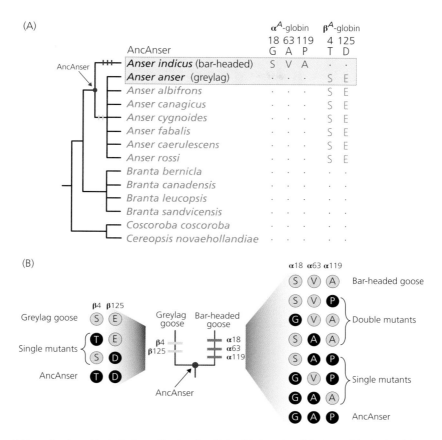

Fig. 9.5. Inferred history of amino acid substitution at five sites that distinguish the major Hb isoforms of the bar-headed goose (*Anser indicus*) and greylag goose (*Anser anser*). (A) Amino acid states at the same sites are shown for twelve other waterfowl species in the subfamily Anserinae. Of the five amino acid substitutions that distinguish the Hbs of *A. indicus* and *A. anser*, parsimony indicates that three occurred on the branch leading to *A. indicus* (αG18S, αA63V, and αP119A) and two occurred on the branch subtending the clade of all *Anser* species other than *A. indicus* (βT4S and βD125E). "AncAnser" represents the reconstructed sequence of the *A. indicus/A. anser* common ancestor, which is also the most recent common ancestor of all extant species in the genus *Anser*. (B) Triangulated comparisons involving rHbs of bar-headed goose, greylag goose, and their reconstructed ancestor (AncAnser) reveal the polarity of changes in character state. Differences in Hb function between the bar-headed goose and AncAnser reflect the net effect of three substitutions (αG18S, αA63V, and αP119A) and differences between the greylag goose and AncAnser reflect the net effect of two substitutions (βT4S and βD125E). All possible mutational intermediates connecting AncAnser with each of the two descendent species are shown to the side of each terminal branch (the sequential order of the substitutions is unknown, so the order in which they are shown on each terminal branch is arbitrary).

Reproduced from Natarajan et al. (2018).

To test the functional effects of the bar-headed goose substitutions, Natarajan et al. (2018) synthesized all possible genotypic intermediates in the line of descent connecting the wildtype bar-headed goose genotype with the most recent common ancestor of bar-headed goose and its lowland relatives. Thus, each of the three bar-headed goose substitutions were tested in each possible multi-site combination. These site-directed mutagenesis

experiments confirmed that αP119A represents a major-effect mutation that significantly increased Hb-O_2 affinity on all possible backgrounds (Fig. 9.7). Analysis of the crystal structure of bar-headed goose Hb revealed that this mutation has very little effect on the main-chain formation and appears to exert its functional effect via the elimination of side chain contacts and increased backbone flexibility. The other two other mutations (αG18S and αA63V)

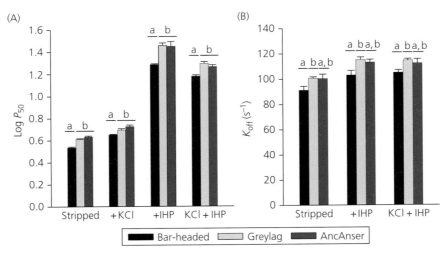

Fig. 9.6. Bar-headed goose evolved an increased Hb-O_2 affinity relative to greylag goose and their reconstructed ancestor, AncAnser. Triangulated comparisons of purified rHbs involved diffusion-chamber measurements of O_2 equilibria (A) and stopped-flow measurements of O_2 dissociation kinetics (B). O_2 affinities (P_{50}, torr; ± 1 SEM) and dissociation rates (k_{off}, $M^{-1}s^{-1}$; ±1 SEM) of purified rHbs were measured at pH 7.4, 37°C, in the absence (stripped) and presence of allosteric effectors ([Cl^-], 0.1 M; [HEPES], 0.1 M; IHP/Hb tetramer ratio = 2.0; [heme], 0.3 mM in equilibrium experiments; [Cl^-], 1.65 mM; [HEPES], 200 mM; IHP/Hb tetramer ratio = 2.0; [heme], 5 μM in kinetic experiments). Letters distinguish measured values that are significantly different ($P < 0.05$).

Reproduced from Natarajan et al. (2018).

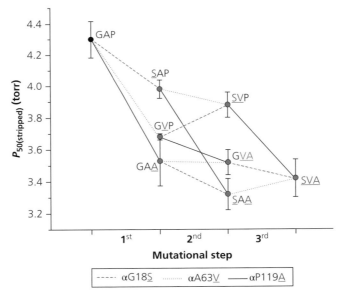

Fig. 9.7. Trajectories of change in intrinsic Hb-O_2 affinity (indexed by P_{50}, torr) in each of six possible forward pathways that connect the ancestral "AncAnser" genotype (GAP) and the wildtype genotype of bar-headed goose (SVA). Derived amino acid states are indicated by underlined letters. Error bars denote 95% confidence intervals.

Reproduced from Natarajan et al. (2018).

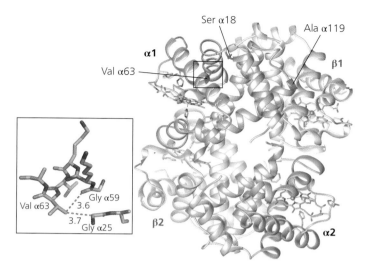

Fig. 9.8. Structural model showing bar-headed goose Hb in the deoxy state (PDB1hv4), along with locations of each of the three amino substitutions that occurred in the bar-headed goose lineage after divergence from the common ancestor of other *Anser* species. The inset graphic shows the environment of the Val α63 residue. When valine replaces the ancestral alanine at this position, the larger volume of the side chain causes minor steric clashes with two neighboring glycine residues, Gly α25 and Gly α59. The distances between non-hydrogen atoms (depicted by dotted lines) are given in Å.

Reproduced from Natarajan et al. (2018). See Plate 14.

exhibited smaller average effect sizes and less additivity across backgrounds. With regard to αA63V, the introduction of the valine side chain causes minor steric clashes with Gly 25 and Gly 59 of the same subunit (Fig. 9.8, Plate 14). This interaction may alter O_2 affinity by impinging on the neighboring His α58 (the distal histidine) that stabilizes the α-heme Fe-O_2 bond.

Although mutational changes in Hb-O_2 affinity were not significantly associated with changes in autoxidation rate or other examined functional and structural properties, the mutagenesis experiments revealed a striking compensatory interaction between the mutations at α18 and α63, residues that are located within 7 Å of one another. The αA63V mutation produced a greater than twofold increase in the autoxidation rate on backgrounds in which the ancestral Gly was present at α18, and the effect was completely compensated by the polarity-changing αG18S mutation (Fig. 9.9). This compensatory interaction suggests that the αG18S mutation may have been fixed by selection not because it produced a beneficial main effect on Hb-O_2 affinity, but because it mitigated deleterious pleiotropic effects of the affinity-altering αA63V mutation.

Alternatively, if αG18S preceded αA63V during evolution of bar-headed goose Hb, then the (conditionally) deleterious side effect of Aα63V would never have been manifest. These findings suggest that compensatory interactions may play an important role in adaptive protein evolution due to trade-offs between different functional properties.

9.5.4 Segregating variants within species are not qualitatively or quantitatively distinct from fixed differences between species

A fundamental question in evolutionary biology concerns the extent to which allelic variation within species can be extrapolated to explain phenotypic divergence between species. The comparative studies of Hb function in high- and low-altitude mammals and birds discussed in Chapter 8 involved pairwise comparisons between closely related species, as well as comparisons between conspecific populations. Thus, the experimental data can be used to test the hypothesis that the mechanisms underlying observed differences in

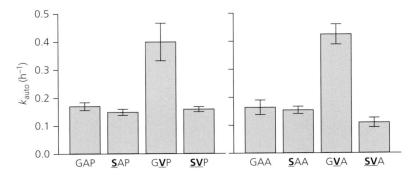

Fig. 9.9. Compensatory interaction between spatially proximal α-chain residues in bar-headed goose Hb. The mutation αA63V produces a more than twofold increase in autoxidation rate (k_{auto}; ±1 SEM) on genetic backgrounds with the ancestral Gly at residue position α18. This effect is fully compensated by αG18S, as indicated by two double-mutant cycles (A and B) in which mutations at both sites are tested individually and in pairwise combination.

Reproduced from Natarajan et al. (2018).

Hb-O_2 affinity are qualitatively different in comparisons between species and between locally adapted populations of the same species. Stern and Orgogozo (2008) hypothesized that segregating alleles that contribute to evolved differences between conspecific populations may generally be characterized by a greater degree of antagonistic pleiotropy compared to alleles that contribute to phenotypic divergence between species. This is because locally adaptive alleles that are conditionally deleterious are unlikely to go to fixation at the species level.

In the case of comparative data for high- and low-altitude birds, a parsing of the data summarized in Fig. 8.12 reveals that altitude-related differences in Hb-O_2 affinity were far more common in comparisons between species than in comparisons between conspecific populations. This is not surprising; at the intraspecific level, local adaptation to different elevational zones will typically be constrained by countervailing gene flow. However, in the few cases in which highland populations evolved higher Hb-O_2 affinities than lowland conspecifics (e.g., HbA and HbD of house wrens [*Troglodytes aedon*], crested ducks [*Lophonetta specularioides*], speckled teal [*Anas flavirostris*], and HbA in the case of cinnamon teal [*Anas cyanoptera*] and yellow-billed pintails [*Anas georgica*]), the nature of the allelic difference in Hb-O_2 affinity was not qualitatively or quantitatively distinct from the patterns observed for interspecific comparisons.

In the case of mammals, the geographically widespread and abundant deer mouse (*Peromyscus maniculatus*) harbors a highly complex, multilocus Hb polymorphism that plays a well-documented role in physiological adaptation to high-altitude hypoxia (Snyder, 1981, Snyder, 1982, Snyder et al., 1982, Snyder, 1985, Chappell and Snyder, 1984, Chappell et al., 1988, Storz, 2007, Storz et al., 2009, Storz et al., 2010a, Natarajan et al., 2013, Natarajan et al., 2015a, Jensen et al., 2016). Complementing experimental studies of Hb function, population genetic analyses of nucleotide variation at the underlying α- and β-globin genes have provided corroborative evidence for a history of spatially varying selection on Hb polymorphism between deer mouse populations that are native to different elevational zones (Storz et al., 2007, Storz and Kelly, 2008, Storz et al., 2009, Storz et al., 2010a, Storz et al., 2012, Natarajan et al., 2015a).

Deer mice native to the high alpine of the Rocky Mountains have evolved significantly higher Hb-O_2 affinities relative to lowland conspecifics from the prairie grasslands (Storz et al., 2009, Storz et al., 2010a, Natarajan et al., 2015a, Jensen et al., 2016). The allelic differences in Hb-O_2 affinity are attributable to the additive and nonadditive effects of amino acid replacements at twelve sites, eight in the α-chain and four in the β-chain (Natarajan et al., 2013) (Figs. 9.10, Plate 15). The affinity-altering mutations are widely shared among geographically disparate populations of *P. maniculatus* from a broad

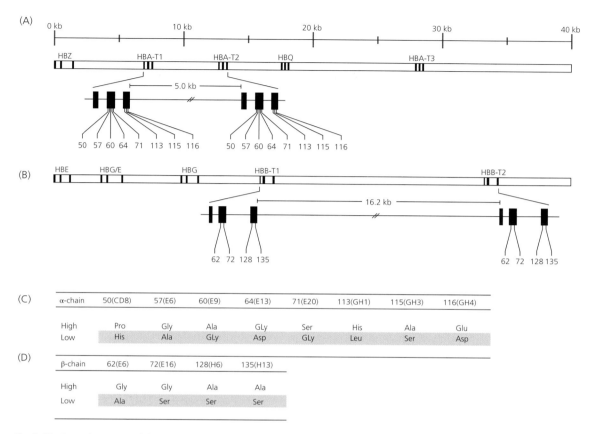

Fig. 9.10. Genomic structure of the α- and β-globin gene clusters of the deer mouse, *P. maniculatus*. (A) The tandemly duplicated adult α-globin genes (*HBA-T1* and *HBA-T2*) segregate 8 amino acid polymorphisms that exhibit significant allele frequency differences between highland and lowland populations. Lines denote nucleotide positions of the 8 replacement changes in *HBA* exons 2 and 3, and numbers refer to the corresponding amino acid positions in the encoded α-chain polypeptide. (B) The tandemly duplicated adult β-globin genes (*HBB-T1* and *HBB-T2*) segregate four amino acid polymorphisms that exhibit significant allele frequency differences between highland and lowland populations. Lines denote nucleotide positions of the four replacement changes in *HBB* exons 2 and 3, and numbers refer to the corresponding amino acid positions in the encoded β-chain polypeptide. In both the α- and β-globin gene clusters, shared polymorphisms between tandemly duplicated genes reflect a history of interparalog gene conversion (Storz et al., 2009, Storz et al., 2010a). (C, D) Amino acid sequences that define the most commonly observed α- and β-chain alleles from high-altitude deer mice in the Rocky Mountains and low-altitude deer mice in the prairie grassland. To see the structural locations of these amino acid replacements, see Plate 15.

Modified from Storz et al. (2010a).

range of elevations, and numerous amino acid polymorphisms are also shared with closely related species (Natarajan et al., 2015a). Consequently, the same amino acid replacements that contribute to variation in Hb-O_2 affinity within the broadly distributed *P. maniculatus* also contribute to divergence in Hb function among different species of *Peromyscus* with different elevational ranges.

Contrary to the predictions of Stern and Orgogozo (2008), data for both birds and mammals reveal no evidence that the allelic differences in Hb function between conspecific populations are qualitatively or quantitatively distinct from fixed differences between species.

9.5.5 Intramolecular epistasis is surprisingly prevalent

"Whatever you do in one part of the molecule affects the molecule as a whole"   **—Perutz (1990a)**

9.5.5.1 Direct measurements of epistasis

Site-directed mutagenesis experiments provide the most decisive means of testing whether particular mutations have different functional effects on different genetic backgrounds. As already mentioned, allelic variation in Hb-O_2 affinity in natural populations of deer mice from the Rocky Mountains and the Great Plains is attributable to the independent or joint effects of twelve amino acid replacements. Site-directed mutagenesis experiments revealed that nonadditive interactions between pairs of residues were surprisingly pervasive. In the presence of physiological anion concentrations, the observed epistasis for Hb-O_2 affinity mainly stemmed from effects on DPG binding. Chimeric Hb tetramers that incorporated alternative allelic variants of the α- and β-chain subunits (high-altitude α in combination with low-altitude β, and vice versa) exhibited suppressed sensitivity to DPG even though all the canonical DPG-binding sites were invariant (Natarajan et al., 2013). Allosteric regulatory capacities of chimeric Hbs were partially restored by reciprocally converting subdomains of each subunit back to the allelic type that matched the other subunit. A comprehensive set of site-directed mutagenesis experiments (Natarajan et al., 2013) and crystallographic analysis (Inoguchi et al., 2017) revealed that the observed epistasis for Hb-O_2 affinity is largely attributable to second-order perturbations of the interdimer $\alpha_1\beta_2/\alpha_2\beta_1$ interface that mediate the oxygenation-linked T↔R transition.

Given the evidence for spatially varying selection on the deer mouse Hb polymorphism that appears to favor different Hb-O_2 affinities in different elevational zones (Snyder et al., 1988, Storz, 2007, Storz et al., 2007, Storz et al., 2009, Storz et al., 2010a, Storz et al., 2012, Natarajan et al., 2015a), the pervasiveness of sign epistasis for Hb-O_2 affinity suggests that the selection coefficient for a given allelic variant may be highly dependent on the allelic composition of the local population. More generally, these results suggest that sign epistasis among segregating amino acid variants could exert a strong influence on the evolutionary dynamics of protein polymorphism in natural populations.

9.5.5.2 Indirect inferences of epistasis

Insights into the prevalence of epistasis and the nature of genetic compensation are also provided by cases where a pathogenic amino acid mutation in a human protein appears as the wildtype residue at the same site in the orthologous protein of one or more nonhuman species (Kondrashov et al., 2002). An example involving the human Hb mutant, Hb Mequon (β41Phe→Tyr), is shown in Fig. 9.11. In such cases, the pathogenic variant is invariably present at low frequency in the human gene pool, but

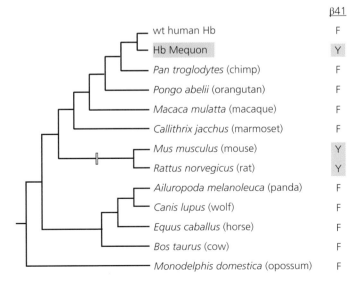

Fig. 9.11. Indirect evidence for genetic compensation is provided by cases where a pathogenic amino acid mutation in a human protein appears as the wildtype residue at the same site in the orthologous protein of one or more nonhuman species. The pathogenic Hb mutant, Hb Mequon (β41Phe→Tyr), provides an illustrative example. Although the Hb Mequon mutation is associated with severe hemolytic anemia in humans, the disease-associated Tyr β41 is wildtype in *Mus* and *Rattus*. In order for the Tyr variant to become fixed in the common ancestor of these rodent taxa, the deleterious effects that are manifest in human Hb must have been compensated by one or more rodent-specific substitutions at other sites in the same protein.

Reproduced from Storz (2018).

the same amino acid is fixed (present at a frequency of 1.0) in the nonhuman species. In order for the disease-associated residue to become fixed in the nonhuman species, its deleterious effects must have been compensated by one or more substitutions at other sites in the same protein or in an interacting protein. This permits an indirect inference of sign epistasis for fitness: the disease-associated residue produces a deleterious effect in the human protein but is neutral in the nonhuman ortholog due to genetic compensation (Kondrashov et al., 2002, Baresic et al., 2010, Xu and Zhang, 2014, Jordan et al., 2015).

A similar approach can be used to infer sign epistasis for biochemical phenotypes without making assumptions about fitness effects (Storz, 2018). This is possible in studies of proteins like Hb where we have detailed information about structure-function relationships. In cases where a particular mutation in human Hb is known to alter a particular functional property, it is often possible to identify Hbs from nonhuman species in which the same amino acid is wildtype and yet the property of interest is not altered in the same way. In such cases, the connection between genotype and biochemical phenotype can be experimentally tested. By contrast, associations with disease states do not generally permit direct insights into the inferred connection between genotype and fitness.

At physiological pH and temperature, the Bohr effect of human Hb is mainly attributable to the oxygenation-linked deprotonation of surface histidines because their imidazole side chains typically have acid dissociation constants, pK_as, in the physiological pH range (Chapter 4, section 4.6.1). The C-terminal histidine of the β-chain, His β146, makes an outsized contribution, accounting for ~60 percent of the Bohr effect in the presence of 0.1 M chloride (Shih et al., 1993, Fang et al., 1999, Lukin and Ho, 2004, Berenbrink, 2006). In deoxy (T-state) Hb, the positive charge on the imidazole sidechain of His β146 is stabilized by formation of a salt bridge with the carbonyl group of Asp β94 in the same β-chain subunit (Fig. 9.12A). This ionization of the His β146 side chain substantially raises its pK_a in the T-state. Consequently, mutational replacements of either His β146 or Asp β94 result in a severely diminished Bohr effect because the Asp β94-His β146 salt bridge in the T-state is replaced by an unionizable hydrogen bond (Fig. 9.12B); consequently, no protons are released in the allosteric T→R transition. Surprisingly, substitutions at these highly conserved residue positions have been identified in the Hbs of several vertebrate species that do not have reduced Bohr effects. For example, human Hb mutants such as Hb Bologna-St. Orsola (β146His→Tyr) and Hb Kodaira (β146His→Gln) exhibit increased O_2-affinities (due to destabilization of the T-state) and severe reductions in the Bohr effect (Shih et al., 1984, Shih et al., 1993, Ivaldi et al., 1999). The same amino acid states are observed as wildtype in the adult Hbs of

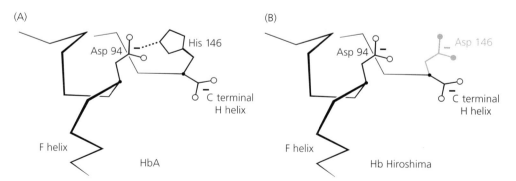

Fig. 9.12. In human Hb, replacement of the C-terminal His of each β-chain subunit (His β146) dramatically reduces the Bohr effect. (A) The positively charged imidazole side chain of His β146 forms an intrasubunit salt bridge with Asp β94, which increases its pK_a. When Hb is oxygenated, the allosteric transition in quaternary structure ruptures the His β146-Asp β94 salt bridge, resulting in the deprotonation of the His side chain (two protons are released per tetramer). (B) In mutants like Hb Hiroshima, where His β146 is replaced by Asp, the C-termini of the β-chains do not form salt bridges in the T-state and therefore do not contribute to the oxygenation-linked release of Bohr protons.

the dwarf caiman (*Paleosuchus palpebrosus*) (Tyr β146) and the golden-mantled ground squirrel (*Callospermophilus lateralis*) (Gln β146), and yet the Hbs of both species exhibit Bohr effects that are undiminished relative to normal human Hb (Revsbech et al., 2013, Weber et al., 2013) (Fig. 9.13). In the case of both caiman and ground-squirrel Hb, the loss of a single key residue with a major effect on pH-sensitivity, His β146, appears to be

compensated by the lineage-specific gain of multiple solvent-exposed, titratable histidines with individually minor effects.

In the case of Asp β94, human Hb mutants such as Hb Barcelona (β94Asp→His) and Hb Bunbury (β94Asp→Asn) also exhibit marked increases in O_2 affinity and concomitant reductions in the Bohr effect due to the disruption of the Asp β94-His β146 intrachain salt bridge. In two different species of

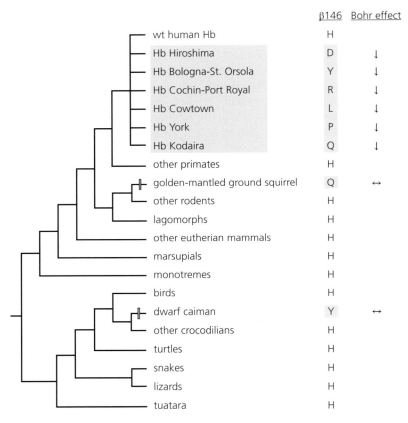

Fig. 9.13. In the Hbs of amniote vertebrates, comparative sequence analysis and experimental data on structure-function relationships reveal intramolecular epistasis for the Bohr effect. In the β-chain subunit of vertebrate Hb, the highly conserved His β146 generally accounts for a major fraction of the Bohr effect. This is well-documented by experimental studies of naturally occurring human Hb mutants which demonstrate that mutational replacements of His β146 (H) with Asp (D), Tyr (Y), Arg (R), Leu (L), Pro (P), or Gln (Q) invariably result in a severely diminished Bohr effect. Surprisingly, however, two of these amino acid states, Q (Hb Kodaira) and Y (Hb Bologna-St. Orsola), occur as wildtype in the Hbs of two nonhuman vertebrates, golden-mantled ground squirrel (*Callospermophilus lateralis*) and dwarf caiman (*Paleosuchus palpebrosus*), respectively, and yet the Hbs of both species exhibit Bohr effects that are undiminished relative to normal human Hb (Revsbech et al., 2013, Weber et al., 2013). In both species, the aggregate effect of other lineage-specific substitutions (e.g., gains of solvent-exposed histidines at other sites in the α- and/or β-chains of the Hb tetramer) may have rendered β146His redundant with respect to oxygenation-linked proton binding, so it could therefore be replaced without unduly compromising the Bohr effect.

Reproduced from Storz (2018).

high-altitude Andean waterfowl (crested duck [*Lophonetta specularioides*] and Puna teal [*Anas puna*]), β94Asp→Glu mutations have contributed to adaptive increases in Hb-O_2 affinity, but the Bohr effect is not compromised relative to wildtype Hbs with the ancestral Asp β94 (Natarajan et al., 2015b). A consideration of the crystal structure of avian Hb provides a clear explanation for this result, as the Asp β94-His β146 salt bridge is not formed in the T-state, so the amino acid state of β94 does not affect the pK_a of His β146. This may also explain why the β146His→Tyr substitution in dwarf caiman Hb is not associated with a diminished Bohr effect relative to the Hbs of other crocodilians, and demonstrates how "major effect" Bohr groups in human Hb may have minor or nonexistent effects in the Hbs of other species. These examples also illustrate how subtle changes in the three-dimensional orientation of highly conserved amino acids (caused by substitutions at other sites) can alter the functional effects of substitutions at those conserved sites (Naoi et al., 2001).

9.5.6 Epistasis is a source of contingency in adaptive protein evolution

"History often precludes useful opportunity; you cannot always get there from here."   **—Gould (1985)**

Compensatory substitutions are central to questions about the role of historical contingency in shaping pathways and outcomes of protein evolution. If the fitness effects of amino acid mutations are conditional on genetic background, then mutations can have different effects depending on the sequential order in which they occur (Weinreich et al., 2005). Consequently, the accumulated history of substitutions in the past will influence the set of allowable mutations in the future, and evolutionary outcomes will be historically contingent on ancestral starting points (Starr and Thornton, 2016, Starr et al., 2017, Salverda et al., 2011, Dickinson et al., 2013, Harms and Thornton, 2014).

The study of Hb evolution in Andean birds (Natarajan et al., 2016) revealed that parallel substitutions that contributed to convergent increases in Hb-O_2 affinity tended to be phylogenetically concentrated, mainly occurring in sets of closely related

species. One possible explanation for this pattern is that function-altering mutations tend to have context-dependent effects, so that identical mutations produce different effects on the divergent genetic backgrounds of different species. To test this hypothesis, Natarajan et al. (2016) focused on replicated substitutions at β83 that contributed to increased Hb-O_2 affinities in multiple high-altitude hummingbird species and one high-altitude flowerpiercer (genus *Diglossa*) (Fig. 9.4C). By conducting site-directed mutagenesis on resurrected ancestral Hbs, the researchers introduced the same β83 mutation into progressively more divergent genetic backgrounds (representing progressively more ancient nodes in the avian phylogeny). The experiments demonstrated that identical mutations at β83 have affinity-enhancing effects in the resurrected Hbs of ancestral hummingbirds and flowerpiercers, but the same mutation has no detectable effect on more divergent backgrounds (e.g., the common ancestor of all birds, "AncNeornithes") (Fig. 9.14). This illustrates a potentially important role for contingency in adaptive protein evolution. In different species that are adapting to the same selection pressure, the set of possible amino acids at a given site that have unconditionally beneficial effects may be contingent on the set of antecedent substitutions that have independently accumulated in the history of each lineage. Consequently, possible options for adaptive change in one species may be foreclosed options in other species.

9.5.7 Epistasis restricts mutational pathways of protein evolution

"If evolution by natural selection is to occur, functional proteins must form a continuous network which can be traversed by unit mutational steps without passing through nonfunctional intermediates."   **—Maynard Smith (1970)**

The adaptive evolution of protein function may often involve the sequential fixation of individual amino acid mutations with each substitution producing an incremental improvement in the selected biochemical property. However, epistatic interactions between mutant sites in the same protein may make some mutational pathways less accessible than others (DePristo et al., 2005, Weinreich et al., 2005, Weinreich et al., 2006, Poelwijk et al., 2007,

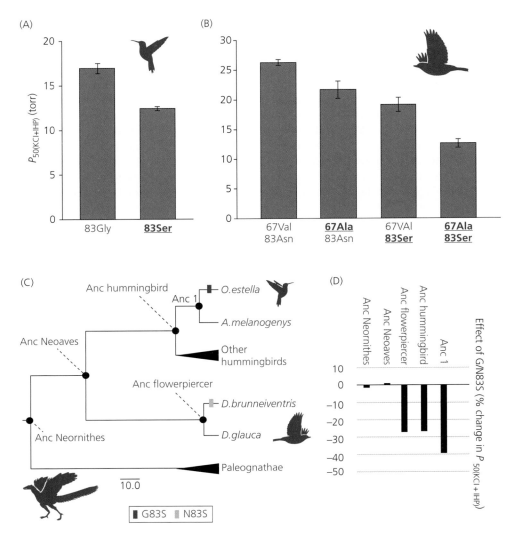

Fig. 9.14. Phenotypic effects of substitutions at β83 are conditional on genetic background. (A) βG83S substitutions occurred in parallel in multiple species of high-altitude hummingbirds, and were consistently associated with derived increases in Hb-O$_2$ affin ty. Consistent with the observed effects in contemporary hummingbird species, the engineered G83S mutation produced a significant reduction in $P_{50(KCl+IHP)}$ (increase in Hb-O$_2$ affinity) in the reconstructed Hb of the hummingbird ancestor. (B) A convergent βN83S substitution also occurred in one high-altitude passerine species, the black-throated flowerpiercer, *Diglossa brunneiventris*, in combination with the α^A substitution A67V. The engineered αA67V and βN83S mutations produced additive reductions in $P_{50(KCl+IHP)}$ in the reconstructed Hb of the flowerpiercer ancestor. Thus, substitutions at β83 produce consistent, affinity-enhancing effects in the Hbs of hummingbirds and flowerpiercers. (C) Diagrammatic tree with time-scaled branch lengths showing internal nodes that were targeted for ancestral protein resurrection. Scale bar denotes 10 million years. (D) N/G83S mutations produced significant increases in Hb-O$_2$ affinity (expressed as reductions in $P_{50(KCl+IHP)}$) in the ancestors of hummingbirds and flowerpiercers but produced no detectable effects on more divergent genetic backgrounds represented by the reconstructed common ancestor of Neoves (Anc Neoaves) and the common ancestor of all birds (Anc Neornithes), which existed ~110 million years ago. These results demonstrate that substitutions at the same site can have different effects on different genetic backgrounds.

Reproduced from Natarajan et al. (2016) with permission from the American Society for the Advancement of Science.

Bloom and Arnold, 2009, Kondrashov and Kondrashov, 2015, Starr and Thornton, 2016). If the sign of a mutation's fitness effect is conditional on genetic background, then pairs of mutations that are individually neutral or beneficial may be deleterious in combination. In such cases, some fraction of all possible mutational pathways connecting ancestral and descendant genotypes will be selectively inaccessible because they include incompatible mutational combinations as intermediate steps. Conversely, pairs of mutations that are individually deleterious may be neutral or beneficial in combination, thereby opening up new pathways through sequence space that previously would have been off limits.

Consider a scenario where positive directional selection produced an evolutionary increase in Hb-O_2 affinity, as might be expected for a lowland species that colonized a high-altitude environment. The change in Hb function is attributable to the independent or joint effects of amino acid substitutions at three sites such that the ancestral, low-affinity genotype ("000") and the derived, high-affinity genotype ("111") are connected by 3! = 6 possible mutational pathways (Figs. 9.15, 9.16). Were each of the possible pathways to the high-affinity/high-fitness "111" triple-mutant genotype equally likely and equally accessible to selection? In some pathways, each successive mutational step is either neutral (Hb-O_2 affinity is unchanged) or beneficial (yielding an incremental increase in Hb-O_2 affinity; Fig. 9.16B,E). By contrast, other pathways may include one or more mutational steps that yield a decrease in Hb-O_2 affinity, representing a potential fitness valley (Fig. 9.16A,C,D,F). This reflects the fact that the sign of a mutation's effect on Hb-O_2 affinity depends on the preceding mutations. At site 3, for example, the 0→1 mutation increases affinity when allele 1 is present at site 1 (Fig. 9.16A,B,C), but the same mutation decreases affinity (Fig. 9.16D) or has no effect on affinity (Fig. 9.16E,F) when allele 0 is present at site 1. Even in pathways with monotonically increasing Hb-O_2 affinity, affinity-enhancing mutations may have deleterious pleiotropic effects on other aspects of protein function (e.g., an increased autoxidation rate; 9.16A,B,E). As this hypothetical example illustrates, there may be epistasis for a mutation's effect on the selected phenotype *and* epistasis for pleiotropic effects, both of which can influence the selective accessibility of mutational pathways to high-fitness genotypes.

Fig. 9.16 also shows two distinct forms of compensatory interaction reminiscent of the pattern described for the bar-headed goose Hb: In Fig. 9.16B, the 000→100 mutation reduces the autoxidation rate, thereby offsetting the subsequent increase caused by the affinity-enhancing 100→101 mutation (this is an example of a permissive mutation). In Fig. 9.16E, by contrast, the increased autoxidation rate caused by the 001→101 mutation is

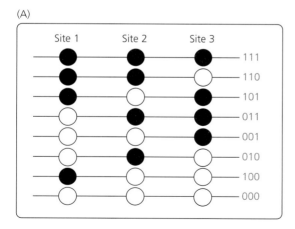

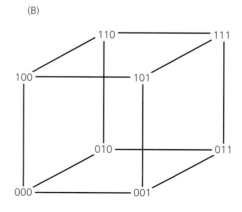

Fig. 9.15. Graphical depiction of genotype space for 8 (=2^3) combinations of bi-allelic variation at each of three sites. Each vertex of the cube represents a discrete three-site genotype, and edges connect genotypes that are separated by a single mutational step.

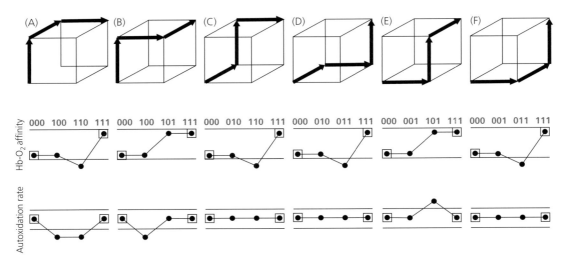

Fig. 9.16. Hypothetical mutational trajectories through sequence space. Of the 3! = 6 possible forward trajectories that connect the ancestral, low-affinity genotype ("000") with the derived, high-affinity genotype ("111"), only two pathways do not involve transient reductions in Hb-O_2 affinity (B and E). In both pathways, the evolutionary increases in Hb-O_2 affinity are produced by mutations to 101 (100→101 [B] and 001→101 [E]). At each site, the phenotypic effects of 0→1 mutations are conditional on the genetic background in which they occur. Even in pathways with monotonically increasing Hb-O_2 affinity, mutations that alter O_2 affinity may have deleterious pleiotropic effects on other aspects of protein function (e.g., an increased autoxidation rate [pathway E]).

compensated by the subsequent 101→111 step in the pathway, which has no effect on O_2 affinity. Experimental reconstructions of mutational pathways can reveal what fraction of pathways pass through conditionally deleterious intermediates, and they can shed light on the form and prevalence of compensatory mutations during the evolution of protein function.

To evaluate how pleiotropic trade-offs and epistatic interactions influence the accessibility of alternative mutational pathways, Tufts et al. (2015) examined the evolution of a derived increase in Hb-O_2 affinity in high-altitude pikas, *Ochotona princeps* (Mammalia: Lagomorpha). By combining ancestral protein resurrection with a combinatorial protein engineering approach, the researchers examined the functional effects of sequential mutational steps in all possible pathways leading from a low-affinity ancestral state to the high-affinity wildtype genotype of *O. princeps* (Fig. 9.17A). These experiments revealed that the effects of mutations on the oxygenation properties of Hb were highly dependent on the sequential order in which they occurred, with the result that only one of six possible forward pathways yields a monotonic increase in Hb-O_2

affinity (Fig. 9.17B). There was no evidence for pleiotropic trade-offs associated with affinity-enhancing mutations, suggesting that in some cases the accessibility of alternative mutational pathways may be more strongly constrained by sign epistasis for positively selected functional properties than by antagonistic pleiotropy.

In another study on this same general topic, Kumar et al. (2017) used a similar experimental approach to determine the fraction of possible pathways connecting ancestral and descendant genotypes that involve nonfunctional intermediates. Specifically, the researchers dissected the molecular basis of an evolved reduction in the O_2 affinity of avian Hb that was caused by the combined effect of four amino acid substitutions, two of which occurred in the highly conserved intradimeric $a_1\beta_1/a_2\beta_2$ interfaces (Fig. 9.18A). Using site-directed mutagenesis, they synthesized genotypes representing all possible mutational intermediates in each of 4! = 24 forward pathways connecting the high-affinity ancestral Hb to the low-affinity quadruple-mutant genotype (Fig. 9.18B). The experiments revealed that half of all possible forward pathways included mutational intermediates with

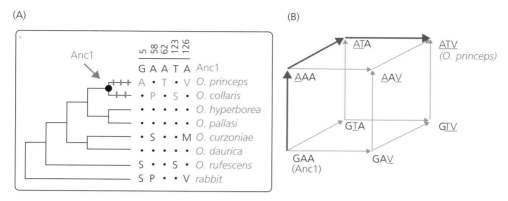

Fig. 9.17. Inferred history of substitution at five sites that distinguish the β-globin genes of high- and low-altitude species of North American pikas, *Ochotona princeps*, and *O. collaris*, respectively. (A) Of the five amino acid substitutions that distinguish the Hbs of the two sister species, ancestral sequence reconstruction revealed that three occurred on the branch leading to *O. princeps*, and the remaining two occurred on the branch leading to *O. collaris*. "Anc1" represents the reconstructed β-globin sequence of the *princeps/collaris* common ancestor. (B) Graphical representation of sequence space, where each vertex of the cube represents a discrete three-site β5-β62-β126 genotype and each edge connects genotypes separated by a single mutational step. The low-affinity ancestral Anc1 and the high-affinity Hb of *O. princeps* are connected by 6 (=3!) possible forward trajectories through sequence space. Due to epistatic interactions, only one of the six possible forward pathways yields a monotonic increase in Hb-O_2 affinity: G5A (first), A62T (second), A126V (third).

Modified from Tufts et al. (2015).

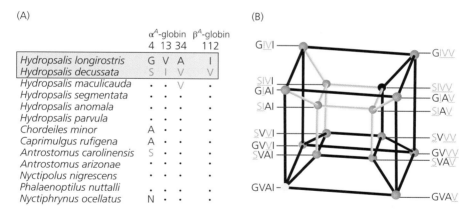

Fig. 9.18. Sign epistasis constrains the number of selectively accessible mutational pathways in the evolution of Hb function in nightjars (Caprimulgidae: Aves). (A) Amino acid substitutions that distinguish the HbA isoforms of *Hydropsalis longirostris* and *H. decussata*, and amino acid states at orthologous sites in other New World nightjar species. At each site, derived (non-ancestral) amino acids are denoted by gray lettering. (B) Four-dimensional depiction of sequence space, showing all possible mutational pathways connecting the high-affinity, ancestral GVAI genotype and the low-affinity, quadruple-mutant SIVV genotype (depicted by white and black nodes, respectively). All sixteen genotypes are represented as nodes (vertices of the hypercube), with edges connecting genotypes that differ by a single point mutation. Putatively inaccessible pathways that pass through the SIAI, SVAV, and SIVI genotypes are denoted by light gray edges; accessible pathways, in which each successive mutational step resulted in an unchanged or reduced Hb-O_2 affinity (increased P_{50}) are denoted by black edges.

Modified from Kumar et al. (2017).

aberrant functional properties because particular combinations of mutations at intradimer interfaces promoted tetramer-dimer dissociation. The subset of mutational pathways with unstable intermediates may be selectively inaccessible, representing evolutionary roads not taken. These experimental results also demonstrate how epistasis for particular functional properties of proteins may be mediated indirectly by mutational effects on structural stability.

These studies illustrate how questions about pathways of evolution can be addressed by using a protein engineering approach to explore the mutational landscape of protein function in experimentally defined regions of sequence space.

9.6 Conclusion and future directions

As the previous examples illustrate, experimental research on functional properties of a well-chosen model protein can be used to address some of the most conceptually expansive questions in evolutionary biology: Is genetic adaptation predictable? Why does evolution follow some pathways rather than others?

The key to a successful investigation of biochemical adaptation is to integrate molecular function with effects on higher-level physiological phenotypes in ecological context. In a statement that still rings true today, Lewontin (1979) lamented that "it has proved remarkably difficult to get compelling evidence for changes in enzymes brought about by selection, not to speak of adaptive changes." The challenge he identified is not about detecting statistical signatures of selection in patterns of DNA sequence variation; it is about characterizing functional differences between the products of alternative alleles and determining how the measured differences impinge on whole-organism performance in a way that affects fitness. The challenge is to determine how such allelic differences translate into differences in trait values, which in turn translate into fitness differences that exceed the threshold at which the effects of selection dominate the stochastic effects of genetic drift. Gillespie (1991) articulated the challenge of investigating fitness effects that fall below the resolving power of experimental methods: "Selection coefficients for single amino acid substitutions as small as 10^{-4} to 10^{-3} are large enough

to dominate genetic drift, yet are refractory to direct experimental investigation. In other words, most of protein evolution could be due to strong natural selection, yet we have no experimental protocol capable of measuring the selective differences."

Perutz (1983) mused on a similar subject with regard to species differences in Hb function: "It could be argued that even an amino acid replacement that produces only a very small shift in the O_2 equilibrium curve may give an animal a selective advantage that would prove decisive over thousands of generations; against this it could be held that homeostatic mechanisms allow organisms to compensate efficiently for quite large shifts in the curve." Case studies of Hb function in high-altitude vertebrates provide compelling evidence that evolutionary changes in Hb-O_2 affinity often have adaptive significance, but Perutz's point highlights the need to investigate how putatively adaptive changes in Hb function revealed by *in vitro* experiments translate into *in vivo* enhancements of red cell function and whole-animal physiological performance. As proof of principle, consider that affinity-altering amino acid variants in rodent Hbs that enhance blood O_2 capacitance have been shown to produce surprisingly dramatic effects on whole-animal aerobic performance (Chappell and Snyder, 1984, Chappell et al., 1988, Shirasawa et al., 2003).

Adaptive modifications of vertebrate Hb provide examples in which phenotypic effects of individual mutations are large enough to measure, and they suggest clear, testable hypotheses about how modifications of protein function translate into changes in more distal physiological phenotypes. The case studies reviewed here have provided hard-won insights into mechanisms of biochemical adaptation and protein evolution. How general are the results? In terms of qualitative and quantitative effects, are the affinity-enhancing mutations in the Hbs of deer mice, hummingbirds, and bar-headed geese generally representative of adaptive substitutions in protein evolution? Regardless of their generality, the value of these examples of biochemical adaptation is that they provide insights into mutational effects that may often fall well below the resolving power of our experimental techniques, and they allow us to gain purchase on otherwise intractable evolutionary questions. As stated by

Gillespie (1991): "They should be cherished as windows into a world of selection that will forever remain beyond the resolution of our techniques."

References

Alayash, A. I., Patel, R. P., and Cashon, R. E. (2001). Redox reactions of hemoglobin and myoglobin: biological and toxicological implications. *Antioxidants and Redox Signaling*, **3**, 313–27.

Arendt, J. and Reznick, D. (2008). Convergence and parallelism reconsidered: what have we learned about the genetics of adaptation? *Trends in Ecology and Evolution*, **23**, 26–32.

Baresic, A., Hopcroft, L. E. M., Rogers, H. H., Hurst, J. M., and Martin, A. C. R. (2010). Compensated pathogenic deviations: analysis of structural effects. *Journal of Molecular Biology*, **396**, 19–30.

Bazykin, G. A. (2015). Changing preferences: deformation of single position amino acid fitness landscapes and evolution of proteins. *Biology Letters*, **11**, 20150315.

Bazykin, G. A., Kondrashov, F. A., Brudno, M., et al. (2007). Extensive parallelism in protein evolution. *Biology Direct*, **2**, 20.

Bellelli, A., Brunori, M., Miele, A. E., Panetta, G., and Vallone, B. (2006). The allosteric properties of hemoglobin: insights from natural and site directed mutants. *Current Protein and Peptide Science*, **7**, 17–45.

Bencowitz, H. Z., Wagner, P. D., and West, J. B. (1982). Effect of change in P50 on exercise tolerance at high-altitude—a theoretical study. *Journal of Applied Physiology*, **53**, 1487–95.

Berenbrink, M. (2006). Evolution of vertebrate haemoglobins: histidine side chains, specific buffer value and Bohr effect. *Respiratory Physiology and Neurobiology*, **154**, 165–84.

Bishop, C. M., Spivey, R. J., Hawkes, L. A., et al. (2015). The roller coaster flight strategy of bar-headed geese conserves energy during Himalayan migrations. *Science*, **347**, 250–4.

Black, C. P. and Tenney, S. M. (1980). Oxygen-transport during progressive hypoxia in high-altitude and sea-level waterfowl. *Respiration Physiology*, **39**, 217–39.

Bloom, J. D. and Arnold, F. H. (2009). In the light of directed evolution: pathways of adaptive protein evolution. *Proceedings of the National Academy of Sciences of the United States of America*, **106**, 9995–10000.

Bloom, J. D., Gong, L. I., and Baltimore, D. (2010). Permissive secondary mutations enable the evolution of influenza oseltamivir resistance. *Science*, **328**, 1272–5.

Bloom, J. D., Labthavikul, S. T., Otey, C. R., and Arnold, F. H. (2006). Protein stability promotes evolvability. *Proceedings of the National Academy of Sciences of the United States of America*, **103**, 5869–74.

Bloom, J. D., Romero, P. A., Lu, Z., and Arnold, F. H. (2007). Neutral genetic drift can alter promiscuous protein functions, potentially aiding functional evolution. *Biology Direct*, **2**, 17.

Bloom, J. D., Silberg, J. J., Wilke, C. O., et al. (2005). Thermodynamic prediction of protein neutrality. *Proceedings of the National Academy of Sciences of the United States of America*, **102**, 606–11.

Bonaventura, C., Henkens, R., Alayash, A. I., Banerjee, S., and Crumbliss, A. L. (2013). Molecular controls of the oxygenation and redox reactions of hemoglobin. *Antioxidants and Redox Signaling*, **18**, 2298–313.

Brauner, C. J. and Wang, T. (1997). The optimal oxygen equilibrium curve: a comparison between environmental hypoxia and anemia. *American Zoologist*, **37**, 101–8.

Bridgham, J. T., Ortlund, E. A., and Thornton, J. W. (2009). An epistatic ratchet constrains the direction of glucocorticoid receptor evolution. *Nature*, **461**, 515–19.

Bunn, H. F. and Forget, B. G. (1986). *Hemoglobin: Molecular, Genetic and Clinical Aspects*, Philadelphia, PA, W. B. Saunders Company.

Chappell, M. A., Hayes, J. P., and Snyder, L. R. G. (1988). Hemoglobin polymorphisms in deer mice (*Peromyscus maniculatus*), physiology of β-globin variants and α-globin recombinants. *Evolution*, 42, 681–8.

Chappell, M. A. and Snyder, L. R. G. (1984). Biochemical and physiological correlates of deer mouse α-chain hemoglobin polymorphisms. *Proceedings of the National Academy of Sciences of the United States of America*, **81**, 5484–8.

Chevin, L.-M., Martin, G., and Lenormand, T. (2010). Fisher's model and the genomics of adaptation: restricted pleiotropy, heterogeneous mutation, and parallel evolution. *Evolution*, **64**, 3213–31.

Couce, A., Rodriguez-Rojas, A., and Blazquez, J. (2015). Bypass of genetic constraints during mutator evolution to antibiotic resistance. *Proceedings of the Royal Society B-Biological Sciences*, **282**, 20142698.

Dean, A. M. and Thornton, J. W. (2007). Mechanistic approaches to the study of evolution: the functional synthesis. *Nature Reviews Genetics*, **8**, 675–88.

DePristo, M. A., Weinreich, D. M., and Hartl, D. L. (2005). Missense meanderings in sequence space: a biophysical view of protein evolution. *Nature Reviews Genetics*, **6**, 678–87.

Dickerson, R. E. and Geis, I. (1983). *Hemoglobin: Structure, Function, Evolution, and Pathology*, Menlo Park, CA, Benjamin/Cummings.

Dickinson, B. C., Leconte, A. M., Allen, B., Esvelt, K. M., and Liu, D. R. (2013). Experimental interrogation of the path dependence and stochasticity of protein evolution using phage-assisted continuous evolution. *Proceedings of the National Academy of Sciences of the United States of America*, **110**, 9007–12.

Fang, T. Y., Zou, M., Simplaceanu, V., Ho, N. T., and Ho, C. (1999). Assessment of roles of surface histidyl residues in the molecular basis of the Bohr effect and of β143 histidine in the binding of 2,3-bisphosphoglycerate in human normal adult hemoglobin. *Biochemistry*, **38**, 13423–32.

Galen, S. C., Natarajan, C., Moriyama, H., et al. (2015). Contribution of a mutational hotspot to adaptive changes in hemoglobin function in high-altitude Andean house wrens. *Proceedings of the National Academy of Sciences of the United States of America*, **112**, 13958–63.

Gillespie, J. H. (1991). *The Causes of Molecular Evolution*, New York, NY, Oxford University Press.

Golding, G. B., and Dean, A. M. (1998). The structural basis of molecular adaptation. *Molecular Biology and Evolution*, **15**, 335–69.

Goldstein, R. A., Pollard, S. T., Shah, S. D., and Pollock, D. D. (2015). Nonadaptive amino acid convergence rates decrease over time. *Molecular Biology and Evolution*, **32**, 1373–81.

Gong, L. I., Suchard, M. A., and Bloom, J. D. (2013). Stability-mediated epistasis constrains the evolution of an influenza protein. *eLife*, **2**, e00631.

Gould, S. J. (1985). *The Flamingo's Smile*, New York, NY, W. W. Norton and Company.

Gould, S. J. (1989). *Wonderful Life: The Burgess Shale and the Nature of History*, New York, NY, W. W. Norton and Company.

Gould, S. J. (1995). *Dinosaur in a Haystack*, New York, NY, Harmony Books.

Harms, M. J. and Thornton, J. W. (2010). Analyzing protein structure and function using ancestral gene reconstruction. *Current Opinion in Structural Biology*, **20**, 360–6.

Harms, M. J. and Thornton, J. W. (2013). Evolutionary biochemistry: revealing the historical and physical causes of protein properties. *Nature Reviews Genetics*, **14**, 559–71.

Harms, M. J. and Thornton, J. W. (2014). Historical contingency and its biophysical basis in glucocorticoid receptor evolution. *Nature*, **512**, 203–7.

Hawkes, L. A., Balachandran, S., Batbayar, N., et al. (2011). The trans-Himalayan flights of bar-headed geese (*Anser indicus*). *Proceedings of the National Academy of Sciences of the United States of America*, **108**, 9516–19.

Hawkes, L. A., Balachandran, S., Batbayar, N., et al. (2013). The paradox of extreme high-altitude migration in bar-headed geese *Anser indicus*. *Proceedings of the Royal Society B-Biological Sciences*, **280**(1750), 20122114.

Hochberg, G. K. A. and Thornton, J. W. (2017). Reconstructing ancient proteins to understand the causes of structure and function. *Annual Review of Biophysics*, **46**, 247–69.

Ingram, V. M. (1956). Specific chemical difference between the globins of normal human and sickle-cell anaemia haemoglobin. *Nature*, **178**, 792–4.

Inoguchi, N., Mizuno, N., Baba, S., et al. (2017). Alteration of the a1b2/a2b1 subunit interface contributes to the increased hemoglobin-oxygen affinity of high-altitude deer mice. *PloS One*, **12**, e0174921.

Ivaldi, G., David, O., Paradossi, V., et al. (1999). Hb Bologna-St. Orsola β146(HC3)His→Tyr: a new high oxygen affinity variant with halved Bohr effect and highly reduced reactivity towards 2,3-diphosphoglycerate. *Hemoglobin*, **23**, 353–9.

Jensen, B., Storz, J. F., and Fago, A. (2016). Bohr effect and temperature sensitivity of hemoglobins from highland and lowland deer mice. *Comparative Biochemistry and Physiology A: Molecular and Integrative Physiology*, **195**, 10–14.

Jessen, T. H., Weber, R. E., Fermi, G., Tame, J., and Braunitzer, G. (1991). Adaptation of bird hemoglobins to high-altitudes—demonstration of molecular mechanism by protein engineering. *Proceedings of the National Academy of Sciences of the United States of America*, **88**, 6519–22.

Jordan, D. M., Frangakis, S. G., Golzio, C., et al. (2015). Identification of *cis*-suppression of human disease mutations by comparative genomics. *Nature*, **524**, 225–9.

Kaltenbach, M., Jackson, C. J., Campbell, E. C., Hollfelder, F., and Tokuriki, N. (2015). Reverse evolution leads to genotypic incompatibility despite functional and active site convergence. *eLife*, **4**, e06492.

Kimura, M. (1983). *The Neutral Theory of Molecular Evolution*, Cambridge, Cambridge University Press.

Kondrashov, A. S., Sunyaev, S., and Kondrashov, F. A. (2002). Dobzhansky-Muller incompatibilities in protein evolution. *Proceedings of the National Academy of Sciences of the United States of America*, **99**, 14878–83.

Kondrashov, D. A. and Kondrashov, F. A. (2015). Topological features of rugged fitness landscapes in sequence space. *Trends in Genetics*, **31**, 24–33.

Kumar, A., Natarajan, C., Moriyama, H., et al. (2017). Stability-mediated epistasis restricts accessible mutational pathways in the functional evolution of avian hemoglobin. *Molecular Biology and Evolution*, **34**, 1240–51.

Lewontin, R. C. (1979). Adaptation. *Scientific American*, **239**, 156–69.

Lozovsky, E. R., Chookajorn, T., Brown, K. M., et al. (2009). Stepwise acquisition of pyrimethamine resistance in the malaria parasite. *Proceedings of the National Academy of Sciences of the United States of America*, **106**, 12025–30.

Lukin, J. A. and Ho, C. (2004). The structure-function relationship of hemoglobin in solution at atomic resolution. *Chemical Reviews*, **104**, 1219–30.

Maillett, D. H., Simplaceanu, V., Shen, T.-J., et al. (2008). Interfacial and distal-heme pocket mutations exhibit additive effects on the structure and function of hemoglobin. *Biochemistry*, **47**, 10551–63.

Maynard Smith, J. (1970). Natural selection and the concept of a protein space. *Nature*, **225**, 563–4.

McCandlish, D. M. and Stoltzfus, A. (2014). Modeling evolution using the probability of fixation: history and implications. *Quarterly Review of Biology*, **89**, 225–52.

Nagel, R. L. (2001). Disorders of hemoglobin function and stability. *In:* Steinberg, M. H., Forget, B. G., Higgs, D. R., and Nagel, R. L. (eds.) *Disorders of Hemoglobin: Genetics, Pathophysiology, and Clinical Management*, pp. 1155–94. Cambridge, Cambridge University Press.

Naoi, Y., Chong, K. T., Yoshimatsu, K., et al. (2001). The functional similarity and structural diversity of human and cartilaginous fish hemoglobins. *Journal of Molecular Biology*, **307**, 259–70.

Natarajan, C., Hoffman, F. G., Lanier, H. C., et al. (2015a). Intraspecific polymorphism, interspecific divergence, and the origins of function-altering mutations in deer mouse hemoglobin. *Molecular Biology and Evolution*, **32**, 978–97.

Natarajan, C., Hoffmann, F. G., Weber, R. E., et al. (2016). Predictable convergence in hemoglobin function has unpredictable molecular underpinnings. *Science*, **354**, 336–40.

Natarajan, C., Inoguchi, N., Weber, R. E., et al. (2013). Epistasis among adaptive mutations in deer mouse hemoglobin. *Science*, **340**, 1324–7.

Natarajan, C., Jendroszek, A., Kumar, A., et al. (2018). Molecular basis of hemoglobin adaptation in the high-flying bar-headed goose. *PLoS Genetics*, **14**, e1007331.

Natarajan, C., Projecto-Garcia, J., Moriyama, H., et al. (2015b). Convergent evolution of hemoglobin function in high-altitude Andean waterfowl involves limited parallelism at the molecular sequence level. *PLoS Genetics*, **11**, e1005681.

Naumenko, S. A., Kondrashov, A. S., and Bazykin, G. A. (2012). Fitness conferred by replaced amino acids declines with time. *Biology Letters*, **8**, 825–8.

Olson, J. S. and Maillett, D. H. (2005). Designing recombinant hemoglobin for use as a blood substitute. *In:* Winslow, R. M. (ed.) *Blood Substitutes*, pp. 354–74. San Diego, CA, Academic Press.

Orr, H. A. (1998). The population genetics of adaptation: the distribution of factors fixed during adaptive evolution. *Evolution*, **52**, 935–49.

Orr, H. A. (2005a). The genetic theory of adaptation: a brief history. *Nature Reviews Genetics*, **6**, 119–27.

Orr, H. A. (2005b). The probability of parallel evolution. *Evolution*, 59, 216–20.

Ortlund, E. A., Bridgham, J. T., Redinbo, M. R., and Thornton, J. W. (2007). Crystal structure of an ancient protein: evolution by conformational epistasis. *Science*, **317**, 1544–8.

Otto, S. P. (2004). Two steps forward, one step back: the pleiotropic effects of favoured alleles. *Proceedings of the Royal Society B-Biological Sciences*, **271**, 705–14.

Pauling, L., Itano, H. A., Singer, S. J., and Wells, I. C. (1949). Sickle cell anemia, a molecular disease. *Science*, **110**, 543–8.

Perutz, M. F. (1983). Species adaptation in a protein molecule. *Molecular Biology and Evolution*, **1**, 1–28.

Perutz, M. F. (1990a). Comments. *In:* Ho, C. (ed.) *Hemoglobin and Oxygen Binding*, New York, NY, Elsevier.

Perutz, M. F. (1990b). Frequency of abnormal human hemoglobins caused by C→T transitions in CpG dinucleotides. *Biophysical Chemistry*, **37**, 25–9.

Petschow, D., Wurdinger, I., Baumann, R., et al. (1977). Causes of high blood O_2 affinity of animals living at high-altitude. *Journal of Applied Physiology*, **42**, 139–43.

Phillips, P. C. (2008). Epistasis—the essential role of gene interactions in the structure and evolution of genetic systems. *Nature Reviews Genetics*, **9**, 855–67.

Poelwijk, F. J., Kiviet, D. J., Weinreich, D. M., and Tans, S. J. (2007). Empirical fitness landscapes reveal accessible evolutionary paths. *Nature*, **445**, 383–6.

Pollock, D. D., Thiltgen, G., and Goldstein, R. A. (2012). Amino acid coevolution induces an evolutionary Stokes shift. *Proceedings of the National Academy of Sciences of the United States of America*, **109**, E1352–9.

Poon, A. and Chao, L. (2005). The rate of compensatory mutation in the DNA bacteriophage phi X174. *Genetics*, **170**, 989–99.

Poon, A. F. Y. and Chao, L. (2006). Functional origins of fitness effect-sizes of compensatory mutations in the DNA bacteriophage phi X174. *Evolution*, **60**, 2032–43.

Povolotskaya, I. S. and Kondrashov, F. A. (2010). Sequence space and the ongoing expansion of the protein universe. *Nature*, **465**, 922–6.

Projecto-Garcia, J., Natarajan, C., Moriyama, H., et al. (2013). Repeated elevational transitions in hemoglobin function during the evolution of Andean hummingbirds. *Proceedings of the National Academy of Sciences of the United States of America*, **110**, 20669–74.

Reeder, B. J. (2010). The redox activity of hemoglobins: from physiologic functions to pathologic mechanisms. *Antioxidants and Redox Signaling*, **13**, 1087–123.

Reeder, B. J. and Wilson, M. T. (2005). Hemoglobin and myoglobin associated oxidative stress: from molecular mechanisms to disease states. *Current Medicinal Chemistry*, **12**, 2741–51.

Revsbech, I. G., Tufts, D. M., Projecto-Garcia, J., et al. (2013). Hemoglobin function and allosteric regulation in semi-fossorial rodents (family Sciuridae) with different altitudinal ranges. *Journal of Experimental Biology*, **216**, 4264–71.

Riggs, A. (1976). Factors in evolution of hemoglobin function. *Federation Proceedings*, **35**, 2115–18.

Rogozin, I. B., Thomson, K., Csueroes, M., Carmel, L., and Koonin, E. V. (2008). Homoplasy in genome-wide analysis of rare amino acid replacements: the molecular-evolutionary basis for Vavilov's law of homologous series. *Biology Direct*, **3**, 7.

Rokas, A. and Carroll, S. B. (2008). Frequent and widespread parallel evolution of protein sequences. *Molecular Biology and Evolution*, **25**, 1943–53.

Rokyta, D. R., Joyce, P., Caudle, S. B., and Wichman, H. A. (2005). An empirical test of the mutational landscape model of adaptation using a single-stranded DNA virus. *Nature Genetics*, **37**, 441–4.

Rollema, H. S. and Bauer, C. (1979). Interaction of inositol pentaphosphate with the hemoglobins of highland and lowland geese. *Journal of Biological Chemistry*, **254**, 2038–43.

Salverda, M. L. M., Dellus, E., Gorter, F. A., et al. (2011). Initial mutations direct alternative pathways of protein evolution. *PLoS Genetics*, **7**, e1001321.

Schechter, A. N. (2008). Hemoglobin research and the origins of molecular medicine. *Blood*, **112**, 3927–38.

Scott, G. R. (2011). Elevated performance: the unique physiology of birds that fly at high altitudes. *Journal of Experimental Biology*, **214**, 2455–62.

Scott, G. R., Hawkes, L. A., Frappell, P. B., et al. (2015). How bar-headed geese fly over the Himalayas. *Physiology*, **30**, 107–15.

Scott, G. R. and Milsom, W. K. (2006). Flying high: a theoretical analysis of the factors limiting exercise performance in birds at altitude. *Respiratory Physiology and Neurobiology*, **154**, 284–301.

Shah, P., McCandlish, D. M., and Plotkin, J. B. (2015). Contingency and entrenchment in protein evolution under purifying selection. *Proceedings of the National Academy of Sciences of the United States of America*, **112**, 7627.

Shih, D. T. B., Luisi, B. F., Miyazaki, G., Perutz, M. F., and Nagai, K. (1993). A mutagenic study of the allosteric linkage of His(HC3)146β in hemoglobin. *Journal of Molecular Biology*, **230**, 1291–6.

Shih, T. B., Jones, R. T., Bonaventura, J., Bonaventura, C., and Schneider, R. G. (1984). Involvement of His HC3(146)b in the Bohr effect of human hemoglobin. Studies of native and N-ethylmaleimide-treated hemoglobin A and hemoglobin Cowtown (β146His→Leu). *Journal of Biological Chemistry*, **259**, 967–74.

Shirasawa, T., Izumizaki, M., Suzuki, Y., et al. (2003). Oxygen affinity of hemoglobin regulates O₂ consumption, metabolism, and physical activity. *Journal of Biological Chemistry*, **278**, 5035–43.

Snyder, L. R. G. (1981). Deer mouse hemoglobins—is there genetic adaptation to high altitude? *Bioscience*, **31**, 299–304.

Snyder, L. R. G. (1982). 2,3-diphosphoglycerate in high-altitude and low-altitude populations of the deer mouse. *Respiration Physiology*, **48**, 107–23.

Snyder, L. R. G. (1985). Low P50 in deer mice native to high altitude. *Journal of Applied Physiology*, **58**, 193–9.

Snyder, L. R. G., Born, S., and Lechner, A. J. (1982). Blood oxygen affinity in high-altitude and low-altitude populations of the deer mouse. *Respiration Physiology*, **48**, 89–105.

Snyder, L. R. G., Hayes, J. P., and Chappell, M. A. (1988). α-chain hemoglobin polymorphisms are correlated with altitude in the deer mouse, *Peromyscus maniculatus*. *Evolution*, **42**, 689–97.

Soylemez, O. and Kondrashov, F. A. (2012). Estimating the rate of irreversibility in protein evolution. *Genome Biology and Evolution*, **4**, 1213–22.

Starr, T. N., Picton, L. K., and Thornton, J. W. (2017). Alternative evolutionary histories in the sequence space of an ancient protein. *Nature*, **549**, 409–13.

Starr, T. N. and Thornton, J. W. (2016). Epistasis in protein evolution. *Protein Science*, **25**, 1204–18.

Steinberg, M. H. (2009). Clinical and pathophysiological aspects of sickle cell anemia. *In*: Steinberg, M. H., Forget, B. G., Higgs, D. R., and Weatherall, D. J. (eds.) *Disorders of Hemoglobin: Genetics, Pathophysiology, and Clinical Management*, 2nd edition. Cambridge, Cambridge Medicine.

Steinberg, M. H., Forget, B. G., Higgs, D. R., and Weatherall, D. J. (2009). *Disorders of Hemoglobin: Genetics, Pathophysiology, and Clinical Management*, 2nd edition, Cambridge, UK, Cambridge Medicine.

Steinberg, M. H. and Nagel, R. L. (2001). Native and recombinant mutant hemoglobins of biological interest. *In*: Steinberg, M. H., Forget, B. G., Higgs, D. R., and Nagel, R. L. (eds.) *Disorders of Hemoglobin: Genetics, Pathophysiology, and Clinical Management*, 2nd edition pp. 1195–211. Cambridge, Cambridge University Press.

Steinberg, M. H. and Nagel, R. L. (2009). Unstable hemoglobins, hemoglobins with altered oxygen affinity, Hemoglobin M, and other variants of clinical and biological interest. *In*: Steinberg, M. H., Forget, B. G., Higgs, D. R., and Weatherall, D. J. (eds.) *Disorders of Hemoglobin: Genetics, Pathophysiology, and Clinical Management*, 2nd edition, pp. 589–606. Cambridge, Cambridge Medicine.

Stern, D. L. and Orgogozo, V. (2008). The loci of evolution: how predictable is genetic evolution? *Evolution*, **62**, 2155–77.

Stoltzfus, A. (2006). Mutationism and the dual causation of evolutionary change. *Evolution and Development*, **8**, 304–17.

Stoltzfus, A. and McCandlish, D. M. (2017). Mutational biases influence parallel adaptation. *Molecular Biology and Evolution*, **34**, 2163–72.

Stoltzfus, A. and Yampolsky, L. Y (2009). Climbing Mount Probable: mutation as a cause of nonrandomness in evolution. *Journal of Heredity*, **100**, 637–47.

Storz, J. F. (2007). Hemoglobin function and physiological adaptation to hypoxia in high-altitude mammals. *Journal of Mammalogy*, **88**, 24–31.

Storz, J. F. (2016). Causes of molecular convergence and parallelism in protein evolution. *Nature Reviews Genetics*, **17**, 239–50.

Storz, J. F. (2018). Compensatory mutations and epistasis for protein function. *Current Opinion in Structural Biology*, **50**, 18–25.

Storz, J. F. and Kelly, J. K. (2008). Effects of spatially varying selection on nucleotide diversity and linkage disequilibrium: insights from deer mouse globin genes. *Genetics*, **180**, 367–79.

Storz, J. F., Natarajan, C., Cheviron, Z. A., Hoffmann, F. G., and Kelly, J. K. (2012). Altitudinal variation at duplicated β-globin genes in deer mice: effects of selection, recombination, and gene conversion. *Genetics*, **190**, 203–16.

Storz, J. F., Runck, A. M., Moriyama, H., Weber, R. E., and Fago, A. (2010a). Genetic differences in hemoglobin function between highland and lowland deer mice. *Journal of Experimental Biology*, **213**, 2565–74.

Storz, J. F., Runck, A. M., Sabatino, S. J., et al. (2009). Evolutionary and functional insights into the mechanism underlying high-altitude adaptation of deer mouse hemoglobin. *Proceedings of the National Academy of Sciences of the United States of America*, **106**, 14450–5.

Storz, J. F., Sabatino, S. J., Hoffmann, F. G., et al. (2007). The molecular basis of high-altitude adaptation in deer mice. *PloS Genetics*, **3**, 448–59.

Storz, J. F., Scott, G. R., and Cheviron, Z. A. (2010b). Phenotypic plasticity and genetic adaptation to high-altitude hypoxia in vertebrates. *Journal of Experimental Biology*, **213**, 4125–36.

Storz, J. F. and Wheat, C. W. (2010). Integrating evolutionary and functional approaches to infer adaptation at specific loci. *Evolution*, **64**, 2489–509.

Storz, J. F. and Zera, A. J. (2011). Experimental approaches to evaluate the contributions of candidate protein-coding mutations to phenotypic evolution. *In:* Orgogozo, V. and Rockman, M. V. (eds.) *Molecular Methods for Evolutionary Genetics*, pp. 377–96. New York, NY, Humana Press.

Streisfeld, M. A. and Rausher, M. D. (2011). Population genetics, pleiotropy, and the preferential fixation of mutations during adaptive evolution. *Evolution*, **65**, 629–42.

Thom, C. S., Dickson, C. F., Gell, D. A., and Weiss, M. J. (2013). Hemoglobin variants: biochemical properties and clinical correlates. *Cold Spring Harbor Perspectives in Medicine*, **3**, a011858.

Tokuriki, N., Stricher, F., Serrano, L., and Tawfik, D. S. (2008). How protein stability and new functions trade off. *Plos Computational Biology*, **4**, e1000002.

Tufts, D. M., Natarajan, C., Revsbech, I. G., et al. (2015). Epistasis constrains mutational pathways of hemoglobin

adaptation in high-altitude pikas. *Molecular Biology and Evolution*, **32**, 287–98.

Turek, Z., Kreuzer, F., and Hoofd, L. J. C. (1973). Advantage or disadvantage of a decrease in blood-oxygen affinity for tissue oxygen supply at hypoxia—theoretical study comparing man and rat. *Pflügers Archiv-European Journal of Physiology*, **342**, 185–97.

Turek, Z., Kreuzer, F., and Ringnalda, B. E. M. (1978a). Blood-gases at several levels of oxygenation in rats with a left-shifted blood-oxygen dissociation curve. *Pflügers Archiv-European Journal of Physiology*, **376**, 7–13.

Turek, Z., Kreuzer, F., Turekmaischeider, M., and Ringnalda, B. E. M. (1978b). Blood O_2 content, cardiac output, and flow to organs at several levels of oxygenation in rats with a left-shifted blood-oxygen dissocation curve. *Pflügers Archiv-European Journal of Physiology*, **376**, 201–7.

Unckless, R. L. and Orr, H. A. (2009). The population genetics of adaptation: multiple substitutions on a smooth fitness landscape. *Genetics*, **183**, 1079–86.

Usmanova, D. R., Ferretti, L., Povolotskaya, I. S., Vlasov, P. K., and Kondrashov, F. A. (2015). A model of substitution trajectories in sequence space and long-term protein evolution. *Molecular Biology and Evolution*, **32**, 542–54.

Varnado, C. L., Mollan, T. L., Birukou, I., et al. (2013). Development of recombinant hemoglobin-based oxygen carriers. *Antioxidants and Redox Signaling*, **18**, 2314–28.

Wajcman, H. and Galacteros, F. (1996). Abnormal hemoglobins with high oxygen affinity and erythrocytosis. *Hematology and Cell Therapy*, **38**, 305–12.

Wajcman, H. and Galacteros, F. (2005). Hemoglobins with high oxygen affinity leading to erythrocytosis. New variants and new concepts. *Hemoglobin*, **29**, 91–106.

Wang, T. and Malte, H. (2011). O_2 uptake and transport: the optimal P50. *In:* Farrell, A. P. (ed.) *Encyclopedia of Fish Physiology: From Genome to Environment*, pp. 893–8. San Diego, CA, Elsevier Science.

Wang, X. J., Minasov, G., and Shoichet, B. K. (2002). Evolution of an antibiotic resistance enzyme constrained by stability and activity trade-offs. *Journal of Molecular Biology*, **320**, 85–95.

Weber, R. E., Fago, A., Malte, H., Storz, J. F., and Gorr, T. A. (2013). Lack of conventional oxygen-linked proton and anion binding sites does not impair allosteric regulation of oxygen binding in dwarf caiman hemoglobin. *American Journal of Physiology-Regulatory Integrative and Comparative Physiology*, **305**, R300–12.

Weber, R. E., Jessen, T. H., Malte, H., and Tame, J. (1993). Mutant hemoglobins (a(119)-Ala and β(55)-Ser)—functions related to high-altitude respiration in geese. *Journal of Applied Physiology*, **75**, 2646–55.

Weigand, M. R. and Sundin, G. W. (2012). General and inducible hypermutation facilitate parallel adaptation in *Pseudomonas aeruginosa* despite divergent mutation spectra. *Proceedings of the National Academy of Sciences of the United States of America*, **109**, 13680–5.

Weinreich, D. M., Delaney, N. F., Depristo, M. A., and Hartl, D. L. (2006). Darwinian evolution can follow only very few mutational paths to fitter proteins. *Science*, **312**, 111–14.

Weinreich, D. M., Watson, R. A., and Chao, L. (2005). Sign epistasis and genetic constraint on evolutionary trajectories. *Evolution*, **59**, 1165–74.

Willford, D. C., Hill, E. P., and Moores, W. Y. (1982). Theoretical analysis of optimal P50. *Journal of Applied Physiology*, **52**, 1043–8.

Wong, A., Rodrigue, N., and Kassen, R. (2012). Genomics of adaptation during experimental evolution of the opportunistic pathogen *Pseudomonas aeruginosa*. *PloS Genetics*, **8**, e1002928.

Xu, J. and Zhang, J. B. (2014). Why human disease-associated residues appear as the wild-type in other species: genome-scale structural evidence for the compensation hypothesis. *Molecular Biology and Evolution*, **31**, 1787–92.

Yampolsky, L. Y. and Stoltzfus, A. (2001). Bias in the introduction of variation as an orienting factor in evolution. *Evolution and Development*, **3**, 73–83.

Yampolsky, L. Y. and Stoltzfus, A. (2005). The exchangeability of amino acids in proteins. *Genetics*, **170**, 1459–72.

Zhang, J. (2003). Parallel functional changes in the digestive RNases of ruminants and colobines by divergent amino acid substitutions. *Molecular Biology and Evolution*, **20**, 1310–17.

Zhang, J. Z. and Kumar, S. (1997). Detection of convergent and parallel evolution at the amino acid sequence level. *Molecular Biology and Evolution*, **14**, 527–36.

Zhu, X., Guan, Y., Signore, A. V., et al. (2018). Divergent and parallel routes of biochemical adaptation in high-altitude passerine birds from the Qinghai-Tibet Plateau. *Proceedings of National Academy of Sciences USA*, **115**, 1865–70.

Zou, Z. and Zhang, J. (2015). Are convergent and parallel amino acid substitutions in protein evolution more prevalent than neutral expectations? *Molecular Biology and Evolution*, **32**, 2085–96.

Index